TAXATION OF MINERAL RENTS

TAXATION OF MINERAL RENTS

ROSS GARNAUT

AND

ANTHONY CLUNIES ROSS

CLARENDON PRESS · OXFORD

1983

Oxford University Press, Walton Street, Oxford OX2 6DP
London Glasgow New York Toronto
Delhi Bombay Calcutta Madras Karachi
Kuala Lumpur Singapore Hong Kong Tokyo
Nairobi Dar es Salaam Cape Town
Melbourne Auckland
and associated companies in
Beirut Berlin Ibadan Mexico City Nicosia

Oxford is a trade mark of Oxford University Press

Published in the United States
by Oxford University Press, New York

British Library Cataloguing in Publication Data
Garnaut, Ross
Taxation of mineral rents.
1. Mines and mineral resources — Taxation
I. Title II. Clunies-Ross, Anthony
336.2′78622 HJ4169
ISBN 0-19-828454-3

Library of Congress Cataloging in Publication Data
Garnaut, Ross.
Taxation of mineral rents.
Bibliography: p.
Includes index.
1. Mines and mineral resources — Taxation.
2. Mines and mineral resources — Government policy.
I. Clunies Ross, Anthony Ian, 1932- . II. Title.
HJ4169.G37 1983 351.72′42 83-2088
ISBN 0-19-828454-3

Set by Hope Services, Abingdon
Printed in Great Britain by
Redwood Burn Ltd, Trowbridge

PREFACE

We began in Port Moresby during 1973 to follow the lines of thought that led to this book. Our stimulus was the government's need to renegotiate the Bougainville Copper agreement without precluding the development of new mines. It seemed to us then that an appropriate tax structure could enable host governments to escape most of the apparent dilemmas over the taxation of mines. We also came to see the importance of maintaining a stable and known system of tax rules. Subsequent events in Papua New Guinea and elsewhere have shown that the government of a small developing country that has a clear and rational policy on mining may be able thereby to act with considerable independence towards large mining companies to the advantage of the people for whom it is responsible.

Our ideas on the matter have been considerably developed by exposure to the writing or conversation of several economists, some of them more or less critical of our position, as well as to practitioners of mineral leasing negotiations on behalf of governments or mining companies. Most important among the economists have been Gordon Goundrey, Mike Faber, Walter Mead, Peter Swan, Michael Sumner, Ben Smith, Peter Lloyd, and Craig Emerson, to all of whom we are grateful. Keith Palmer gave us invaluable service by reading a large part of the draft text of the book and making extensive comments and suggestions on both content and presentation. We are indebted also for helpful comments on parts of the text and for information to Alex Kemp and to a senior executive of a UK oil company. We have learnt much that has been useful from many advisers to governments, especially from Roland Brown, and from many company officials, especially from John Ralph. For genial and efficient typing of many drafts we are grateful to the secretaries in our universities, in particular Margaret Tulloch, Jean Nelson and Eleanor Riches at Strathclyde, and Margaret Dietze at the Australian National University.

We also most heartily thank our own and each other's families, especially Jayne and Morag for (among much else) accommodating the intercontinental visits that have helped so much in getting the text written, and for providing part of the incentive for those visits.

Canberra and Glasgow
November 1982

CONTENTS

1

INTRODUCTION AND SYNOPSIS

This book is intended to be of help to those who have to play a part in devising tax systems for mining, especially in developing countries, and also to advance economic knowledge about mineral rents and their relevance to economic development. For various reasons, mining and the charges applied to it have become highly topical and controversial over the last twenty years. A number of governments that had no previous experience of dealing with mining were obliged to take decisions on the subject that were of enormous importance to their peoples. Others, which had taken for granted an established approach to the industry, found existing policy challenged by public opinion or by developments of a completely new order for which the old rules seemed inadequate.

The capacities and structures of governments and the characters of the mining industries and mining companies that confront them are too diverse to make it possible to propound a single set of rules. Nevertheless, certain forms of taxation are intrinsically much more satisfactory than others for reasons that we hope to make apparent, and we believe that a broad model of a tax system can be defined which provides a useful starting point on which variations can be made to meet special circumstances.

Many people believe that the only important characteristic of a tax is how much it takes. This is very far from true. The form of the tax may have extremely weighty effects in encouraging some activities or discouraging others. It is easy to assume, as governments often seem to have done in meeting the question of taxing mining companies, that there is a simple dilemma between heavy taxation, which discourages mining, and light taxation, which yields little in the way of revenue. On the contrary, provided that the form of the tax regime is chosen prudently, it is possible to improve the trade-off considerably between encouraging mining projects on the one hand and reaping large revenues from those that are actually undertaken on the other. To a great extent it is possible to have one's cake and eat it too. The reason for this is not very difficult to understand, and we hope that we manage to explain it. The lesson is a vital one, especially in poor countries, since the revenue from mining is often capable of making possible substantial improvements in living standards if the government that gets hold of it uses it wisely and responsibly. We know that this condition will not always be fulfilled. Not every government that seeks to maximize its revenue from minerals will be a wise and responsible one. But we believe that a wise and responsible government will be right in wanting (so far as consistent with certain other conditions) to maximize its revenue therefrom.

The book describes the operation and analyses the strengths and weaknesses of various forms of taxes on mining that have been suggested or applied in practice. We find it convenient to apply the terms 'tax' and 'taxation' to any charges through which governments collect revenue from their mineral endowments, even though some of the charges may not be covered by the normal usage of these terms. Most of the 'taxes' involve payments that are conditional on the volume, value, or profitability of mining operations. One, the fixed fee, involves payments that are fixed in advance of investment and so are independent of the outcome. This broad distinction between *conditional* and *prior fixed* forms of taxation is of considerable importance to the analysis of several parts of this book.

Beside maximizing mining revenue, a responsible government will also want to stabilize the flow of the revenue or at least to stabilize its effect on the economy. We discuss some approaches to this objective.

The basic concepts developed in this book are applicable to all minerals, including petroleum, which is sometimes considered to be a special case. However, the analysis in this book makes it clear that special characteristics of an industry, including the degree of competition among investors and the typical scale of mining operations or profitability, may argue for variations in tax treatment. For example, the large scale and high profitability of many petroleum mining operations introduce some special requirements for taxation of this industry.

This book inevitably draws on the language and method of economics. We hope, however, that it will be of use to concerned parties who know little of economic analysis and to others who have some acquaintance with economic thinking but do not have the time or patience to follow through a fairly abstract argument. At the same time, we want to be able to answer, for those with economics training, the questions that they are likely to ask. Consequently, there are some sections where algebraic or geometric methods are used to explain or to establish some point.

Thus, not everyone interested in policy on mining taxation will be advised to read the book from beginning to end. To help any reader to discover quickly its theme and tendency, we provide in this introduction a synopsis of the argument of each of the next eleven chapters and a summary of the information and appraisal contained in the six chapters that deal with case studies.

We suggest that any reader should read the synopsis in this introduction first and then the conclusion (chapter 19). He or she will then probably be able to decide which of the chapters it is worth reading in full; but chapters 4 (on the supply price of investment), 6 (on the implications of monopoly or competition in investment), and 7 (on the main kinds of tax and a summary of their characteristics) will probably be necessary if the main line of the argument is to be properly understood. Anyone interested in the operation of mineral tax systems might also at least skim through the case study chapters (13–18) if not already familiar with the countries covered there. Working through chapters 3, 5, 8, and 9 may be more the part of the specialist who wants to test the assertions in the synopsis, but chapters 10–12 should be accessible to the general reader.

Synopsis

Chapter 2 Minerals and the Economy

The *economic rent* of a mineral resource is to be understood as the value of the product when all necessary costs have been deducted. There are good grounds for governments to aim at appropriating the rent of any mineral resource because such rent represents the value of the resource, which is commonly taken to belong to the community at large, and because the alternative is commonly to hand much of the value of the resource to foreign or local investors, to private owners of land under which minerals are found, or to mine workers, and none of these groups has a moral claim to the resource itself.

Since, in order to exploit their resources efficiently, states are normally obliged to rely on private mining companies which have a substantial equity in any mine that they manage, taxation has to be used in order to extract the value of the resources. Basing the tax upon the economic rent is consistent in principle with raising substantial revenue without distorting production decisions.

While international equity in distribution of the value of minerals is presently out of the question, except for deposits in non-territorial sea or in the Antarctic Continent, identifying the rent for tax purposes is a necessary first step to any such fair distribution. Individual countries may also reasonably be thought to have an obligation to the world at large of extracting minerals in an efficient sequence, and this will broadly be favoured by taxes that do not distort production decisions, and hence by taxes based on economic rent.

The aim of maximizing government revenue is not of course the same as that of maximizing social income or welfare. Nevertheless, if (i) mineral exploitation is by private firms, (ii) externalities are compensated, (iii) the tax system is otherwise neutral, in the sense of not distorting production decisions, and (iv) the government spends wisely and responsibly, then it is proper for the government to seek to maximize revenue from mining.

Chapter 3 Economic Rent and Neutrality in Taxation

The economic rent of any asset is the reward that the owner of the asset can expect purely on account of ownership and without any personal sacrifice on his part. The rent of a piece of agricultural land per unit value of output depends on the difference between the costs per unit value of output on that land and those on the least productive land on which production is worthwhile. The rent of a mine can be thought of as depending on the difference between the present value of the net proceeds of the mine accruing over the whole of its life and the present value of the expected net proceeds of the least profitable mine that will ever be in production. Exclusive rights of use over resources make possible not only the extraction but also the existence of the economic rents of those resources: a free-for-all would be socially inefficient because it would tend to destroy the real social income represented by the economic rent.

In practice, the economic rent of a mine can be taken as including windfall profits or losses, which are added to, or subtracted from, the expected rent.

The economic rent of a mine should be distinguished from various *quasi-rents*, such as the returns required on exploration or mine development. Such returns may be taxed away from an existing project without affecting current production, but to do so is likely to halt any future exploration or mine development.

Chapter 4 The Supply Price of Investment

The reward required for the investment of capital must be treated as part of the cost of a project for purposes of determining the economic rent. This required reward is called the 'supply price of investment'. Since the cost of investment per unit is the cost of time, the supply price of investment is often measured by the discount rate or interest rate for time which the investor applies to future expected cash flows when assessing whether or not a project has a positive present value (and hence is worth undertaking).

The supply price of investment, at least under competitive conditions, will not be less than the interest rate on riskless borrowing, but those who venture the capital on a mine will require an additional reward for the risk that they accept (since the risk naturally adds to the cost of the time for which they tie up their capital). The risk arises from the investors' ignorance or uncertainty about the outcome of the investment. An important element in it is a political risk or a *sovereign risk*, which springs from uncertainty over future treatment by the state.

Different investors, and different participants in any investment decision, may respond to risk in different ways. We suppose that the most important response is to add a *risk premium* to the discount rate applied in evaluation of investments, which may take the form of the addition of a few percentage points to the interest rate that would be applied to riskless borrowing. But investors may require a risky investment to pass tests other than the expectation of positive present value at the risk-adjusted discount rate. The special concerns of managers and bankers may prevent large-scale investment if there is a substantial probability of failure (*total risk aversion*). The special concerns of bankers may prevent large-scale investment unless all outlays seem likely to be recouped within a short period (*short pay-back period*).

Thus, compensation of the cost of riskless capital and for risk are at the heart of the concept of the 'supply price of investment'. There may also be *monopoly rents*, which particular investors can claim either as a premium for their exclusive possession of know-how or because of other forms of exclusive access or rights.

The supply price of investment may include all these elements, but the form of tax treatment, by affecting the probable distribution of outcomes and the likelihood of repayment within any given period, may itself alter (i) the equity-holders' risk and hence their risk premium, (ii) the impact of any requirements for a short pay-back period, and (iii) the impact of any total risk aversion. Government contributions to equity may also reduce the private investor's risk premium by reducing the scale of the investor's commitment to a single project. The general character and stability of fiscal policy may, in addition, affect the political element in the risk premium, and rules for dealing with investors, together with government bargaining strategy, may remove some or all of the monopoly rent from the supply price of investment.

Thus the supply price, and hence the amount of economic rent available to the state, can be influenced by government policy. (See figure 4.4, p. 52, for a representation of the fixed and the alterable elements in the supply price.)

If the supply price of investment in any mineral industry is to be estimated in practice, it can be taken as substantially above international borrowing rates currently incurred by major governments, but no higher, and probably lower, than the rate that mining executives say their companies require, and no higher (translated into real terms) than the average real rates that investors have realized in similar situations in the recent past.

Chapter 5 Public Investment in Resource Development

It will often be worthwhile for governments themselves to invest in resource projects because the supply price of public investment may be lower than that of private investment, at least for part of the relevant range of scale of private and public expenditure. The lower public discount rate derives mainly from the fact that governments do not need to worry about sovereign risk, and sometimes from governments' access to concessional finance from abroad. Use of the latter can lead to distortions and inefficiencies, but the incidence of these can be reduced greatly by appropriate administrative arrangements. Governments can increase the economic rent of natural resources by encouraging consortia of private investors when this has the effect of lowering the supply price of private investment. Governments can also increase the economic rent by themselves investing up to the point where, at the margin, the supply prices of public and private investment are equal, provided that the private share of the investment is at no less than the minimum level necessary to engage the private partner's interest.

Where there is government investment, the appropriate mixture within it of expenditure on exploration, undifferentiated purchase of equity, and provision of infrastructure depends on the factors that may cause the net effective supply price of private investment to differ according to the investment's form. For example, if exploration is regarded by the private investor as especially risky, there may be a case for the government to concentrate its investment on exploration. The appropriate amount and form of government investment and any charges made for services provided by government are best determined case by case, but subject to certain clear principles, which limit both the tax and the subsidy element and which are designed to reduce uncertainty about government policy.

Access of a government to concessional loans for resource investment demands special disciplines to ensure that such loans are not of negative value to the nation receiving them.

Chapter 6 Relationships between Governments and Investors

Models of the relationships between governments and mineral investors range between two extremes, here labelled as the 'bargaining' or 'North American' model and the 'resource rent' or 'Commonwealth' model. The first model supposes that the investor is supplying his investment monopolistically, the second that there is a competitive market and thence a market supply price. The one logically emphasizes the importance to the government of bargaining tactics, including the appeal to what is normal in the relevant part of the world at the time, and consequently of renegotiation as the standard of what is normal changes; the other emphasizes the importance of discovering and applying the minimum supply price of the potential investment.

Each model probably approximates to the truth for certain industries and in certain particular circumstances. General knowledge of the industries and experience recounted in later chapters, however, suggest that competition is present, at least potentially, in a number of the mineral industries. Where it is not present in fact, this is often because a government has conceded to a particular firm a moral claim to exploit a resource before the fiscal conditions have been fixed. The need for such a commitment can be avoided if the fiscal rules for each industry are 'general' and thus known to the investor in advance of exploration, or alternatively if the government itself undertakes exploration and can therefore negotiate over rights or auction them at a stage when much is already known about the deposits, without prior obligation to any particular investor.

Rules that are 'general' and therefore known in advance of exploration and more probably stable have the further advantage that they greatly reduce the investor's uncertainty, and it will probably be worthwhile having them even if the government undertakes exploration itself. But there is one condition that must be observed if much revenue is to be expected under general rules: the form of the tax used must be such as to adjust its burden closely to the realized profitability of the particular investment.

Chapter 7 Types of Rent Tax and Criteria of Evaluation

In this chapter six broad types of tax that have been used or proposed for extracting mineral rents are considered, together with some variants that have a rather similar effect but simultaneously confer a title to output or to equity. The types of tax are the fixed fee (FF), specific or *ad valorem* duty (SAVD), higher rates of proportional income tax (HRIT), progressive profit tax (PPT), resource rent tax (RRT) and Brown tax (BT). These are considered in 'general', in negotiated, and in auctioned, form.

Each of the six types of tax is described in the text. Briefly, the FF requires the investor to pay a lump sum for access to the resource, usually upon being granted the right to explore or to mine. Specific duties involve payment of a specified sum per unit of output or of sales, and *ad valorem* duties, payment of a specified proportion of the value of production or sales. HRIT requires payment of normal corporate income tax at a higher rate than would be applicable to non-mining income. The PPT applies progressively higher rates of corporate income tax as the rate of return on invested capital rises, with the rate of return calculated on an annual basis and the rate of income tax accordingly varied year by year. RRT is a tax on realized net present value, assessed by accumulating net cash flows with interest over the life of a project, and taxing positive accumulated value. BT involves payment of a proportional subsidy on annual cash 'losses' and an equivalent tax on annual cash 'profits'. The BT is financially roughly equivalent to the government's having fully contributed equity in the mining enterprise.

The types of tax are judged comparatively on nine interrelated criteria: (i) neutrality (how far each can be made to avoid biasing investor decisions); (ii) stability (how far the rules of the system seem likely to persist under changing circumstances); (iii) intrinsic investor risk (how far it reduces the risk to the investor); (iv) maximum rate limit (how much revenue it can be made to collect on a project still viable at the maximum rate of tax consistent with incentives to efficiency); (v) adaptability (how far tax revenue responds to profitability); (vi) revenue risk reduction (how far it can be made to reduce the dispersion of possible government revenue outcomes for a given expected revenue); (vii) reduction of delay in revenue; (viii) imposition (how easy it is to determine appropriate tax rates and other parameters); and (ix) administration (how easy it is to apply effectively). The first of these is independently important, but the first five are all relevant to the capacity of the taxes to realize revenue.

Rough ordinal rankings of the six types of tax on these criteria are given with the relevant assumptions in tables 7.1–7.3 (pp. 108–12). These results are not easily summarized, but in general, the taxes lower on the list of six given above perform better on neutrality, stability, intrinsic investor risk, adaptability, and imposition than those higher up, but worse on revenue risk reduction, revenue delay, and ease of administration. The criteria on which they perform better are generally much more important than those on which they perform worse.

No conclusion can be drawn as to which type of tax is best in all circumstances, but the PPT, and even more the RRT and BT, are capable of meeting the requirement stated in the previous chapter that their burden should be related to realized profitability. Succeeding chapters suggest that the relative merits of the HRIT, PPT, RRT, and BT depend to some extent on the circumstances, but they provide a strong presumption in favour of the BT (if the government finds the necessary risk to its own revenue acceptable) and of the RRT. In practice, there are generally advantages in adding such taxes to the regular company income tax prevailing in the country, and often advantages in combining them with an auctioned fixed fee, which serves in certain circumstances to allocate the extraction rights.

Chapter 8 The Neutrality of Rent Taxes

If we put aside for the moment the effect of sovereign risk, strict neutrality in all respects is achieved by the auctioned FF ('cash bonus bidding') or (if its effects in *reducing* risk are ignored) by the BT. Strict neutrality cannot be expected from any of the other types of tax. The auctioned FF will probably be especially subject to political (sovereign) risk. The BT may also incur a higher element of sovereign risk than some of the other taxes. The presence of high degrees of sovereign risk is likely to distort investment decisions in the direction of deterring some investments with long-term pay-offs and accelerating the rate of production from those mining projects that qualify to be undertaken, and independently, in the case of the FF, to reduce revenue yields.

Among the taxes other than the BT imposed in general form, it will be possible to set parameters of comparable expected revenue-raising capacity on projects actually undertaken, so that the departures from neutrality of the HRIT, PPT, and RRT are much less important than those of the FF (in general form) or the SAVD; those of the PPT less important than those of the HRIT; and those of the RRT less important than those of the PPT, political risk again being ignored. The bankers' requirement of quick pay-back, and the fact that managers will plan for a safety margin over the minimal expected requirements of equity-holders, both tend to favour the approach to neutrality of the RRT and the BT over that of the rest.

If neutrality is so interpreted that political risk entails a departure from it, the effect of political risk is to favour the approach to neutrality of the RRT and PPT (with suitable parameters) over that of the rest, even perhaps that of the BT. The RRT, with suitable threshold and rate of tax, thus appears to have considerable advantages in terms of neutrality. Whether the RRT is superior to the BT in practice on grounds of neutrality interpreted in this way will depend on whether investors see the political environment as likely to keep the rules of the BT stable (which will favour the BT) and whether the investors' discount rate is reasonably consistent and easy to assess (which will favour the RRT).

Promoting neutrality may well be served best either by (i) the BT alone, or (ii) subsidizing exploration (or providing it by government enterprise and then auctioning licences for a fixed fee) and taxing mine development and mineral recovery by the RRT with a threshold slightly higher than the investors' discount rate for those activities.

Chapter 9 Government Revenue Maximization and the Degree of Revenue Uncertainty

The revenue that can be raised by any system of taxation is assumed to be limited first by the maximum marginal rate of tax on positive cash flow that is consistent with economizing behaviour on the part of the investor (a limiting rate which is assumed to be the same across the profit-related taxes), second by the risk entailed in the particular form of tax applied, and third by the readiness with which the tax burden adapts to profitability. It is supposed in the comparisons made here that top marginal tax rates are set no higher than this maximum 'economical' rate, and that the 'threshold' rate (for those types of tax that depend on a threshold, namely the PPT and RRT) is set at such a level as to approach as closely as possible to neutrality.

Then, if the taxes could be negotiated, ideally, project by project, or where appropriate auctioned, the profit-related taxes could be expected to raise more revenue than the others (with the possible exception of the FF, where sovereign risk was low and other risks could be spread); there is a fairly strong presumption that the BT, RRT, and HRIT would raise more revenue than the PPT Mark II without, or the PPT Mark I with or without, depreciation allowances; and there is a somewhat less strong presumption that the BT and RRT would raise more revenue than the HRIT.

If, on the other hand, the tax regimes were to be established uniformly within each mineral industry (as is generally advocated here), and (subject to the conditions on thresholds and maximum tax rates set out above) rates were set so as to maximize expected revenue for each type over a whole range of potential projects, the revenue-raising capacity of those taxes in which yield to the revenue was closely dependent on profitability would, relatively speaking, be greatly enhanced: the FF would probably come out worst; the PPT might well be a better revenue-raiser than the HRIT, the BT and RRT even more probably so; and the BT and RRT would have a further advantage over the PPT.

On the other hand, the tax types that are better for revenue-raising tend to be worse from the viewpoint of their capacity to deliver a steady, early, and predictable stream of revenue to government. Some countries, particularly those with chronic budget or balance of payments problems, may judge these considerations to be important, but large losses in expected revenue yields will follow if they emphasize these considerations in their choice of fiscal regime. For many countries, however, especially if they are mindful of the means available to stabilize the economic impact of fluctuations in mineral revenues, the tendency to irregularity and delay may not be important except in the case of the BT, under which there is the possibility of absolute loss to the public revenue.

Thus, except in the circumstances (low political risk, an opportunity to spread other risks) that favour the revenue-raising capacity of the auctioned fixed fee, the revenue maximization criterion tends to favour the same tax types as the neutrality criterion, namely the BT and RRT, with the advantage of one or the other depending on the same conditions as favour neutrality of one or the other, as set out above in the synopsis to chapter 8. Any aversion on the part of the government to risk or irregularity over its flow of revenue will mitigate the advantage of these two types of tax; but it may be that this consideration is important only over the risk of absolute loss of revenue entailed by the BT, and if that is so it may merely enhance the case for the RRT. A government concerned for early and certain revenue on a single large project might choose the PPT or HRIT.

The maximization of government revenue may sometimes be served better by a combination of auctioned fixed fee and one of the superior forms of conditional payment than by reliance on one of the conditional taxes alone. This combination is favoured the more strongly when neutrality and revenue maximization are considered together as objectives of policy. This follows because the addition of the auctioned fixed fee to the conditional tax makes it possible for a given net value of revenue to be collected with lower marginal rates of tax, and hence greater incentives to economizing behaviour. The presence of conditional taxes, as against the auctioned fixed fee alone, not only increases expected revenue by reducing risk, but also reduces the net present value that must be collected through a fixed fee, and so increases the number of companies whose financial strength would allow them to participate in an auction to establish an appropriate FF. The combination also provides the opportunity for the fixing of an appropriate compromise between high expected revenue on the one hand and early and assured revenue on the other.

Chapter 10 Imposition and Administration

The various problems discussed in this chapter on administration and imposition of rent taxes lead to the following judgements.

Auctioning tax parameters is most appropriate in the case of the fixed fee paid in advance of detailed exploration, and the auctioned fixed fee itself is normally appropriate only in certain circumstances, and most reliably so in combination with a profit-based rent tax. When conditional tax parameters are auctioned, winning bids may involve offers of fiscal regimes that are so severe that they deter investments that would be made in the absence of tax or remove necessary incentives to economization on expenditures, or deter full economic recovery from established mines. Only auctioned SAVD will have the third of these effects, and this is an especially distorting form of tax. Problems of choosing appropriate rates, if they are either to be negotiated or to be imposed generally across an industry, tell very much in favour of profit-based taxes and most in favour of those furthest to the right in tables 7.1–7.3. (The choice between negotiated and general rates is explored in chapter 6.)

The rates of rent taxes should be set no higher than a rate at which unduly large incentives to unnecessary outlays set in. This may be higher with projects where rent is very high in relation to necessary costs than with others.

Clarity in definition is required between those activities that will be treated as extractive for tax purposes and other activities.

If rates and thresholds are negotiated by project, the project should be the tax unit. If, on the other hand, there are to be general rules, arguments can be made for either the company or the project as a basis for tax. Various compromises are possible to mitigate the disadvantages of each, such as adopting the project method but either offsetting all of a company's exploration expenditure in the country against receipts in the company's first successful project, or else 'trading' exploration losses among projects. The difficulties relate particularly to unsuccessful exploration and may be mitigated by the government's undertaking or subsidizing exploration.

There are ideal methods by which, with perfect knowledge, appropriate combinations of tax rates and thresholds could be devised for PPT and RRT so as to maximize the revenue consistent with avoiding incentives within the project to over-investment. In practice, however, sets of rates and thresholds will probably best be fixed for each industry in advance of discussions over particular projects; and because of uncertainty over, and variation in, investors' discount rates there may be a case for using more than one threshold and rate of tax as a means of maximizing revenue and approaching neutrality.

Simple adjustments can be made to all the profit-based taxes to take account of exchange-rate variation and of inflation in the currency in which investors are interested. The purpose of these adjustments is to reduce uncertainty for the investor without increasing it for the government. The available methods include indexation, for the value of the relevant currency and also for the rate of exchange, and connection of thresholds to standard interest rates.

Creditability for favourable international tax treatment generally favours the use of profit-based taxes rather than specific or *ad valorem* royalties or fixed fees, and it sets certain technical requirements on any profit tax.

Creditability may dictate that even those rent taxes that are not designed as addenda to a conventional income tax on companies should be combined with such a conventional income tax, whichever tax is collected first being treated as deductible for purpose of assessment of the other. Where creditability does not require this, however (because the mineral rent tax is itself creditable against the investor's home government's tax), there may be good grounds for not applying the conventional income tax to mining projects, especially in countries for which investment in mining comes principally from overseas.

Purely administrative problems of assessment and collection of tax favour the cruder types such as the fixed fee and the specific or *ad valorem* duty, but it will not as a rule make sense for a government that can buy expertise on the international market to eschew profit-related taxes on administrative grounds alone. Among the profit-based taxes, administrative considerations may possibly favour the HRIT and PPT where an income tax system is already firmly established but favour the RRT and BT, which are intrinsically simpler, where it is not.

If a new system of rent taxes is to be introduced where there are mines already in operation, attempts to impose taxes, such as the BT, RRT, and PPT, which are generally the most desirable for new projects, are likely to lead either to administrative problems with old ones or to problems over equity between new and old. More or less acceptable compromises are possible, but in certain circumstances it may be necessary for the authority to declare at an early stage in discussion that a purchase price paid for mining rights by one company to another will not be recognized as an investment or deductible outlay in the assessment of rent tax on the purchaser.

Chapter 11 The Dissipation and Diversion of Mineral Rent

Governments do not always receive the full potential mineral rent into their revenues, because it is *dissipated* in less than optimal patterns of production, or *diverted* to other parties.

Diversion of rent commonly occurs when private rights holders, regulatory agencies, and tiers of government other than that which has the prime responsibility for managing the mineral resources can negotiate some of the terms on which mining can proceed, or when investors or workers hold veto powers over smooth production at the mine.

Poor co-ordination of the demands made by monopolistic suppliers of labour and investment or by various governments and regulatory agencies (such as occurs in tax conflicts between federal and state or provincial governments) may also dissipate rent in the attempt to divert it or to prevent its diversion.

Dissipation of rent also often results from government regulation aimed at objectives other than maximizing revenue; for example, requirements for exploration or for processing may impose extra costs on the investors that are not matched by their own additional receipts or by external benefits or by both together.

An important special case of diversion and dissipation of rent occurs when producers use a cartel to raise mineral prices. There is then some loss in mineral rent through the fall in production. This loss may be offset by an increase in monopoly rent. The extent to which the monopoly rent is available to governments is likely to be greater the more competition there is in the supply of investment. Such competition may be present if it is governments themselves that are enforcing the price cartel.

Chapter 12 Economic Stability and the Timing of Mineral Investment, Production, and Revenue

The most efficient rent taxes from the point of view of neutrality and revenue maximization are likely to generate highly unstable revenue flows. These create problems for the wider economy. The best solution will *not* as a rule be to use administrative controls or changes in rent charges as the means of maintaining steadier expenditure levels. The problems are generally managed best by varying the level of government fiscal reserves and indebtedness.

Chapters 13–18 Case studies

The case studies show various approaches to the extraction of mineral rents. They were chosen for the authors' familiarity with them and because together they exemplify the workings, the weaknesses, and the strengths of the various types of rent tax discussed in the book.

The six studies illustrate a wide range of forms of rent tax, degrees of stability in arrangements, general and *ad hoc* rules, government equity participation, and dissipation and diversion of mineral rent. Some of these differences would seem to result from differences in the objective circumstances of the countries considered, and others from differences in historical experience and in perceptions of policy-makers.

The US federal petroleum policies provide the world's principal example of competitive bidding for FF, which is combined with more or less distorting forms of conditional tax (most commonly SAVD, but also HRIT). The UK has also briefly auctioned petroleum leases for a fixed fee, but the current UK system relies on a complex hybrid of SAVD, HRIT, PPT, and RRT. Australia illustrates well the case of heavy reliance on SAVD.

Indonesia (over oil), Papua New Guinea, Zambia, Tanzania, and the UK have all employed special profit-based taxes on mining and, like the USA and Australia, have made mineral income subject to normal corporate income tax as well. In Papua New Guinea since 1976 the main additional tax for new projects has been an RRT, while on the only large project upon which terms were settled by the new national government before that a PPT was applied. The additional tax in Indonesia is an HRIT, and in Tanzania an RRT. For Zambia it is a tax with some of the characteristics of PPT. In Botswana the main additional levy, which takes the form of 'free equity', operates in its financial effects as an HRIT. In most of these cases there are *ad valorem* royalties, although often at low rates, in addition to additional profit-based taxes.

Instability in fiscal arrangements has been present at times over the past decade in all of the countries covered by the case studies, although Papua New Guinea is notable for the high degree of political consensus about the mineral taxation arrangements since the adoption of an RRT-based regime in 1976. The costs of instability in fiscal arrangements seem to have been considerable in Zambia, and probably in the UK and Australia. The effective neutrality of Zambia's system has clearly suffered from the frequency of fiscal changes between 1966 and 1973 and from other elements conducive to political uncertainty. This has compounded the effect of the fact that the PPT-type threshold had been set too low. The frequent changes in the UK petroleum tax system and its excessive complexity seem to have resulted from an insufficiently clear view of what is intended, coupled with a mistaken first step in the historical development of the additional tax. The instability in the Australian arrangements partly derives from their arbitrary nature (there having been no attempt to articulate a rationale for the rates or forms of mineral taxes in operation), which is especially damaging under the federal constitution, which grants relevant taxing powers to both federal and state governments.

Indonesia (over oil), the UK (over oil), Papua New Guinea, and Zambia have all sought to apply fixed 'general' rules, which is the approach generally favoured in this book. Indonesia (for minerals other than oil), Botswana, and Tanzania have, by contrast, followed a project-by-project approach, within more (Botswana) or less (Indonesia) clear guidelines. Botswana, over its most important mineral (diamonds), and Tanzania, over uranium, have provided examples of the gains that may in suitable circumstances be made from this policy. Conditions were probably right for bargaining because of the monopolistic element in the investment market for diamonds, and because of certain special features in the particular investor's demand for Tanzania's uranium. The approach is less obviously appropriate for Botswana's other mineral industries or for Indonesia's, in both of which there is potential competition for investment rights. The US federal government has applied general rules, but with more than usual scope for flexibility within these rules because of the important auctioned FF component of the arrangements, and with some experimental variants since 1978.

Zambia has been open in principle to making a majority government contribution to investment in any new mines, as has Tanzania, but there is no provision in the new Tanzania uranium agreement for majority government contribution to failed investment. Papua New Guinea has so far arranged minority government contribution. Indonesia has insisted on formal national ownership of oil production within production-sharing arrangements, but the economic reality is that private investors take risks and receive rewards in much the same way as with conventional direct investment. Some Indonesian production of other minerals is undertaken by state corporations, and some by private foreign companies. The UK has nationalized its longstanding coal industry; for some years its retained the right to buy 50 per cent of equity in new oil fields through its state company, BNOC, which was also independently an active operator and had by agreement the right to buy 50 per cent of oil output from any licensee at the current market price, though in November 1982 BNOC's oil operations were to be put on the market; it also has a substantial share in BP which is one of the international oil 'majors'. The Botswana government has not contributed financially to mining except through geological survey work; nor has Australia, apart from certain of its state governments.

Exploration has not been actively undertaken by the governments considered here beyond geological surveying, except that in the UK BNOC, and in Indonesia Pertamina, have explored in the same way as if they were private oil companies.

Papua New Guinea's experience over the Ok Tedi lease in 1975–6 provides an interesting test case of the degree of competitiveness potentially present in the market for investment in a major copper and gold mine. A government that took the risk of treating the market as competitive found it to be far more so than much conventional wisdom in the industry and in academia allowed.

Australia, through its uncoordinated combination of state and federal demands, provides among the countries considered perhaps the greatest possibilities for dissipation and diversion of rents. Diversion of rents to petroleum consumers was important in several of the larger petroleum producers after the OPEC oil price increases of 1973–4 (Indonesia, the USA, and Australia), but it remains in high degree only in Indonesia.

Instability in mineral revenues has been of major importance to several of the developing countries. Explicit mineral revenue stabilization arrangements have helped Papua New Guinea to avoid transmitting this instability to domestic economic activity. Zambia provides an extreme case in which instability in mineral revenues has led to severe dislocation in the domestic economy. Sudden expansion of mineral production in the 1970s contributed to major long-term structural problems in Australia, but appears to have generated only short-term problems of a similar kind in the labour-surplus Indonesian economy[1].

Chapter 19 Conclusion

The theme of the book is summarized, and the reader is directed to pages of the text that expand major points in the summary.

[1] Note that the case-studies are not necessarily up to date beyond 1980. This is particularly relevant in the case of Australia.

2

MINERALS AND THE ECONOMY

This book is concerned with the taxation of income from the exploitation of mineral resources. It demonstrates the effects of alternative methods of taxation on the amount and distribution of mineral income, and discusses some special problems of economic development in regions and nations within which mineral deposits are located.

'Minerals' here include all non-renewable natural resources. There are certain features of the mineral industries that demand special treatment in taxation. Some of the same features are present in the renewable resource industries such as agriculture, hydroelectricity, forestry, and fishing. Each of these, however, has special characteristics that are different from those that need to be considered in the taxation of minerals. Thus, though much that is said here will be relevant to them, they are not specifically considered.

The production and use of minerals play important roles in all economies. Since demand for minerals expands more than proportionately with the growth of per capita incomes up to relatively high income levels, it is concentrated most heavily in the advanced industrial countries and the rapidly industrializing developing countries. Mineral deposits, on the other hand, are scattered more or less at random over the earth's surface, and the location of the deposits, within certain economic limits, determines the place of production. The disparity is reinforced by the earlier depletion of the most accessible resources adjacent to the major industrial centres.

It follows that there is a large international trade in minerals (including petroleum) representing between one-third and one-half of the total value of trade in commodities. The trade is principally from poorer or more sparsely populated countries to wealthier, more heavily industrialized, and more densely populated countries. Minerals often contribute an overwhelming proportion of exports, and a major proportion of production and of government revenue, in the former.

The mineral extraction industries commonly contribute a much larger proportion of the value of a country's production than of the labour force employed. This characteristic derives partly from the presence of economic rent in many extractive activities, and often partly from high capital intensity in production. Here we define the economic rent of a mineral resource, or the *mineral rent*, as the value of its produce minus all the costs of production (apart from costs deriving purely from claims of ownership). The costs of production include the minimum returns to capital that are necessary to induce investment, and these minimum returns include several elements, as discussed in chapter 4. Mineral rent is analysed in chapter 3.

These characteristics are important for the structure of economies in which mineral production plays a prominent role and are a major subject of the present inquiry.

Mineral Rent and Taxation

Mineral rent is commonly thought to be a suitable base for the collection of public revenue, for three reasons. First, in many countries mineral rights are treated as belonging to the state, which is presumed to act on behalf of the public. Taxation or additional charges on private parties who are granted the right to remove minerals can thus be seen as the price of exploitation of a public asset. Second, the levying of taxation or charges on economic rents holds out the possibility of raising substantial amounts of public revenue without distorting decisions on production, consumption, or trade. The first and second reasons lead to similar approaches to mineral taxation despite their different philosophic origins, since the economic rent that is the possible base of neutral taxation or other charges is the same as the property value of the mineral resource owned by the state.

The third reason is that in the popular view the concentration of income from mineral production in large, often foreign, corporations appears to establish a case in equity for high taxation. The case is not necessarily valid in general, but there are many instances where higher charges on mineral income would lead to more equitable income distribution between nations and within mineral-producing countries. To the extent that the additional charges are confined to economic rents, the improved equity may perhaps be achieved without loss of efficiency in resource allocation.

The charges on mineral rent that are applied in practice to collect public revenue take many forms. They may be lump sum charges payable at the time when exclusive rights to exploit a mineral deposit are granted to an investor, or they may be conditional payments, the amounts of which depend on the volume or value of mineral production, or upon annual profits, or upon net present value realized through mining investment. The greater part of this volume is devoted to analysing the properties and the relative merits of various forms of conditional payments but there is some treatment of lump sum charges, for example in chapter 7. We argue that governments seeking to collect as much revenue as possible from the mineral industries will generally be wise to rely more heavily on conditional than on prior lump sum payments, but that there will usually be some place for both in an efficient fiscal regime. Although the term 'taxation' is in some usage applied only to certain forms of conditional charges, we find it convenient to use the term more loosely to include all mechanisms for raising public revenue.

Let us now examine each of the characteristics of mineral rent that are thought to make it a suitable base for the raising of public revenue, before concluding with a brief survey of the issues that are developed in the following chapters.

Minerals and Public Rights

Many governments in countries in which mineral rights are vested in the state put the minerals under a certain area of land or sea at the exclusive disposal of a private person or company. The good reason for allowing a private agent to exploit the deposit is that those who make the decision for the state doubt that it has the technical and managerial capacity to produce the minerals as efficiently (that is, to generate as much domestic product) as the private operator.

This is not always the case, and many governments use state-owned corporations to exploit some or all mineral deposits within their jurisdiction. But, even where a government is committed for ideological reasons to state ownership of mineral projects, it frequently finds that it must still employ a private company as manager, and if, as is generally prudent, it requires the managing firm to have some financial interest in the success of the enterprise, the problem of distributing the proceeds between the private firm and the state are much the same as if the former were the sole entrepreneur. The circumstances in which it is useful on economic grounds to have private investment in mineral production, and the proportion of private ownership that is likely to be suitable to various objectives of and constraints upon the state, are discussed in chapter 5 below.

Many of the issues discussed in this book are relevant only when private firms have some financial interest in the success of mining enterprises, but some of the problems that are addressed arise as well when all mineral-producing enterprises are owned wholly by the state. The latter is true, for example, of some stabilization and investment timing problems discussed in chapter 12.

The reason for giving exclusive rights over a piece of land or sea to one enterprise as operator rather than allowing a free-for-all is the need to prevent the wasteful over-investment that might arise if several firms were racing each other to get as much ore, coal, or petroleum from the one site as possible. Just as no economic rent is likely to be generated in fisheries to which there is uncontrolled access (Gordon, 1954), so the whole of the economic rent in mineral deposits would be dissipated in unproductive duplication of expenditure if they were treated as common property resources. This is obvious when several firms extract petroleum simultaneously from a single pool, or minerals from a single ore body. However, similar dissipation of potential economic rent can occur when there is competition to find a mineral deposit, even when exclusive rights to production are the reward for the first discovery. The latter possibility is discussed further in chapter 11.

There is another reason for granting exclusive rights to explore for or to exploit mineral deposits within a given area: free competition in these circumstances creates major difficulties for the maintenance of order in society. In the past this has often been the compelling reason for granting exclusive exploration or production rights to private individuals or firms. Much current mining legislation around the world still carries marks of its origin in the authorities' concern for the maintenance of law and order in a 'gold rush' (Faber, 1977:50).

It is now common for the state to issue licences to explore for minerals and also separate leases to extract them. The 'exploration licence' generally covers a larger area and provides less secure tenure than a 'mining lease'. Holders of 'exploration licences' usually have some implicit or explicit right of first refusal on the issue of the 'mining leases' within the licence area (Brown, 1977:3–8).

The special problem of taxation for mineral projects is that of setting an appropriate price on the piece of resource that is handed over to the private firm to exploit. The case is rather different from that of land. If an isolated piece of state agricultural land were sold to a private operator for his exclusive use, either for a certain period or for ever, the appropriate price for the land or for its use could be set according to the current price of that kind of land in the market. There is not in the same sense a market or a market price for mineral deposits. They are less divisible and are bought and sold less frequently, and it is far more difficult to determine in advance of exploitation the quality of a mineral deposit than to determine the quality of a piece of land. Thus the expedient of charging a known market price in advance simply will not serve. Yet a valuable public asset is being transferred whenever an area of mineral deposit is handed over to a private firm that wants to exploit it, so that it is important to find a way of determining an amount of payment that will measure the asset's real value.

Mineral Rent and Neutral Taxation

It was mentioned above that the property value of a mineral resource is the economic rent that it can be expected to generate. The economic rent is also income that can in principle be taxed away in a neutral fashion – that is, without affecting decisions on production, consumption, or trade. This concept lies at the heart of the present study. It is expounded at length in chapter 3.

Mineral Rent Taxation and Equity

We noted above that mineral production often represents a higher proportion of exports and economic output in relatively poor countries than among the rich. The operators of mineral projects in these countries are often large foreign corporations. With modern techniques of production, the total amount of value-added in mineral production that is provided directly by labour may be rather low, very much lower than the value of mineral output. This may mean, especially in a developing country, that the share of the value of the output of a mineral project that accrues to domestic factors of production through ordinary commercial processes is low. If there is to be a substantial contribution of the project to the national income of the country in which the project lies, this must generally come largely through government revenue.

In an ideal world the rights to minerals and to other natural resources might be regarded as belonging to the whole world community, to be distributed among its members according to need. It is possible that, as the social net proceeds from mineral endowments become more clearly identified and the advantages to any country of spreading its mineral receipts over time come to be recognized, some deliberate international redistribution of the benefits may occur, just as it is possible that pressures will arise for wider international redistribution of income. Even in the 1970s, some of the countries that suddenly became exceptionally rich as a result of mineral discoveries or mineral price rises turned over some of their receipts into aid to much poorer nations, though in the unsystematic and semi-equitable fashion by which international aid is usually given. For the present, however, mineral deposits are vested in the state that is sovereign in the area in which they are located, except, of course, in the potentially important case of extra-territorial oceans. Mineral wealth must be seen mainly as a bonus for those countries lucky enough to be endowed with it. If properly managed it can be a lifeline for the poor, enabling them to make infrastructural investments and to improve health and education facilities in the broadest sense, and it can also be a very useful bonus for the rich (such as Australia or Britain), allowing them to refashion their institutions while still paying off expressed or implied promises to various interests within their borders. It is also true that access to mineral wealth can easily be mismanaged so that the welfare of the community is not enhanced at all, or arguably is even reduced, by the development of the mineral deposits.

Whatever the inequities of the international distribution of mineral wealth and the possibility of its misuse by governments, there can be no moral case for handing the whole value of any such resource to the mining company that happens to be used to exploit it. The activities of the mining company and the capital it uses will need to be rewarded according to market prices which reflect their scarcity, natural or artificial. Beyond that, the claim of a mining company is no greater than that of any other person. Governments may or may not act as good stewards of their people's wealth, but this is at least a role that they profess to accept. Mining companies, mine workers, and the private owners of land under which mines are dug accept no such function. It may be necessary or convenient for the government to reward or to compensate these groups in various ways out of the gross value of what a mine produces, but their association with mining gives them no moral claim in general to the value of the deposits, except (it may be argued) where there is a recognized legal claim (as in the USA) by those who have purchased mineral rights or land with mineral rights. Where purchase prices bear little relationship to the present value of the resource at the time of purchase, even such legal claims may appear to have a dubious moral grounding. The principles of deciding what is the necessary or expedient reward for mining companies, and what can reasonably be regarded is the value of the resource to which they are given access, will form a large part of the subject matter of chapter 3.

It was suggested above that there is a strong moral claim on the part of a government to retain the value of the natural resources of its nation, if not for the use of the people of the world, at least for that of its own public. The economic rent of any natural resource has been presented as the best equivalent of the value of the resource. Distribution considerations, therefore, suggest the desirability of attempting to collect as much as possible of the economic rent as government revenue. The analysis of allocation also suggests that collecting a tax which confines itself to economic rent may be neutral in the sense of not introducing any unintended interference in decisions on production, consumption, or trade. Thus, distribution and allocation considerations alike suggest the desirability of devising a tax on mineral projects that will go as close as possible to being a tax on economic rent alone.

A government's policy towards the mining of a particular product may conceivably affect other producing countries by altering the world price of the product or the readiness of investors to mine in those other countries. It may affect consumer countries by altering the prices that they must pay. It may affect the distribution of benefits between present and future generations of its own citizens by altering the timing of the exploitation of its deposits. Finally, it may affect distribution among the present generation of its own citizens by the way in which it disposes of or dissipates the rent.

Rules of appropriate conduct over the various kinds of action that might affect world distribution of mineral proceeds would be difficult to draw up in detail and would probably not be observed if they were drawn up. Yet a few broad principles may be proposed, on the supposition that it is desirable to maximize total world welfare.

One is that the *sequence* of extraction of deposits of any particular mineral over the world should as far as possible be efficient; that is to say, deposits that are cheapest to extract with existing techniques should be mined first and each area of deposit should be mined or drilled to the point where the marginal cost of the product equals its current marginal market value. This rule dictates that an individual government should not, by subsidies for example, try to alter its place in the queue and that it should not take measures that encourage premature abandonment of mines or oil wells. In general, a tax falling purely on rent would not err in either of these directions. It would not of itself tend to raise or to lower producer prices from the level that they would have in the absence of tax. If the tax took a large proportion of the rent, policy could not be accused of engaging in a competition to attract investors who might otherwise mine elsewhere. It would not encourage premature abandonment of ore bodies or oil wells.

A second broad rule that might be widely accepted is that the interests of the poorer are to be weighted more highly than the interests of the richer. This applies between countries, between generations in the same country, and between members of the one generation in one country. This would mean that governments of producing countries should not combine to raise the price of product unless they had a strong tendency to be poorer than the consuming countries,[1] and that, if they managed to raise the price by this means, they should compensate the poorer countries that would otherwise suffer thereby. The rule would, in a country whose real national product per head is growing and is expected to go on growing, justify some preference of present over future proceeds from natural resources; but, where the total known and suspected deposits of an exploitable kind are likely to be limited and their proceeds form a significantly large part of the national income, there may be a case for deliberately delaying the opening of new ore bodies or oil or gas fields to exploitation. Furthermore, the rule might dictate decisions or guiding rules for directing public expenditure permitted by the fiscal proceeds of mining projects differentially to the poor within the nation itself.

[1] A discussion of the possibilities of such international combinations to raise the prices of commodities in the interests of developing countries appears in Radetzki (1976) and in chapter 11 below.

Finally, governments conscious of the enormous and swiftly changing international inequalities that endowments of natural resources entail might make at least tentative moves toward internationalization, or at least towards more equitable international distribution, of the proceeds of the natural resource industries.

In all this, a tax on mineral rent has little positive part to play except in ensuring the state's right to dispose of the value of the resource. Control over the timing of projects in the interest of inter-generational equity might require restrictions on the rate at which resources are released to exploitation. However, the development of a single or more or less uniform system for taxing away the economic rents of minerals would make those rents more visible and identifiable, which might be an early precondition for moves to redistribute some of the proceeds internationally in a more equitable fashion.

Mineral Rents and Public Policy

The above rather simple exposition of the theory of mineral rent raises a number of questions for public policy that are mentioned briefly here and expounded in more detail in later chapters.

As we shall show, it is difficult in practice to design a tax that is strictly neutral and collects the whole of the mineral rent. Thus in practice there may be some trade-off between the amount of mineral rent tax on each project undertaken and the total value of mineral production (and hence the total amount of mineral rent).

This raises the question of whether, in fixing rules for mineral taxation, policy-makers should be trying to maximize government revenue, domestic product, national product (income), or wage employment.

Modern approaches to cost-benefit analysis, which have been faced with an analogous question, suggest that we should not be trying to maximize any one of these measures, as usually computed, on its own (United Nations, 1972). To transfer their methods to the present problem, we might say that some more fundamental concept of social income should be adopted as the maximand and that the part of income that is channelled to the poorest of the population should be weighted more heavily than other income for purposes of calculating the aggregate to be maximized.

To maximize social income in this sense will not be the same as maximizing government revenue. Nevertheless, measures taken to keep the costs of mineral production (including the supply price of investment) as low as possible, and (as far as is consistent with that objective) to collect as much as possible of the economic rent, will tend to increase domestic product, the national income associated with a given domestic product, and (provided the government disposes of the revenue equitably and efficiently) the share of national income going to the poor and the wage employment associated with a given national income. Furthermore, it will be suggested in later chapters of this book that tax policies over minerals that can be expected over the longer term to maximize government revenue will depend on rent-extracting taxes, which, because they approach neutrality in their impact, also tend to maximize the real value of domestic product.

Thus, while simply maximizing revenue may not in general be an acceptable objective in itself, government revenue will usually serve for purposes of this book as a sufficient surrogate for the less easily definable maximand of social income.

3

ECONOMIC RENT AND NEUTRALITY IN TAXATION

An accepted ideal in any general system of taxation is that it should as far as possible be *neutral*. Taxation by its nature removes disposable income from private into state hands, and the reduction in disposable income will necessarily alter the pattern of private spending. The ideal of neutrality is that, beyond the changes that would arise from the reduction of disposable income in itself, the tax should not alter decisions on consumption, production, or trade. While this ideal can seldom, if ever, be completely fulfilled in practice, its maintenance as a criterion by which tax systems should be judged is important. Assessments need to be made about the relative significance of the various departures from the neutrality ideal that different systems of tax may entail.

The quest for neutrality does not imply the view that the market provides an ideal allocation. It does not exclude the use of special taxes to correct what are called 'externalities' that is to say, cases in which the market does not of itself tend to provide the best use of resources. The use of taxes on particular activities to charge for, or to pay for the cost of removing, environmental pollution, for example, is by no means inconsistent with the search for a *general* system of taxation that is as far as possible neutral. We do not want a tax system, without good reason, to induce consumers to switch from tea to coffee or investors to switch from picks and shovels to bulldozers.

It has been recognized that the quest for complete neutrality in taxation very often reduces itself to a quest for finding ways of extracting no more and no less than what is called the 'economic rent'. Mineral taxation is an area in which this is true in a very clear and practical sense. The concept of economic rent also provides a way of defining the property value of a mineral deposit, the extraction of which on behalf of the community is, as we have suggested, a proper objective of government when it is faced with the task of distributing the revenues generated by mining. Hence we proceed to some discussion of the concept of economic rent and to how it can be defined so as to have practical relevance.

The Classical Concept of Economic Rent

Economic rent is the excess of total revenue derived from some activity over the sum of the supply prices of all capital, labour, and other 'sacrifical' inputs necessary to undertake the activity. The idea of economic rent was developed and promoted by the classical economists. In essence, it referred to the reward that a landowner could derive by virtue simply of being a landowner and without exerting any effort or making any sacrifice (Ricardo, 1911: chapter 2). By analogy, it could be taken to refer to the reward accruing purely and simply to the possession of any property, and Ricardo (1911: 67–8 and chapter 3) specifically extended it to the rewards of owning a mine. Since the rent in this sense covered that part of the reward of landowner or mine owner that could be had without effort or sacrifice and purely because of his good fortune in possessing claims on a valuable bit of nature, a government could in principle tax away the rent without affecting the rewards that anyone received for effort or sacrifice. The reduction of the income of the landowner or mine owner might induce him in another capacity (as labourer or capitalist) to do more or to take different risks. But this would be purely what is called an 'income effect'. Such a tax on rents might encourage some productive activities because it reduced the disposable income of certain people. It would not discourage any productive activities through reducing the rewards payable to effort or sacrifice at the margin. It would not encourage a change in the mix of activities or consumption as a means of avoiding tax.

In Ricardo's exposition, rent on agricultural land derived from the fact that the amount of labour and capital that must be expended to provide a given amount of produce at the place where it was in demand increased as the margin of cultivation was extended to less fertile and more distant land. In other words, the marginal cost of agricultural production (let us say it is corn) rose with any increase in the amount of land under cultivation, as described in figure 3.1.

$MC\ MC^1$ represents an aggregate marginal cost curve, or supply curve, of corn from a given territory. If the amount of production is marginally greater than zero, the unit cost of production is OMC. If the amount of production is OQ, the unit cost of production on the land in production that is least favoured with respect to fertility and location is QR. Figure 3.1 is drawn on the assumption that the price of corn is independent of the amount of production. Such an assumption might be justified by a small size of the territory, or by the availability of perfect substitutes for corn at a fixed price.

The margin of cultivation will be extended to the point where the price of corn is equal to the marginal cost of production. At price OP, OQ of corn is produced. The total cost of producing OQ is given by the area $OMCRQ$, which is the supply price of the total amount of labour and capital required to produce OQ of corn.

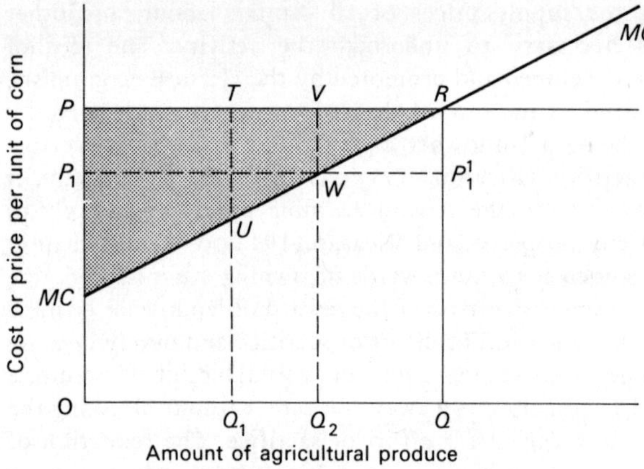

Fig. 3.1 The determination of rent on agricultural land

The excess of revenue over the supply price of sacrificial productive factors is the shaded area *MCPR*. This is the economic rent of agricultural land, which could in principle be taxed away without affecting the allocation of resources to productive uses.

Several simple but important points about the determination of rent can be made from figure 3.1. Economic rent is specific to a particular piece of land, and depends on the amount by which the unit cost of production is below that on marginal land. It is possible, even in a closed economy, for all the land to earn rent; but, under conditions of competitive enterprise, that could not be a permanent state according to traditional analysis. This conclusion relies on the assumption that 'land' in agricultural uses has no opportunity cost, in the sense that there is no non-agricultural use to which it could be put, so that there is no sacrifice on the part of the owner in allowing it to be used, however intensively, for agriculture. A further implicit assumption is that land, being under private ownership, is not used in such a way as to impair its fertility. If there is temporarily a rent accruing to the owner of the least fertile acre of land, then, according to the analysis, investors will individually have an incentive to increase the intensity with which land is worked, in such a way as to increase production. Though each owner individually will have an incentive to allow his land to be thus used, because otherwise the rent he receives will be less than that from land of similar qualities on which the more intensive use is allowed, the more intensive use will tend to raise the 'extra' or marginal cost of the last corn, that is, the extra cost of the 'last' unit; and in addition it is possible that the total effect of the increased output that will result may be to lower the price of the product. It will pay individual farmers to increase production in this way until the poorest land in production earns no rent. (If price is lowered beyond this point, of course, the poorest land goes out of production.) The amount of rent depends on the price of corn as well as on the fertility and location of land: if the price fell from OP to OP_1, the rent of the most favoured land would fall from PMC to P_1MC.

The amount $MCPR$ could be collected as taxation without diminishing total output, but only if it were collected in the correct amount from each piece of land. Income from the most fertile land could support taxation at the rate PMC without the land's being withdrawn from production; land corresponding to production at point Q_1 could support TU; and land corresponding to production at point Q could not be taxed at all without effects on production decisions.

If an attempt were made to collect an amount of tax revenue equal to the pretax economic rent by means of a uniform *ad valorem* tax on sales, say at rate PP_1/OP, rather than by means of a tax that was tailored to the profitability of production of each piece of land, the less fertile land would be withdrawn from production. Total production would fall to OQ_2, and the tax would yield only the much smaller amount of revenue PP_1WV.

The Rent of Mines

The concept of economic rent is applicable directly to mineral deposits, but only if we think in terms of total mineral output over long periods, rather than of annual output as in the case of agricultural land. The marginal mine is not the highest-cost mine in production, but the highest-cost mine that can be expected ever to be in production.

Classical economists familiar with the concept of agricultural land rent were puzzled by the apparent truth that at any time even the highest-cost mines in some mineral industries appeared to generate rent. This led to the formulation of a doctrine that the income of a mine comprised two parts: a 'royalty', which was compensation for reducing the capital value of a mine through the removal of minerals, and 'rent', which was a payment for differential costs of production or location determined in the same manner as land rent. The mine owner would not allow removal of minerals without payment of the 'royalty'. This view was expounded by a long line of authors, including Ricardo (1911, chapter 2), Marshall (1920), and Orchard (1922), and was the dominant position of economists, at least until the 1930s.

At the same time, a directly contrary view was put by a few economists including Ricardo in a special chapter on the rent of mines (1911, chapter 3) which made no reference to the earlier statement of the other position. The alternative position was that the income of mines is determined in exactly the same way as land rent, and represents a payment of favourable costs of mineral production or location. The position of Ricardo in chapter 3 was refined and developed by Taussig (1933) and Gray (1914), without affecting very much the course of economic opinion.

The various classical views on the rent of mines can be reconciled through more detailed examination of the factors that determine the rate of exploitation over time, or *depletion*, of mineral deposits. The reconciliation favours the opinions of Ricardo in the chapter on the rent of mines, especially as developed by Gray (1914).

To develop the ideas that are necessary to reconcile the conflicting classical views, it is assumed that property rights (perhaps vested in the state) are held in each mineral deposit, so that there is no 'free-for-all' which leads to dissipation of the economic rent and premature exhaustion of mineral deposits. It is assumed for the moment that the distribution and quality of mineral deposits and future demand for minerals and costs of production are known with certainty, that there is competition in factor markets, and that the present value of unit production costs from a particular mineral deposit is constant over time. We will refer to the 'income' of mines, and put aside for the moment the sometime classical distinction between 'rent' and 'royalty'.

On these assumptions, the price of mineral deposits discounted to present values is constant over time; if it were expected to be otherwise, owners of mineral rights would adjust their activities until the expectation was removed.[1] Similarly, the income per unit of output from mineral deposits discounted to present values would be constant over time and independent of the time at which mines were established to exploit them.

It is not possible to apply figure 3.1 to the problem of determining mineral rent. Whereas the figure as applied to agricultural land refers to price and marginal cost of production over some relatively short period (say one year), in its application to the analysis of mineral income it must be taken to refer to the whole of future time. This concept is a little difficult to comprehend, although it occurs in any calculation of the present value of a permanent income stream. It may aid comprehension, and be not too unrealistic, to think of owners of mineral deposits and users of minerals as having a finite 'time horizon', T, that is a time beyond which economic outcomes are of no interest to them and therefore do not in any case enter into their present decisions.

Under the assumptions described above, the individual owner of mineral rights will be indifferent to the time at which deposits are

[1] This is demonstrated formally in Hotelling's (1931) pioneering paper on optimal depletion.

exploited. Given this indifference, it may happen that at any time the whole of world mineral production is from clearly intra-marginal mines, say from deposits the unit cost of which falls in the range OQ_1 to OQ_2 in figure 3.2. In this situation, the highest-cost mine in production *at that time* would earn rent. Even under free competition and the other conditions assumed, the rent on the marginal project might be prevented from falling to zero because a ton of ore now is not available for use later, and a unit of output will not be worth producing unless the price received covers this element of 'cost' as well as actual production costs. Provided there are exclusive property rights, the exploitation of the non-renewable resource, unlike that of agricultural land, is not costless to its owner. There is an 'opportunity cost' of extracting the mineral now, and that cost is the value of extracting it later.

It is this phenomenon that led to the confusion about the rent of mines in classical economic thought. It is clear that all of the net surplus of the mine can be regarded as 'rent' in the sense that it is part of the property value of the mine and its removal will not in itself alter decisions on the extent or techniques of production. The reduction in the property value of the mine associated with reduction in the amount of minerals in the ground is equal to the amount of the potential rent that is realized through the depletion of the mine. There is no 'royalty' in addition to this rent.

In figure 3.2, OP is the present value of price, which is constant over time. OQ corresponds to the amount of mineral that is demanded at price OP between the present and some distant time T. If at some time all minerals happen to be produced from deposits corresponding to costs of production between Q_1 and Q_2, the lowest-cost mines in production will earn rent of R_1P per unit of output, and the highest cost mines, R_2P per unit of output. The unit income from each mineral deposit is determined by the extent to which costs of production are below those at the highest-cost mine expected to be in production by time T.

The successive relaxation of the restrictive assumptions upon which the above analysis has proceeded modifies some results but preserves the main conclusions. Here we examine the implications of changes over time in the present value of unit production costs from given deposits, of uncertainty about mineral endowments and demand for minerals, and of insecure property rights.

If we allow unit production costs to rise at less than the discount rate, the tendency for the present value of the flow of mineral rents to remain constant over time will cause the present value of product price to fall over time (Herfindahl, 1974:66–7). If present values of unit costs are expected to fall at the same proportional rate for

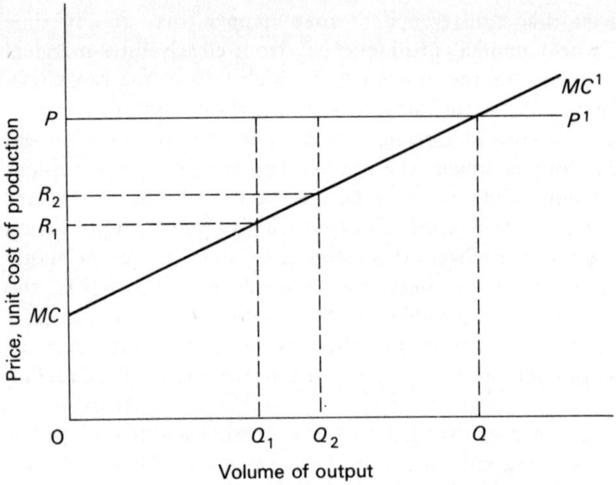

Fig. 3.2 The present value of mineral price and unit marginal cost of production

lower-cost as for higher-cost mines, this expectation of falling unit costs introduces a positive incentive to bring low-cost (high-rent) mineral deposits into production earlier and high-cost mines later than if the present value of unit costs were expected to be constant (Solow, 1974:4). The intuitive explanation of the latter phenomenon is that, if the present value of the unit cost of producing minerals is falling at a uniform rate over time for all deposits, the present value of the cost advantage of superior over inferior deposits is also falling over time. By contrast with the case described in figure 3.2, the owner of a mineral deposit is no longer indifferent to the time at which it is exploited; any delay in the mining of a superior ore body will lead to the diminution of the present value of rents. The conclusions drawn from figure 3.2 are otherwise unaffected by the relaxation of this assumption. (Of course, the less likely possibility that production costs rise at the same rate, but one *greater than* the discount rate, for all potential mines, so that the present value of unit production costs rises, has an opposite effect, causing the present value of product price to be higher in later years and introducing an incentive to delay exploitation of superior mineral deposits and to bring forward exploitation of inferior ones.)

There is probably some tendency for technological change in mineral production to be biased towards more rapid reduction in unit production costs from lower-quality than from higher-quality deposits, since it is the low-grade and otherwise inferior deposits that become economically more important with the passage of time, thus encouraging the development of cost-reducing technological change. This tendency would reinforce the effects of declining present value of unit production costs in favouring earlier development of higher-quality deposits.

Uncertainty about future demand and price for minerals or about mineral endowments causes prices and rents to be determined on the basis of subjective expectations. The uncertainty itself may raise the private discount rates that are applied to future cash flows, with complex effects on the rate of production and price of minerals and the rent of mineral deposits (see chapter 12). In general, uncertainty accelerates the depletion of known deposits, and reduces the rate at which new mineral deposits are discovered. Where expectations formed in uncertain conditions turn out to be unrealistic, the owners of mineral resources will experience windfall gains or losses.

Insecure tenure on mineral deposits, like high private discount rates, tends to accelerate the rate of depletion of known deposits and to deter the search for new deposits.

The Concept of Rent in General

The term 'rent' can also be applied to any profits in any kind of enterprise that exceed those whose prospect the investor would have required to induce him to invest in the enterprise. This part of profit is referred to as 'pure profit'. Such profits might or might not be expected. The necessary condition is that they should not have been required as a condition for making the investment appear attractive. If we regarded a mine owner as an investor who has bought the rights to the mine for nothing, then the rent of his enterprise would be the same as the rent of mine ownership, as it was discussed above. Charging the investor the rent of mine ownership as a price for the right to mine would be exactly equivalent to taxing away the pure profits of the mining enterprise.

It may be useful here to defend the relevance of the classical concept of rent in its application to mineral exploitation against a tendency in modern neo-classical thought that may have the effect of minimizing its importance. This is the habit of thinking of economic rent as either *windfall profits* (which are by definition unexpected) or *transfer rent*, attributable simply to barriers to entry into the activity concerned. The rationale of this approach is that, if investors know of the advantage of one form of investment over another, they will, unless prevented by some obstacle, shift resources out of the latter into the former until returns at the margin in both forms are the same. Hence, unless there is some artificial barrier, economic rents, other than strictly unexpected windfall profits, can not persist.

This approach to rent has a particular ideological slant in that it tends to obscure the distribution issues raised by property inheritance and by genetic and environmental advantages in a market economy. In the discussion of mineral rents, it would tend to suggest that such rents are exceptional, or that they depend simply on artificial obstacles that could and perhaps should be removed, or else that they are really *quasi-rents* (discussed in the next section of this chapter), whose removal would eventually dry up, for example, the flow of investment in exploration or in knowledge or divert it to other countries, or that they merely appear to be rents and are in fact part of the necessary expected reward for investment in an unusually risky industry. The possibility that what appear to be rents are really quasi-rents, so that they cannot be removed without reducing economic efficiency in future production, has to be taken seriously, and is discussed later in this chapter.

There is a sense in which true mineral rent depends either on unexpected outcomes or on barriers to entry, but this should not be interpreted so as to diminish its importance. If returns to investment in a mine turn out to be higher than an acceptable minimum, whether or not the high returns were expected, there is no way by which other investors may enter that particular project except on terms laid down by the holders of the mining rights. This restriction may be due in part to artificial obstacles, but, as we have seen, those obstacles are partly imposed for sound reasons of efficiency. High returns throughout the industry may after long delay inspire additional production from mineral deposits that have been rendered intra-marginal by a high mineral price. But, in the absence of tax, returns much higher than expected (and higher again than the minimum tolerable), or much lower than expected may persist indefinitely in some mines.

If there are pure profits, one might call them 'windfall profits', but the term could give a wrong impression since such profits do not necessarily come as a surprise. One might also call them 'transfer rents', but again the term might be deceptive since the barriers that prevent new entrants from quickly removing such profits are often in fact either 'natural' obstacles that follow from the technical character of the industry, or obstacles imposed for sound social reasons. With lumpy investments, highly imperfect knowledge, and long delays between market signal and response, the real situation in the industry differs substantially from one of market equilibrium brought about under conditions of perfect competition. But even if there were perfect competition, mineral rents would remain so long as exclusive rights to exploit each mineral deposit (or part thereof) were held by one or other economic agent. The 'barrier to entry' that gives rise to what might appear to be 'transfer rent' is the institution of property rights itself. Exclusive property rights are necessary to the emergence of mineral rent in the same way as they are to land rent.

Quasi-rents

We observed above that the excess of income over the minimum necessary to hold some factor of production in an activity is known as rent, and that its extraction can be regarded as neutral. However, as Marshall observed, the earnings from past and irreversible investments are in some ways like economic rent and in some ways not. Here we enter the world of quasi-rents, which are of considerable importance to the present problem.

Marshall used the example of the earnings from a machine to illustrate the concept of quasi-rents, and it is also commonly applied to earnings due to extended training in valuable skills. In the short run, training can be treated as fixed in supply, and increasing or reducing the rewards that it earns cannot affect the number of trained people. In the longer run (and how much longer depends on the skill concerned), the supply can be expanded, and removing the reward might eventually eradicate the skill.

Whenever a case arises in which it appears appropriate to tax away a rent, one has to be careful that the apparent rent is not a quasi-rent, that is to say, that its payment is not in the longer term providing some incentive to a particular allocation of resources which in the absence of the payment could not continue (Marshall, 1920:62–3, 301, 341, 349).

It is useful to think in terms of three separate marginal cost or supply curves in the mineral industry corresponding to short-term, medium-term and long-term costs. The short-term marginal cost curve relates to the variable cost of extracting ore from established mines. The medium-term marginal cost curve relates to the total cost of producing ore from new mines based on mineral deposits that are known. The long-term marginal cost curve relates to the total cost of mineral production, including the cost of finding deposits.

Taxes that have a heavy incidence at the margin of production, such as the *ad valorem* royalties discussed in relation to figure 3.1 above, are likely to affect production from established mines (that is, to deter full economic resource use) even if applied at a low rate. However, applied at low rates they are less likely to affect new mine development and exploration, which are affected relatively more by the expected amount of taxation per project and its effect on the probability of unfavourable outcomes. In figure 3.3, an *ad valorem* tax at the rate P_1P_2/OP_2 would reduce total output from OQ_1 to OQ_2 and dissipate RSP_1' of rent, but would not affect exploration and new mine development; while a lump sum tax designed to remove a much larger sum ($P_1MCP_1' + OMC_D MC_D'Q_1$) might be strictly neutral with respect to production decisions but a complete deterrent to new mine development and exploration. Too much should not be read into the particular shapes given to the

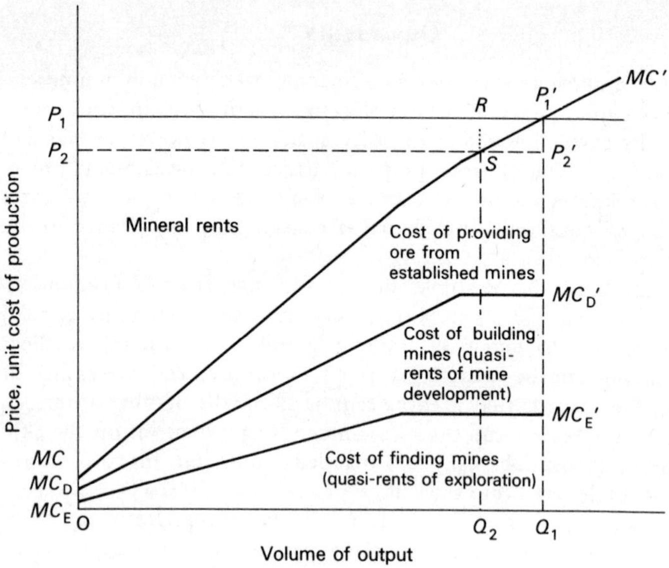

Fig. 3.3 Rents, quasi-rents, and the present value of unit marginal costs of exploration, mine development, and production: $MCMC'$, overall marginal cost of providing ore; MC_DMC_D', marginal cost of finding mines and building them; MC_EMC_E', marginal cost of finding mines.

curves in figure 3.3, but the idea that a tax may be neutral at the margin of investment decisions but not at the margin of production, and vice versa, is important.

In the exposition of the present chapter we have abstracted from many of the problems that arise in practice in the definition of rent for taxation purposes. The more severe problems relate to the valuing of private investment supplied to production. The fact that there are different degrees of uncertainty about the outcomes of expenditure at the three margins (production, new mine development, and exploration) compounds these problems. Most of the following chapter is devoted to analysis of the supply price of investment.

4

THE SUPPLY PRICE OF INVESTMENT

The supply price of private investment is part of the cost of mineral production. If the concept of economic rent is to be used in practice as the base for taxation, it is necessary to establish the full private costs of a mine, including the supply price of investment.

The cost of labour, materials, and equipment are of course declared for income tax purposes and raise no particular problems in principle. For purely practical reasons, however, some price has also to be put on capital; that is to say, some reward must be given to people for being willing to tie up claims on resources in mining. To say this does not imply that capital should be thought of as a factor of production with some natural equilibrium price. But potential investors will require a return on their investment, and without the prospect of such a return will apply their funds elsewhere. We shall refer to this required return as the supply price of investment.

The supply price of investment is a behavioural concept. It refers to the returns required by an investor to justify a decision to invest. For a host government's purposes, the supply price of investment means the returns required by the least demanding investor. Since there must always be uncertainty over what the actual returns will be, it is more precisely defined as the probability distribution of returns after tax, as assessed by the investor, that will just meet the investor's requirements.

In a later section of this chapter we discuss in detail the form in which investors' requirements for expected returns are commonly expressed. For the moment, we simplify the discussion by considering the most important single requirement for an investment to be acceptable: we consider the supply price of investment to be expressed as the discount rate that would give a zero expected net present value to some mineral investment that was just marginally acceptable.

We use the term 'rate of return' loosely, aware of the problems that may arise in actually measuring the internal rate of return of an investment, and aware also that the internal rate of return is not an adequate measure for ranking projects with a view to selecting them. The 'required rate of return' is taken here to mean simply that discount rate which would provide zero net present value for an investment that the investor regards as marginal, that is to say, on the dividing line between the acceptable and the unacceptable. This discount rate can be taken to be the one applied by the investor to evaluate time in the assessment of investments and it will be referred to also as the 'investor's discount rate'. We shall also sometimes refer to the 'returns' required of an investment rather than to an expected rate of return or discount rate applied to expected outcomes. This is not just carelessness but springs from a view that the investor's requirements under conditions of uncertainty can not simply be summarized in an expected rate of return or in a rate of discount applied to expected outcomes. We believe, as we explain below, that the pattern of the probability distribution of outcomes and the timing of outlays and receipts may also be important, and for this reason we use a looser and less specific term.

The supply price of investment for any particular project will be determined in part by the returns that are expected on other projects within industry in general, in part perhaps by the scarcity of expertise and of any other entry requirement for the particular industry (which may give some monopoly power to particular investors), and in part by the special risks of investment in the industry or the project concerned. It follows that the supply price of investment for any particular investor may vary between industries and between projects in the one industry. It may also vary between investors, since one investor may differ from another in the assessment of the riskiness of the project and of the rate of return that can be expected in the industry and in industry in general. The supply price of investment, as applied by a particular investor to a particular project, may also vary at different stages in the development of the project or otherwise over time.

If we knew what the supply price of investment was for a particular investor and a particular project at a particular time, and we used this price to discount the net cash flows of the project, the resulting net present value could be regarded as the economic rent of the project for that investor. To collect the amount of that net present value as tax would be to collect the economic rent.

The propositions that have been advanced here assume a certain model of the way in which the capital market works. To suppose that the expected rate of return required in, say, mining is determined in part by the returns available within other industries implies that capital may move freely between the mining industry and others. To suppose that the expected rate of return required from a project is likely to be increased as the riskiness of the project increases supposes what is called 'risk aversion' on the part of the investor. To say that any investor is 'risk-averse' means that a more risky project is less attractive to him than a less risky one with the same expected net present value (when discounted at the same rate). This implies that the investor requires a reward for taking risks. We regard both elements of this model as fairly realistic. If executives of mining companies deny that their companies are risk-averse, they are generally using the term in some other sense than this. Executives of mining companies *are* generally ready to say that their companies require extra rewards to induce them to take extra risks. This implies precisely that they are risk-averse in the technical sense in which the term is used here.

For reasons that are discussed below, the degree of investors' risk aversion—that is, the extent to which a given increase in the dispersion of the probability distribution of outcomes raises discount rates—commonly depends also on the scale of investment. The 'risk premium' is likely to be greater for a large than for a small investment.

Sources of Uncertainty

The main problem of using rent as the base for mineral taxation results from unavoidable ignorance: ignorance on the part of government and investor about the technical and commercial prospects of any particular project, and ignorance of each about the other's motivations and likely behaviour. It is ignorance that makes it difficult to establish what is an appropriate supply price of investment.

If both government and investor knew fairly clearly what the outcome of any investment in strict commercial terms would be, the government knew what the investor's discount rate was and the investor completely trusted the government not to make additional demands once the project had begun, then the economic rent of the project could be simply calculated, and sums with a present value equal to the rent could be levied at once or in instalments as a kind of licence fee for permission to undertake the project. Investment and production would then proceed at the same pace, and with the same techniques, as in the absence of tax. But these conditions are far from being fulfilled in practice, and it is necessary for us to explore the sources and the implications of uncertainty about the outcome of mineral investment.

Neither the investor nor the government knows how on sheer technical and commercial grounds any project is going to turn out: precisely how big and of what quality the deposit is, what the real costs of extraction will be, and what the course of real prices for the product will be over the life of the project. When the decision to explore is being made, the degree of ignorance is of course even greater.

The ignorance of a government about these matters is likely, if anything, to be greater than that of the investing firms' executives. It will not know, even within very wide limits, what is likely to happen; it will not know what the executives of the firm are expecting to happen; and it will not know for sure the limits to the returns on investment that the investor will accept, and hence would not know what terms would be minimally acceptable to the investor even if it knew the forecasts about facts on which the investor was working. For all these reasons the government is in a very weak position to know in advance what the value of the mining right is to the firm, that is to say, how much the firm is prepared to pay for it. But in a sense the firm itself does not know the value of the mining right. This cannot be known until after the event. If forced to decide in advance the maximum price it would be prepared to pay, it must, because of risk aversion, fix a price that is below the expected value of the project. This is one major disadvantage in any kind of fee, fixed in advance either by bargaining or by auction, as the sole means of collecting the value of the deposit.

The investing firm has a further ground for uncertainty, what is called 'political risk' or 'sovereign risk'; that is, the risk that the government may take action in unpredictable ways to reduce the firm's cash flow–for example, by renegotiating terms, or by various forms of arbitrary action ranging to expropriation without compensation. An empirical study of investor motivation has suggested that the perceived risk of expropriation does not have to be very great to deter investors altogether from seriously considering a project (Aharoni, 1966).

In general it is probably true that political risk, as generally perceived, increases greatly as one considers cash flows farther and farther into the future. Thus the effect of political risk may be to increase substantially the rate at which time is discounted, not just for the usual reason that risk must be paid for, but because prospective benefits seem to become rapidly less and less certain as they move farther from the present.

There is reason to believe that political risk depends in part on the dispersion of the probability distribution of after-tax outcomes as they stand under existing legislation. This is because a serious chance that returns may be very high under current rules makes it more likely that the rules will be changed. Hence political risk partly depends on the taxation regime that is used to collect resource rents. It is common for governments to change laws relating to tenure over mineral property or to fiscal charges, or otherwise to vary laws in ways that reduce after-tax returns from mineral investment. Sometimes these changes bear little relation to realized returns from mineral investment, rather having their origin in developments within the political processes of the host countries. But such changes are more likely when mineral developments turn out to be exceptionally profitable and are subject to relatively low rent charges; these circumstances give rise to community perceptions that the terms upon which investors have been granted access to mineral resources are unfair, and to calls for renegotiation of fiscal arrangements (Brown and Faber, 1977; Smith and Wells, 1975:23; Herfindahl, 1974:242–6; Gillis *et al.*, 1978:105). It follows that sovereign risk is reduced to the extent that rent charges are concentrated on income that, *ex post*, represents high rates of return on investment.

Reaction of Investor to Uncertainty

In order to deal in practice with the supply price of investment under conditions in which the outcome of the investment is uncertain and the investor is risk-averse, it is necessary to have some general picture of how the investing company will take account of uncertainty in its investment decisions. The main line of our argument assumes that investors' main response to risk is to add a premium to the discount rate that they apply in evaluation of investments. We believe, from widespread contact with managerial employees of mining companies and from contributions of mining company executives to the literature (Ralph, 1971), that this is the most important corporate response to a risky environment. However, mining investors require a risky investment to meet other tests as well.

In a company's decision over what returns it should require of a project, there are three classes of people whose objectives are important —bankers, equity-holders, and managers—and these may respond to risk in different ways. The effects of their various characteristic responses approximate to those that would follow from the application of risk premia, but in some circumstances we must consider other implications of their motives.

Bankers we assume to avoid risk as far as possible in the sense of setting an upward limit upon it and consequently to take little account, in the rates that they charge, of the risk of the project to which they are lending. We describe this behaviour to risk as approaching 'total risk aversion'. Where there is political or commercial risk over the outcome of the project on which they are lending they may, however, stipulate early repayment, which may have the effect of making greater demands on the project than it would otherwise be required to fulfil. This second aspect of the banker's approach we describe as the 'short pay-back period'. They may thus act indirectly to impose a kind of risk premium, although with an even heavier discounting of returns that are distant in time.

Equity-holders we assume, in conventional manner, to apply a premium for risk to the requirements that they impose upon a project. The premium for risk can be thought of as an addition of a certain number of percentage points to the expected rate of return that they would otherwise require or to the discount rate that they would otherwise apply to expected cash flows. It may be argued that equity-holders, who can spread their risks over a variety of assets, are irrational to make larger demands on more risky holdings. It may also be argued, consistently with modern risk analysis, that if larger demands are to be made on more risky holdings, it would be more sensible for an investor to require a significantly positive net present value at the riskless discount rate than to apply a risk premium. Yet, as already mentioned, mining men certainly claim that those who invest in mining place heavy emphasis on the risk premium approach, and we shall for the present purpose assume that it is so, though perhaps the apparent evidence for it may be explained instead by the managerial behaviour which we consider below. Some of the consequences of equity-holders' requiring significantly positive net present value are discussed later in the content of the choice between the fixed fee and profit-based taxes.

The interest of the *managers* in the decision can not be so neatly covered.[1] Following the general tendency of the theory of managerial capitalism, we assume that managers of mining firms set a very high importance on avoiding loss, or at least on preventing returns from falling below levels that equity-holders would find just barely acceptable.[2] This will probably mean that, in an industry in which outcomes are highly uncertain, managers will have a safety margin between the minimum expected returns that they require in advance and the returns that, *ex post*, they will find just acceptable. We shall call this gap the 'managerial safety margin'. It may cause the expected return required by managers to rise above that required by equity-holders and may thus set a further margin over any risk premium and know-how premium applied by the equity-holders. Like a risk premium, the managers' requirement will tend to be greater the greater the risk. Yet we deliberately distinguish it from a risk premium because we believe that its implications for taxation and rent extraction are different, for reasons that we shall try to explain. This approach to risk has much in common with that part of the bankers' approach which we described as 'total risk aversion', although managers may not share that part of the bankers' concern which we called 'short pay-back period'.

To illustrate total risk aversion, suppose that in the absence of tax the probability distribution of outcomes of a project is symmetrical with a mode at the expected value (for example, that it is normal; see figure 4.1). If the minimum rate of return that will when realized be acceptable to the equity-holders is i, the managers may set their requirements for the expected value in such a way that, say, only 5 per cent of the tail of the probability distribution of outcomes falls below i. If r

[1] For further treatments of the effects of managerial attitudes to risk on mineral investment, which follow similar lines of argument to that presented here, see Herfindahl (1974:240–2); Gillis *et al.* (1978:102–3); Ralph (1971); Gutman (1973); and especially Mikesell (1975: chapter 1).

[2] This view and its implications are interestingly explored by Aharoni (1966) in discussing a survey of US executives whose firms have contemplated investing in developing countries.

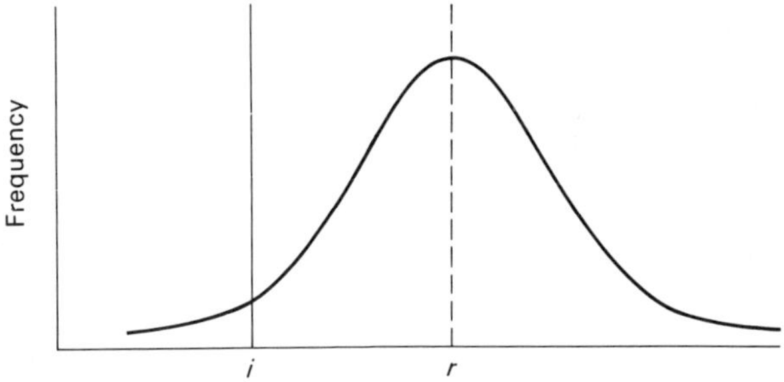

Fig. 4.1 Expected and minimum rates of return

then is the lowest value that the managers will accept for the expected rate of return, there will be a gap between i and r, and the gap will be greater the greater the dispersion of the probability distribution of oucomes.

If we now suppose that a tax regime is imposed on the same probability distribution of outcomes (which in the absence of tax would just satisfy the managers' requirements for a safety margin above i) and that this tax regime is such as to reduce to a rate of return $i + \Delta i$, negligibly above i, all outcomes that in the absence of tax would have been above that level, but to leave unchanged all outcomes that would have been equal to or below that level, then the post-tax distribution of outcomes would still satisfy the managers' requirement that no more than 5 per cent of the probability distribution should fall below i. The expected value of the post-tax distribution would fall between i and $i + \Delta i$ (see figure 4.2).

If it is true that the managers' sole requirement is that no more than a certain proportion of the probability distribution should fall below i, then, for certain purposes relevant to taxation, the supply price of investment may be regarded as approximating to i, rather than to r.

This treatment of course greatly over-simplifies the considerations that would operate in practice. The hypothetical tax imposed has a 100 per cent marginal rate, which would not be efficient in practice. The managers may in fact set not one requirement but several on the probability distribution of post-tax outcomes. (They may hate falling below a 15 per cent discounted cash flow (DCF) return and allow only a 5 per cent possibility of doing this, but hate like hell falling below a 0 per cent DCF return and allow only a 1 per cent possibility of that. The effect for our purposes would be discovered in the same way.) And it should be stressed that the managers may not *consciously* make

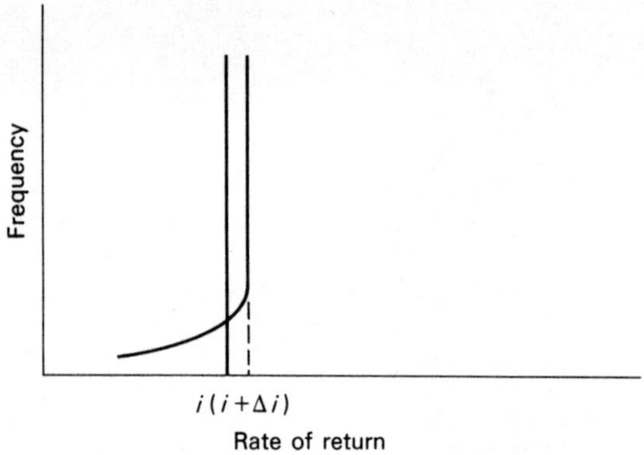

Fig. 4.2 Expected and minimum acceptable rates of return with rent tax

their decisions in this form but may merely act *as if* some rule like this were being applied. This discussion, however, may be taken to provide some rationale for supposing that there will be some safety margin applied for reasons of managerial motivation. This margin will set a further constraint on the expected returns required from investment *before tax* but will not impose a similar constraint on the requirements for after-tax returns that are actually *realized*, and will hence allow scope for a suitably structured tax system to collect rents that are not in any real sense unplanned or abnormal. In the example given, a particular form of tax allows the government to 'collect' the managers' safety margin without making investment unacceptable to the managers.

It is possible that the behaviour of shareholders corresponds better to the view that they are applying a safety margin than to the view that they are applying a risk premium, or that, independently of the managers, they would apply both. The important conclusion of the immediately preceding paragraph in the text will hold if any party is applying a safety margin. The paragraph immediately following would become redundant, however, if no one were actually to apply a risk premium.

To represent the allowances made by shareholders for risk by a risk premium measured by a simple addition to the discount rate no doubt over-simplifies the account of the behaviour of investors, but it will be accurate enough for many of our present purposes, since it will show correctly the direction in which any change in the degree of risk will affect the supply price of investment. Nevertheless, from time to time in subsequent chapters we find it useful to draw attention to certain consequences of the possibility that some of the actors may respond to some kinds of risk in ways not readily predicted by a simple view of premia for risk. While the analysis of this book proceeds on the basis that the risk premium approach is the main response to risk, it takes account of other approaches when those seem to affect practical outcomes in important ways, for example in relation to the effects of *sovereign risk*.

Special Reactions to Sovereign Risk

Sovereign or political risk is in some ways similar to and in some ways different from other sources of uncertainty in its effects on the supply price of investment and on the amount of economic rent. It also introduces some complex influences on the distribution of economic rent that are not associated with other sources of uncertainty. It is necessary to draw out some implications of these differences between sovereign risk and other sources of uncertainty here, and to clarify some definitions so as to simplify the subsequent discussion.

The supply price of investment has been defined above as a behavioural concept that is related to the investor's subjective perceptions of outcomes. Thus it is the taxation system that the investor expects to apply to future mineral income, rather than the system in place at the time of investment or the system that the government expects to apply in future, that is relevant to the determination of the supply price of investment. A taxation system is in an important sense the system that is in force at the time of investment together with anticipated changes in that system.

It follows from this definition of the taxation system that there may be windfall gains or losses to investors whenever governments' true patterns of behaviour with respect to taxation vary from those anticipated by investors. If the government increases fiscal charges after an investment has been made by less than the investor has anticipated, part of the economic rent is left with the company. If the government increases fiscal charges by more than the investor has anticipated, the government appropriates part of the normal return to capital as well as the economic rent. We can expect that variations of anticipated from realized taxation outcomes will affect perceptions about the incidence of taxation in future, but in the meantime each party may experience windfall gains or losses.

In the following discussion we seldom refer specifically to the possibility of such windfall gains and losses, but when we refer in the remainder of this book to the effect of a taxation system on the availability of pre-tax rent, we should be understood to mean the system as perceived by investors, which may incorporate their beliefs about future variations in that system.

Even if anticipated fiscal arrangements were to remain in force over the life of investments, and no windfall gains and losses were associated with unanticipated developments in the fiscal regime, the presence of sovereign risk could still have important effects on the availability and distribution of mineral rent. The expectation that the fiscal regime may change affects both the mean and the shape of the subjective probability distribution of after-tax outcomes. Historical experience teaches investors that revisions of fiscal arrangements are much more likely to reduce than to increase returns to investors, so that recognition of this reality and a natural tendency to 'fear the worst' are likely to cause uncertainty about the future fiscal regime to lower expectations about the mean of after-tax outcomes and to extract a high penalty in terms of what an investor is prepared to offer as payments for access to a mineral deposit. The tendency to 'fear the worst' probably also causes sovereign risk to widen the dispersion of the *ex ante* subjective probability distribution of outcomes, thus reducing still further the payments to which a risk-averse investor is prepared to agree.

Here we discuss two important mechanisms through which the choice between prior fixed payments and conditional payments might be affected by the presence of sovereign risk. First, as discussed in the next paragraph, the choice of fiscal regime may affect the investor's perception of the stability of the arrangements and with it the subjective perception of the probability distribution of an investment's after-tax outcomes, and so the amount that the investor is prepared to offer by way of fiscal payments. Second, as analysed in the second following and subsequent paragraphs, the presence of sovereign risk may reduce the mean of expected after-tax outcomes, by different amounts for different types of announced fiscal arrangements, or may affect the choice of fiscal regime by leading the investor to set an especially low value on prospective receipts that are likely to be delayed.

The literature suggests that very successful after-tax outcomes for investors are the most likely to incite pressures for changes in the fiscal arrangements. For this reason, fiscal regimes that compress the probability distribution of after-tax outcomes are more likely to be stable over time and more likely to be seen to be stable. This reality is a major focus of chapter 6. Systems of charges that are sensitive to realized rates of return have this effect. This causes the presence of sovereign risk to introduce some presumption in favour of the suitability of conditional payments rather than fixed fees payable in advance of investment.

Suppose now a given perceived probability of changes in the fiscal arrangements, and consider the implications of the several characteristic responses to risk defined above. When total risk aversion and short pay-back periods are important, this seems to establish an even stronger case than with a simple risk premium for exacting revenue as late payments rather than as early payments, given that in either case the system is designed to yield revenue of similar expected net present value at the discount rate applied by investors.

Where *risk premia* are applied as the only response to risk, sovereign risk is likely to raise the discount rates applied by investors. This reduces the amount that the investor is prepared to offer in fiscal payments. This effect is greater when there is relatively heavy reliance on early payments and less when there is relatively heavy reliance on late payments, because of the differential effect of a higher discount rate on early and late payments.

When there is *total* risk aversion, sovereign risk is likely to be more damaging to the incentive to invest, and therefore to the amount of revenue that the investor is prepared to offer, when revenue is collected as early payments than when it is collected as late payments. The reason for this is that there is more scope for additional demands to be made within a regime that depends on early payments, since in this case all the cash flows generated within the project after a change in fiscal arrangements would be at risk, whereas within a regime of late payments some would already have been claimed for the revenue.

Finally, *sovereign risk* is one major reason for the adoption of *short pay-back periods* by investors. Where this response is important, *ceteris paribus*, the investor will offer more expected present value of revenue as payments that will fall partly or entirely outside the pay-back period than as payments that will fall entirely or largely within the period.

Scale, Risk, and the Supply Price of Investment

The risk premium applied by investors is likely to be greater for large than for small investments. This follows from the bahaviour of bankers and managers.

Bankers avoid risks, and so generally do not finance investments if there is a significant chance that they may not be recovered. Thus, investments with this characteristic, such as those in the early stages of exploration, must be financed completely from the investor's cash flow or from new contributions by equity-holders. Only relatively small investments can be financed completely from cash flow, so that the more severe requirements of bankers are more likely to affect large than small investments.

In addition, managers are likely to apply greater risk premia to large than to small investments, even among investments financed in similar proportions by bank loans. The risks of small investments are more easily diversified within the corporation. In addition, the consequences of failure to meet the requirement imposed on managers by equity-holders—that returns should not fall below some more or less clearly specified minimum—are likely to be more severe for large than for small investments. Equity-holders may be prepared to accept returns that they consider to be inadequate on some small investments, and in the mining business to accept as inevitable that some exploration expenditures will lead to losses. But losses from large investments, for example, loss of the whole development cost of a new mine, could possibly threaten the continued existence of a large corporation. Inadequate returns from large and highly visible investments are likely to damage the careers of company managers. It can be expected that they will sometimes be avoided even at the cost of passing over investments of which the expected returns and dispersion of probability distribution of outcomes would have rendered them eligible had they been smaller in scale.

Figure 4.3 maps the relationship between the supply price of investment, risk (the dispersion of the probability distribution of outcomes), and scale of investment. Each of the 'indifference curves' traces the combinations of risk and scale of investment that are just acceptable with various supply prices of investment, D_1, D_2, and D_3. The discount rate applied by investors is lower on indifference curves closer to the origin O (D_1) than on indifference curves further from the origin (D_3). For any given scale of investment, the supply price of investment is higher the greater the risk. For any given risk, the supply price of investment is higher the greater the scale of investment (though there may be some minimum scale of investment M below which it is not worthwhile for an investor to outlay the overhead expenses of committing resources to a project).

The Supply Price of Investment to Exploration, Mine Development, and Production

The uncertainty associated with the outcomes of investment differs among exploration, new mine development, and mineral production, commonly being greatest for the exploration and least for the recovery.

The scale of investment also varies across these activities. It is commonly the case that investments for mine development are large and for exploration small.

The relationship between the supply price of investment at the three stages of mineral development thus depends on the interrelationship between risk and scale considerations. The supply price of investment will generally be relatively low at the mineral recovery stage. It will be higher at the exploration and mine development stages, in which uncertainty of the outcome of any decision is greater. Uncertainty is especially prominent in exploration, but the greater scale of units of mine development investment has an opposite effect on the supply price of investment. It is probably in development expenditures

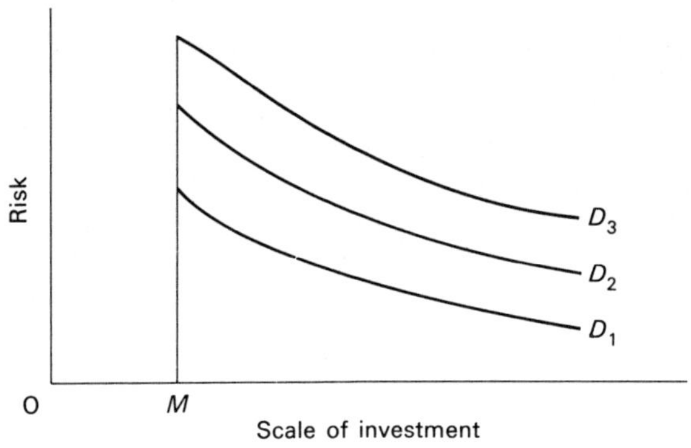

Fig. 4.3 Risk, scale, and the supply price of investment

that managers will be most wary about realizing inadequate returns.

At the time when investors are contemplating expenditure on mine development, much of the exploration expenditure is already water under the bridge; it is not strictly relevant to the current decision. We noted in chapter 3 that returns to exploration investment are in the nature of a quasi-rent. Once exploration investments have been made successfully, host countries might possibly induce companies to invest for the prospect of returns that do not include a return on exploration. But in the long run this must cause private investment in exploration to cease.

Clearly, a firm in the mining industry must find some way of covering costs of both successful and unsuccessful exploration. It is not satisfactory to treat prior exploration costs associated with a particular project, accumulated at the relevant supply price of investment in exploration, as constituting the contribution of exploration to the total project cost. Whether such exploration expenses are large or small has no real bearing on the eligibility of the project once those expenses have been incurred.

It is necessary here to point out that, even if the combined effect of risk and scale leaves the supply price of investment the same for exploration as for mine development, the greater dispersion of exploration outcomes requires higher returns on successful investments. This higher return is required not as an additional risk premium, but simply to compensate for the inevitable 'dry holes' of exploration. (With the same mean but more zeros, positive values must be higher.) The managerial caution about mine development failures makes the recoupment of losses in successful ventures rather less important at the mine development stage; failures are inevitably less common.

One possibility, then, is that firms will seek to recover exploration expenditures in general through the application of an additional premium on returns acquired from mine development.

Alternatively, the investor may seek means of ensuring that at least some investments in mine development or expansion generate returns substantially in excess of the minimum required to induce investment of this type.[3] The former procedure is less satisfactory from the point of view of efficient resource allocation since it would lead to the exclusion of some investments that were expected in themselves to be profitable. However, the supply price of investment is a behavioural concept, and when firms adopt such practices, governments should take them into account in the identification of rent.

Rents Associated with Scarce Technology and Monopoly

A further charge, in the nature of a quasi-rent or transfer rent which might be included in the supply price of investment for the present purpose, is the scarcity price of the technical and managerial know-how that a mining or oil company has assembled. While this capacity is scarce, its services will command a rent that a company will expect to recover through its investments, and which we shall term a 'know-how premium'. In the short run it is conceivable that host countries by concerted action might eliminate the know-how premium from the supply price of investment, since it may be regarded as a quasi-rent. In the longer run, however, the elimination of these charges would not be consistent with the continuing existence of mining companies' expertise.

It sometimes happens that one company or a small number of companies has for a period exclusive access to technology or to markets that are necessary to the commercially successful exploitation of some mineral deposit. An investor or cartel of investors in these circumstances is in the position of a monopolist, who may receive a monopoly rent, and who may also bargain for part of the 'resource rent', which cannot be realized by the owner of the resource without the co-operation of the particular investor or cartel. The investor may thus be able to secure returns in excess of the supply price of investment, as discussed in chapter 6, and in another context in chapter 12.

Means of Reducing the Cost of Investor Uncertainty

Uncertainty, as we have seen, tends to increase the supply price of investment by raising the risk premium, and perhaps also the managerial safety margin, and through the introduction of requirements that outlays be recouped within a relatively short period. Since what can be collected as rent is in principle increased as the supply price of

[3] See also the discussion in Mikesell, Bartsch and Bahrman (1971:35–7).

investment is reduced, it is important to explore the means by which the investor's uncertainty may be reduced.

First, there are measures that the investor himself might take with or without state assistance. An example is the conclusion of long-term contracts of sale and other measures to stabilize product prices (Drysdale, Nakatani, and Smith, n.d.). Further, the scale of investment to any one firm can be reduced by the formation of a private consortium.

Second, there are measures taken by governments in their fiscal dealings with investors. The government can reduce the scale of private investment by contributing capital itself, as discussed in chapter 5. Another such measure is to give investors security of tenure at as early a stage as possible in exploration and to inform investors of the fiscal terms of exploitation as early as possible to create an atmosphere of greater certainty about the behaviour of government.

The supply price of investment can also be reduced by the government's adopting (and maintaining over time) a tax structure within which charges fall much more heavily on cash flows that represent high returns than on those that represent low returns, and so reduce the dispersion of the probability distribution of after-tax outcomes for a given expected tax revenue. Similar considerations make it advisable for the government to ensure that rent charges fall more heavily on later than on earlier income, again for a total *expected* revenue that, discounted at the government's rate of interest, is the same. The former of these procedures reduces the supply price of investment directly by reducing the dispersal of after-tax outcomes, and both of them reduce it indirectly through their effects in reducing sovereign risk, as discussed above.

These measures will all, prima facie, tend to increase the share of the proceeds of investment that a government is in principle able to collect. The relevance is examined more fully in chapter 6.

Summarising Components of the Supply Price of Investment

Figure 4.4 illustrates the components of the supply price of investment. The point of this diagram is that the supply price of investment with which a government would be faced can be affected by policies. It is not meant to imply that the lower limit of the supply price of investment and the corresponding upper limit to the mineral rent available can be attained. But appropriate policies may enable the authorities to *reduce* the actual amount of the equity-holders' risk premium and the bankers' requirements (by modifying the taxation regime in such a way as to reduce risk, as explained above and in chapters 7 and 9, or by government investment, as shown in chapter 5); to *reduce* the

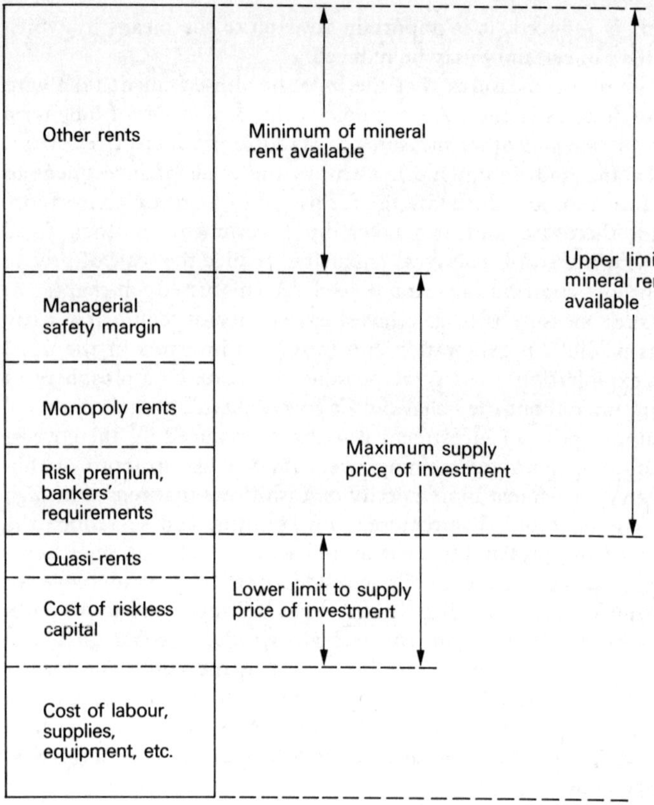

Fig. 4.4 Rents and components of the supply price of investment

monopoly rent that the investor is able to extract (through bargaining or altering the investors' terms of access–see chapter 6); and to *reduce* the managers' post-tax safety margin (again, by modifying the tax regime so as to reduce risk post-tax, as explained in this chapter).

In the diagram the know-how premium might be divided between quasi-rents and monopoly rents.

The important point of the division between quasi-rents and monopoly rents here is that the part of receipts represented by monopoly rents may conceivably be reaped by the state if it can break the monopoly by making entry to the industry freer. Quasi-rents, on the other hand, are the scarcity rewards of some resource that has been acquired at a cost and whose supply is costly to maintain. The state may capture quasi-rents in the short term, but in the longer term the costs that they represent must be covered.

Action taken to reduce the investors' monopoly rent where this depends on a monopoly of marketing may also reduce the price of the product and hence total receipts. Therefore the proceeds of such action may not all accrue to the state.

Determining the Supply Price of Investment

Any satisfactory attempt to collect mineral rents *ex post*, that is to say by assessing profits that have actually been realized, will have to make judgements about the supply price of investment. There is no very satisfactory way of doing this. The best that can be done here is to emphasize the elements that have to be taken into account.

We have suggested that the supply price of investment in mining might be thought of as a discount rate composed of the rate of interest on near-riskless lending plus a premium for risk and a premium for know-how (in a rather broad sense), which itself might be a combination of quasi-rents and monopoly rents and possibly other monopoly charges. The monopoly charges, like the risk premium, are not necessarily unalterable; yet it remains true that the rate we want must be substantially above the current long-term bond rate in major international markets. The long-term bond rate would set a lower limit. On the other hand, it cannot in real terms be much higher than rates of return after tax commonly received on that kind of mining project in the past. An upper limit is less easy to set than a lower limit, but a suggested discount rate that was higher than the rates of return realized after tax on, say, half the projects in the industry concerned that had been started in a comparable political environment between, say, then and twenty years before might be judged too high. Comparisons would have to be made between like situations. In the USA, returns on past US projects would be relevant. In an African developing country, returns on past African projects might need to be considered. Discussion with mining executives of the after-tax returns that they require would also be relevant, but it must be remembered that, apart from any professional bias, which would presumably tend towards overstatement rather than understatement on this question, managers would for good reasons tend to require a safety margin between their expected returns and what we have regarded as the true supply price of investment.

Apart from the elements already listed that may reduce the supply price of investment by reducing uncertainty, there are other, more measurable, changes that may be reflected in it. Of these the most obvious is the long-term bond rate. If, because of a change in the rate of inflation or for other reasons, the long-term international bond rate alters, it is reasonable to suppose that the supply price of investment for any mining industry in any one country will change in the same direction and by a similar amount.

While we are hesitant to give much credence to a particular set of numbers when the discount rate applied by investors varies so much with the circumstances, it may be useful to present a few indicative numbers drawn from our own experience of the mining industries. In the 1960s, when interest rates and inflationary expectations were low, discount rates of about 15 per cent were often applied in evaluation of normally risky large-scale metal mining investments in developing countries that were judged to be politically stable. This corresponded to a 'real rate of return' of a little over 10 per cent, or to a premium of about 10 percentage points over prime long-term lending rates, each measured in one of the major international currencies (say, the US dollar). The 'supply price of investment' would have been a couple of points lower in those advanced industrial countries, like Australia, whose mining sectors drew heavily on foreign investment, and lower again in the main source of direct foreign investment in mining, the USA. It was a good deal higher in developing countries with a record of political instability or of instability in fiscal, mineral tenure, or foreign investment policies.

Investors in petroleum mining commonly applied discount rates around 5 percentage points higher than those mentioned in the previous paragraph for metal mining, presumably to compensate for the fact that a higher proportion of total investment was in high-risk exploration.

Higher inflationary expectations and interest rates caused discount rates applied to the typical 'normally risky large-scale metal mining investments in developing countries judged to be politically stable' to rise to about 20 per cent in the mid-1970s. This still represented a 'real rate of return' of about 10 per cent, and a spread of about 10 percentage points over prime long-term lending rates.

It is our impression that there has been a variety of responses to the experience of unprecedentedly high interest rates in 1980 and 1981, with some metal mining investors rather unrealistically still seeking a 10 percentage point margin on long-term bond rates, but rather more seeking to maintain the established relationship to inflation rates, or a 'real rate of return' of about 10 per cent.

5

PUBLIC INVESTMENT IN RESOURCE DEVELOPMENT

This chapter discusses the circumstances in which the amount of mineral rent, and of government mineral revenue, can be increased through public provision of investment funds.

The issue arises particularly because the government will often apply a lower discount rate than private investors to possible future cash flows in the evaluation of mineral projects, for reasons introduced in chapter 4 and discussed further below. When the government applies a lower discount rate than private investors, the public provision of part of the investment in a resource project reduces the amount of investment that needs to be rewarded at the private investor's high rate and thus increases the amount of mineral rent.

The statement of these propositions raises the question of why private investment is desired at all. Given its requirement of higher rewards, private investment would not be useful unless it could generate gross returns that were higher than those that could be generated by public investment. It will often be the case that a specialist private firm will be able to bring together and manage the various inputs to a successful mining operation more efficiently than a public corporation. This is very likely to be the case in a developing country with relatively little experience of mineral production. The case for private investment depends mainly on the superior efficiency of private management, and is strong only if this superiority is present. To enlist the direct interest of a managing company in the commercial success of the venture, it may be necessary for the company to have no less than a certain share in the equity. Given that its participation is desired, the company may well require a somewhat larger share of equity than this to make the project worth its consideration. Whichever is the larger of these shares is regarded as the minimum necessary private investment.

In some developing countries, private foreign investment may be useful also in reducing the probability of public loss and the scale of any loss associated with public investment in mineral development. Here we take as given the case for management by a private firm with an equity interest, and examine the case for public investment alongside the private investment.

Public and Private Discount Rates

Public discount rates can generally be expected to be lower than private discount rates applied to resource projects for reasons related to sovereign risk and, especially in developing countries, to the availability of concessional finance, and in some cases perhaps also to a government's superior capacity to spread risks across activities. This general expectation may not be realized in countries that do not have access to international capital markets, either because of a record of international insolvency or because chronic balance of payments problems suggest the likelihood of future insolvency.

Sovereign risk (the risk owing to the possibility of adverse action by the government) contributes to the raising of private investors' discount rates but not of course to the raising of the sovereign's own discount rates. Uncertainty about future government policy is important in all jurisdictions, but more so for foreign investors in developing countries. In this very important case, sovereign risk alone is likely to cause private investors' discount rates to be substantially higher than those applied by the government.

It was observed in chapter 4 that private investors' sensitivity to risk and uncertainty is likely to be greater for large investments, where the scale of possible loss is greater. This effect is compounded by the tendency for large-scale investments to attract more political attention than smaller investments, and so to be more vulnerable to sovereign risk. Thus, even when two possible investments have the same variance, and the same probability of very favourable outcomes on purely commercial considerations, the supply price of private investment may be higher to the larger than to the smaller project. At the other end of the range of investment scale, there will be some minimum size of investment below which the absolute size of possible profits is too small to interest major firms. The likely relationship between the discount rate applied by an investor and the scale of investment and shape of probability distribution of possible outcomes was described in figure 4.3.

The existence of the relationship described in figure 4.3 between discount rates and scale of investment opens the possibility that, beyond some scale of private investment, the discount rate applied at the margin by private investors exceeds the government's discount rate, even if the private investor's discount rate is no higher than that of the government for moderately sized investments.

It is commonly suggested (for example, by Brown and Faber, 1977:70) that sovereign risk is reduced by government investment, especially equity investment, in a resource project. The connection is that, because the public investment lowers the supply price of private investment by giving the investor greater confidence that the fiscal rules will not be changed unilaterally, it will increase the rents available to be taxed by the state. This can be regarded as increasing the reward to the public investment or alternatively as lowering its cost. This is a further reason why, at the margin, the effective supply price of government investment might be regarded as lower than the supply price of private investment, at least over part of the range of the two investment supply functions.

Another reason why public discount rates may be lower than private rates is that the supply price of government investment to resource projects is affected considerably by developing countries' access to concessional funding from foreign governments and the international lending agencies. Though the terms upon which most governments of developing countries borrow on commercial international financial markets are not very different from those offered to large multinational corporations, and though it is usual for total government borrowing to equal or to exceed the total amount of any concessional funding that is available, so that the marginal cost of foreign funds to the government for most purposes is the commercial rate, it is often the case that additional concessional funding is made available to finance government investments in resource projects. Of course, the concessional element in funds provided by foreign governments will sometimes be reduced by the excess cost of procurement-tying, and the use of concessional funds may limit the manner in which the government can invest in resource projects. The origins and some implications of these various biases in international capital markets are examined later in this chapter. For the moment, it is sufficient to note that they do not upset the general observation that the availability of concessional funding often lowers the supply price of government investment.

The amount of concessional financing that is available to a developing country for resource investments is not likely to be unlimited, so that the supply price of government investment can be expected to be higher for very large investments than for resource investments of more moderate scale.

Although we have suggested reasons why we might expect governments to apply lower discount rates than private investors to resource projects on account of a lower aversion to risk, there are important exceptions to this general tendency. Some governments may be less averse to risk than the investors with which they have to deal, but the opposite relationship may hold when a large investor is contemplating investment in a small country. Many governments of smaller countries may rationally react to risk in similar ways to private corporations, raising their discount rates for very large investments above those that might apply to commitments on a more modest scale even where the projects have identical shapes of probability distribution. Where this response occurs, it increases the likelihood that the supply price of government investment at the margin will be higher than that of private investment at the margin when government commitment to a project is large, even if the supply price of government investment is lower than that of private investment when government commitment is small (see figure 5.1). The costs of funds to the internationally insolvent government may in some cases be so high that the government's discount rate is higher than that of the typical mining investor at all levels of government commitment to a project. There is no 'economic' case for public investment while such a situation continues. A government in this situation would stand to realize large financial gains by removing the insolvency through adjustments in macroeconomic policies.

It is most common for government investment in mining development to take the form of purchases of equity or of the provision of infrastructure. The two forms of investment may be provided, at the margin, at different supply prices, because of the different degrees of risk associated with them. The following two sections discuss the respective characteristics of equity and infrastructure investment. There then follows an analysis of some special considerations relating to government exploration investments, which may or may not be rewarded with equity in mining enterprises. The chapter then addresses the question of how much government investment is desirable to maximize mineral rent.

Government Equity Investment

Here we are concerned with the case in which the government pays a commercial price for the equity and accepts the risk of the project in the same way as private investors do. Thus we distinguish *government equity investment* from the *carried-interest* and *production-sharing* arrangements discussed in chapter 7. Although the latter arrangements may take the legal form of government ownership, they are economically equivalent to an additional tax, the effects of which depend on the extent and timing of their impact on cash flows.

Nevertheless, production-sharing and carried-interest arrangements may be a preferred form of tax in circumstances in which the presence of *de jure* government ownership is thought to reduce private investors sovereign risk and so their discount rates.

The government may sometimes pay for its equity by providing infrastructure to the project. Such arrangements may allow the government to use concessional funds, which otherwise would not be available, in order to finance its equity investment. This situation is analytically equivalent to the case in which equity is purchased, so long

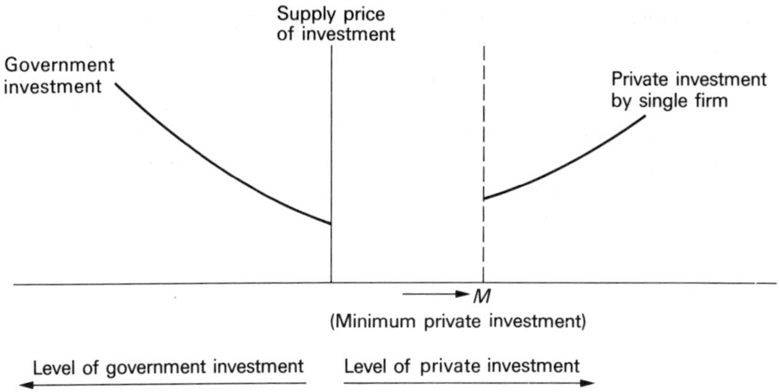

Fig. 5.1 Supply price of government and private investment to projects of similar shape of probability distribution of outcomes

as the value of infrastructure equals that of the equity purchased by the government.

Government equity investment reduces the scale of private investors' exposure in a project, but does not affect the mean value or the shape of the probability distribution of rates of return on investment. The government carries some risk of absolute loss in the same way as private investors. It also receives a share of any after-tax rents, which are potentially important given the difficulties of complete resource rent collection in practice.

Government Provision of Infrastructure

Government provision of infrastructure, like equity investment, reduces the scale of private investment and with it the average supply price of investment to the project.

However, the provision of infrastructure differs from equity investment in several ways. Investment in infrastructure is more easily financed on concessional terms than simple purchase of equity, although, as observed above, it is often possible to convert provision of infrastructure into equity. In situations in which government equity ownership is thought to reduce sovereign risk, the provision of infrastructure without equity may do less to reduce the supply price of private investment.

The economic effect of government provision of infrastructure without the acquisition of equity depends on whether charges are applied and if so whether they are applied in such a way as to recoup its cost to the government, or to recoup its value to the private investor.

User charges may be levied at rates that introduce any desired degree of subsidy or additional taxation. Here we will consider three benchmark rates of user charges: zero; rates designed to generate a stream of revenue with a present value equal to the present value of government outlays on infrastructure at the government's discount rate; and rates designed to generate a stream of revenue with present value equal to the present value of government outlays on infrastructure at the private investor's discount rate.

There are thus four cases to be compared:

A no government provision of infrastructure;
B government provision of infrastructure with no user charge;
C government provision of infrastructure with user charge covering costs at the government's discount rate;
D government provision of infrastructure with user charge covering costs at the investor's discount rate.

In comparing the effects on government revenue and government uncertainty of these various methods, we assume:

(i) that government management of infrastructure provision, if it plays any significant role, is as effective as private management;
(ii) that the government's discount rate is lower than the investor's;
(iii) that the tax system yields more revenue the higher the economic rent;
(iv) that the tax system taxes economic rent at less than a 100 per cent marginal rate.

As we move from B to C to D, revenues derived from charges are progressively higher. As we move from A to D to C to B, however, the discount rates applied by the investors are progressively lower and there is progressively more post-charge rent to be taxed on this account. This is additional to the fact that, as we move from D to C to B, there is of course progressively more post-charge rent to the extent that charges are progressively lower.

It is thus impossible to tell *a priori* which of the methods B, C, and D will yield most revenue when rent taxes as well as charges are taken into account. There is however a strong presumption, on the assumptions made above, that C and D will yield more revenue than A. This is because in C the government investment, after covering its full cost from the government's own point of view, gives the investor the service of the infrastructure more cheaply than he could provide it for himself and reduces his risk, and hence his discount rate, tending on both grounds to increase the surplus available to be taxed, while in D the government more than covers its own cost with the charges and still possibly leaves a greater rent to be taxed, because it tends also to lower the discount rate below what it would be if the government provided no infrastructure.

There would of course be some risk to the government incurred by B, C, or D, less by C or D than by B since user charges would have a high priority in claims on the investor's cash flow.

Government Investment in Exploration

It was observed in chapter 4 that the return on successful private exploration expenditure must be very high, to make up losses on unsuccessful exploration and also to allow recoupment of a premium for risk. These exceptionally high realized returns on exploration investment, which of course are inseparable in practice from returns on the rest of the operation, can attract criticism and calls for renegotiation of fiscal arrangements. The possibility of this response raises further the supply price of private exploration investment and, where high-risk exploration represents a high proportion of total investment and in countries in which fiscal arrangements are judged to be highly unstable, may make this part of the industry unsuitable for private investors.

These considerations have led some experienced commentators on public policy in the minerals sector to argue that governments should undertake directly the early high-risk exploration (Brown and Faber, 1977; Herfindahl, 1974). Herfindahl argues in addition that there are economies of scale in basic geological mapping which make it suitable for public investment.

Public investment in mineral exploration is now facilitated by the availability of finance from a United Nations 'Revolving Fund', and by World Bank willingness to make concessional finance available to governments for petroleum exploration, as exemplified by loans to Tanzania negotiated between 1979 and 1981. The Brandt Commission recommended in favour of public international financing of mineral exploration in developing countries (Brandt *et al.*, 1980:231). However, after the publication of the Brandt Report there was some loss of momentum in what had been a growing movement towards the provision of funds by international organizations for mineral exploration in developing countries, especially after the election of the Reagan administration in the United States. An expert group set up by Commonwealth heads of government (Arndt *et al.*, 1980), recognizing this loss of momentum, suggested greater use of levies such as the resource rent tax to reduce the deterrent effect of sovereign risk on private foreign investment in exploration. By 1981 there had been no movement from the pre-Brandt position, in which highly concessional funding is provided bilaterally by the governments of a few industrial countries (especially Japan and the Federal Republic of Germany) to support the exploration activities in developing countries of their own national corporations, but otherwise only by international organizations on semi-commercial terms.

Where governments are able to organize effectively to undertake mineral exploration, the payment of high exploration risk premia to private investors can be avoided by public investment, whether or not concessional funding is available. In addition, the supply of exploration investment is maintained in circumstances in which private investment would not be forthcoming in economic amounts because of sovereign risk.

The Amount of Government Investment

Provided that over some relevant range the supply price of government investment is less than that of private investment, and provided that there is at least the minimum scale of private investment necessary to interest the private investor and to ensure efficient management, the government is able to increase the economic rent of a mineral deposit by investing up to the point where, at the margin, the supply price of government investment equals the supply price of private investment.

Figure 5.2 examines graphically the problem of determining the desirable level of private investment, describing as two separate cases the two situations that are likely to arise in practice. In both case 1 and case 2, *AB* represents the amount of investment required by the project and *OM* the minimum necessary scale of private investment. In case 1, the supply price of private and public investment is equalized at a price OP_1, with the sum of *OA* government investment and *OB* private investment equalling the required amount of investment *AB*. But in case 2, the total investment requirements *AB* are insufficiently large to allow both the equalization of the supply prices of public and private investments and the utilization of the minimum necessary scale of private investment. The latter situation is probably a common one.

The supply price of private investment in figure 5.2 is described for a single firm. The supply function for private investment by a consortium of several firms to a project is likely to be lower and flatter than that described in the figure, although the higher transactions costs of investment decision within consortia may have an effect in the opposite direction on the supply price of investment. In this situation, the private discount rate for the project would be reduced, with a reduced optimal level of government investment, if a consortium of investors were organized to provide private investment to the project.

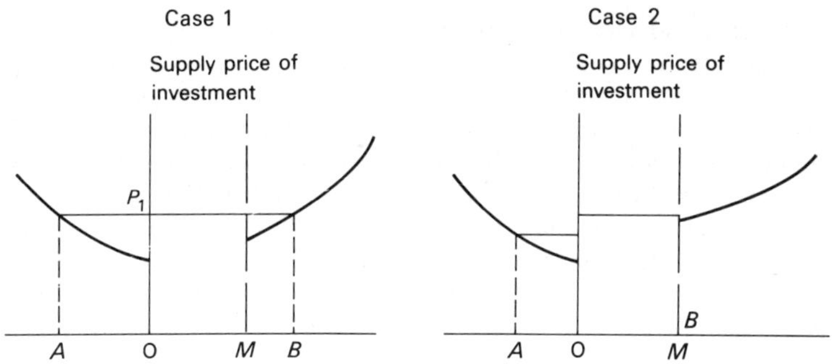

Fig. 5.2 Alternative cases of optimal levels of government and private investment in projects with similar shapes of probability distribution of outcomes: *OA*, amount of government investment; *OM*, minimum private investment; *OB*, amount of private investment

The Optimal Form of Government Investment

The amount of government investment that maximizes economic rent is that amount which equalizes the supply prices of private and public investment. The form of government investment that maximizes economic rent depends upon the supply prices of various forms of public and private investment. It could profitably include some combination of public exploration investment, equity, infrastructure with user charges, and infrastructure without user charges, depending upon the various influences, described above, on the supply price of these various forms of public investment and, in the case of exploration, on the supply price of alternative, private, investment.

A further question arises, parallel to the need to choose between *ad hoc* and general rules of rent taxation discussed in chapter 6: should the authorities announce the principles upon which they will invest in resource projects, or should they handle the question case by case?

Government investment has the potential of raising the attractiveness of a resource development to an investor. Prior knowledge of government intentions may thus reduce investor uncertainty, reduce the supply price of investment, and increase the amount of economic rent. These are the clear advantages of general rules. (Of course, government exploration expenditures are known to private investors at the time of subsequent private investment decisions.)

On the other hand, negotiated arrangements tailor-made to each project could be used to offset possible deterrent effects on investment of the inevitably imperfectly neutral taxation system. These are the advantages of *ad hoc* rules.

There is no clear basis for preferring in general either *ad hoc* or general rules for public investment. However, there would seem to be advantages in a hybrid approach, with the amount and form of investment being settled case by case, but, where it was possible to give the announced rules sufficiently clear operational content, within two announced principles. The first principle would be that a clear upper limit, perhaps zero, should be placed on the tax element in future government equity investment in resource developments. The second principle would be that public provision of infrastructure or other public contribution to investment should be used to raise the private profitability to investors of projects that could be shown to have fallen below the threshold of attractiveness because of non-neutrality in the taxation system, but not otherwise to subsidize resource investments. These principles would contribute to considerable reduction in the perceived probability of unsatisfactory outcomes from private resource investments, without removing useful flexibility.

Effects of Concessional Finance

Policy on public investment in resource developments is likely to be influenced considerably by the availability of concessional finance. Much of the concessional finance will be allocated to specific purposes and in the case of bilateral aid may be tied to purchases from particular countries of origin. These realities introduce a potential for distortion of domestic development priorities and for wasteful use of concessional finance.

It is obvious that aid is used most efficiently in terms of the priorities of the recipient countries' governments (provided those priorities are clearly enunciated and translated into decisions) if it is untied by source and made available in support of the total development programme of the recipient country. Unless the donor can be presumed to have better information and judgement on the recipient country's affairs than that country's own government, aid is used inefficiently to the extent that it promotes activities that would not have survived a listing of the recipient's priorities within a single budget constraint. It is also used inefficiently even when it is allocated to genuinely high-priority programmes if source-tying causes it to be used to purchase relatively high-cost goods and services. Aid for resource developments poses all of the usual problems of source-tied and project-tied aid, although often in an extreme form because of the scale of the projects.

Although aid for specific purposes and source-tied aid is likely to be less efficient than untied aid in support of a recipient's general development programme, it will often appear to be better than nothing. It is indeed likely to be better than nothing, possibly much better, when its availability reduces the supply price of private investment and increases government revenues from mineral rent. However, there are circumstances in which it can be worse than nothing, and it is always difficult for a recipient government to judge whether the alternative really is nothing at all rather than some less restrictive form of aid.

There are three circumstances in which the acceptance of aid to resource projects might bring negative net benefits to a recipient country. The first occurs where the availability of concessional funds from a particular source leads to the inefficient use of a valuable resource, perhaps because the funds are tied to the exploitation of the project by a relatively inefficient firm of the donor's nationality. The second circumstance is one in which the recipient government's net financial commitments to a project exceed the concessional element ('grant-equivalent') in the foreign funding, and where the value of the taxation and other revenue from the rents generated by the investment is less than this shortfall. This situation could arise where public provision of infrastructure caused a sub-marginal project to be undertaken by private investors. The third circumstance is one in which the resource project, with effective concessional financing, itself brings positive net benefits, but where the acceptance of aid that is specific to resource investments reduces the amount of aid from the same source that is available for other, higher-priority purposes. In this case, the true opportunity cost of the concessional finance might exceed its nominal cost. Alternatively, even if there were no prospect of raising the quality of aid from the relevant donor, there might be some danger that the acceptance of restrictive conditions of aid from one donor would weaken the commitment of other donors to provide aid in less restrictive forms.

It is always difficult for an aid recipient to identify this last circumstance in practice, and high orders of political judgement are required in the administration of policy.

A few relatively simple devices in the systems within which aid is administered in recipient countries could remove the major distortions that are associated with project aid and source-tied aid in the mineral sector, without requiring radical changes in the formal conditions that are usually applied to government-to-government assistance in the financing of resource investments.

There would be less chance that government-to-government aid transactions would lead to the waste of resources if three administrative rules were followed. First, donors should undertake to provide a specified level of total aid over a period of years. Second, the funds represented by this commitment should be made available to support a specified range of activities, including major resource investments where government expenditure on these activities survived a listing of the recipient's priorities. The aid would be 'portable', or transferable across the listed activities. Third, the commitment should be drawn down *selectively* against purchases from sources to which the donor expected its aid to be tied.

Such arrangements would work most effectively if they were supported by all major donors. The recipient government would add the commitments of aid to its estimates of resources available to support public expenditure in general, and adjustments would be made to intended future levels of general borrowing and total public expenditure. The recipient government would allocate this total level of expenditure among programmes through its normal budgeting and planning procedures. Decisions to allocate expenditure to the purchase of equity or the provision of infrastructure for resource investments would be made on the merits of these expenditures relative to others, with the contribution of these expenditures to future revenue growth being weighed against the economic and other returns from alternative programmes.

The development programme having been defined without reference to the purposes for which aid was to be allocated, a range of activities could be selected from the total programme for support by foreign donors. For the system to work efficiently, it would be necessary for total expenditure on activities that a particular donor was prepared to support to be very much larger than the amount of the donor's aid commitment.

Goods and services required by each aid-supported activity would be purchased competitively, through the normal supply procedures of the recipient government. When a source of goods and services to which some aid was tied was revealed to be the most efficient supplier, the aid commitment of the corresponding donor would be run down.

The system would clear all aid commitments so long as each donor country was a competitive source of certain goods and services required by activities that it was prepared to support in an amount that substantially exceeded the amount of its aid commitment. If aid commitments cleared automatically, there would be no distortion in priorities associated with the project orientation of aid, and no excess cost of tying. For the donor aid might appear to be project-oriented and source-tied, but this would carry no distortions in expenditure and no excess costs for the recipient.

The system might fail to clear all aid commitments if the priorities of the recipient allowed few activities of types a donor was prepared to support, or if a donor was uncompetitive as a supplier of goods and services required within the activities that it was prepared to support.

As discussed above, the financial returns to a government on funds allocated to resource investments can be very high, so long as the projects operate within an efficient system of rent taxes. It is likely that genuinely profitable expenditures of this type would find their way into a recipient's development programme on their own merits, unless the recipient placed unusually low value on revenue growth. The largest donors who restrict the use of their aid narrowly to resource investments are major industrial countries that are competitive suppliers of the machinery and materials required in large mining investments. These realities would assist in the clearing of aid commitments through the suggested processes.

Nevertheless, it is possible that a donor would be prepared to make more funds available in support of expenditure on resource investments than could be used efficiently within activities that survived a conscientious listing of the recipient's priorities. One advantage of the suggested administrative system is that it would make variations from allocation according to the recipient's priorities, and the excess cost of aid-tying, clear for all to see.

Some bilateral 'donors' make aid available readily for resource investments because they see advantages for themselves in the increase of world-wide supplies of resource-based products. This is most importantly the case with concessional finance from the Federal Republic of Germany and Japan in recent years. There are some circumstances in which it is against the interests of resource-exporting countries, taken as a whole if not individually, to have world-wide mineral supplies increased by these means. The forced development of otherwise submarginal projects does not generate rents and, through its effects on world markets for resource-based commodities, can reduce economic rents in established mines. Recipients of concessional finance have a strong interest in channelling concessional funds into intra-marginal projects, where the reduced supply price of investment is reflected in increased revenue from mineral rents.

Summary

It will often be worthwhile for governments to invest in resource projects because the supply price of public investment is lower than that of private investment, at least for part of the relevant range of scale of private and public expenditure. The lower public discount rate derives mainly from the fact that governments do not need to worry about sovereign risk, and from governments' access to concessional finance. Use of the latter is associated with possible distortions and inefficiencies which, however, can be reduced greatly by appropriate administrative arrangements. Governments may be able to increase the economic rent of natural resources by encouraging consortia of private investors if this has the effect of lowering the supply price of private investment. Governments may increase the economic rent also by themselves investing up to the point where, at the margin, the supply prices of public and private investments are equal, provided the private share of the investment is at no less than the minimum level necessary to engage the private partner's interest. Where there is government investment, the appropriate mix of expenditure on exploration, equity, and infrastructure depends on the various factors that may cause the net effective supply price of government investment to differ according to its form. The appropriate amount and form of public investment is best determined case by case on clear principles designed to keep as low as economically possible the probability, as perceived by investors, of unfavourable outcomes.

The opportunity of access to concessional lending for resource investment demands special disciplines to ensure that such finance does not, through raising costs, encouraging sub-marginal extraction, or distorting national priorities, prove of negative value to the recipient population.

6

RELATIONSHIPS BETWEEN GOVERNMENTS AND INVESTORS

The amount of mineral rent depends on the supply price of investment. In practice, however, investors are often able to secure an element of monopoly rent, which then has to be treated as part of the supply price of their capital (see figure 4.4). The timing and character of a government's offers to investors and its negotiations with them have a considerable bearing on the extent to which this is likely to occur. Hence this chapter explores the manner in which governments come to terms with investors on the taxation of mineral rent, and the relationship of this process to the degree of competition in the supply of investment. It first examines the considerable literature that has developed on these questions over the past two decades. This leads on to the identification of two general models of government–investor relations that are explicit or implicit in most contributions to the literature. The two models are distinguished by the presence in one of competitive and in the other of monopolistic supply of private mineral investment. There is then some discussion of how the degree of competition in the supply of investment depends upon both the organization of firms in the mining industry and the security of tenure over mineral properties at the time at which rent charges are fixed.

The remaining sections of this chapter discuss three specific issues with normative implications for government's management of relations with investors: (i) the manner in which governments' relations with investors are influenced by whether rent charges are announced in advance of, or in the course of, investment; (ii) the relative merits of stable rules, which apply for the life of an investment, and flexible rules, which are varied by negotiation or unilateral government decisions from time to time; and (iii) the choice between general rules, which apply in a consistent way to all investments, and *ad hoc* rules tailored to each investment. A concluding section evaluates the various models of government–investor relations, and suggests some generalizations about the way in which these relations can be arranged so as to maximize government receipts from resource rents.

Models of Government–Investor Relations

The recent literature includes case studies of conflict between governments and investors; positive theorizing about the usual course of government–investor relations; normative theorizing about how these relations can best be structured to promote the mutual interests of the parties; and, in the older works, a little moralizing about how governments should behave independently of their perceptions of the national interest.

When the historical evolution of ideas about government–investor relations is examined, two general and overlapping approaches to this issue can be identified. The first, which we refer to as the 'resource rent' or 'Commonwealth' approach, follows some of the lines of reasoning introduced in the early chapters of this book. The contributions of Penrose (1959), Herfindahl (1974), Faber and Potter (1971), Faber (1977), Hughes (1975), the present authors (Garnaut and Clunies Ross, 1975, 1979) and the recent contributions of Mikesell (1975, 1979) fall mainly within this tradition. This approach rests on the view that the government should seek to tax away as much as possible of, and no more than, the resource rent generated in a mining enterprise. Complications and constraints associated with uncertainty about the supply price of investment, with the political unacceptability of very high returns to the private investor (though these high returns, if they were acceptable, might nevertheless induce optimal levels of investment in some circumstances), and with monopoly in the supply of investment are admitted by most of these authors. The need for bargaining over some fiscal parameters is generally admitted. But it is nevertheless judged that some concept of 'resource rent' should guide mineral taxation, and that it is generally possible for a government to collect through one means or another the greater part of any resource rent that can be identified.

The second approach, which we refer to as the 'bargaining' or 'North American' model of government–investor relations, had its origin in attempts to explain the history of concession agreements in the postwar period, particularly those concluded in Latin America during the 1960s. The presence of resource rent is not denied in this approach, but it is supposed that the amount of revenue that can be collected from mining enterprises depends mainly on power relationships. Resource rents, quasi-rents, monopoly profits, and normal returns to investment are distributed between governments and investors through a highly political bargaining process. The government's receipts out of this process are indeterminate: a wide range of influences affect the bargain that is struck, and agreements are renegotiated frequently in line with changes in any of these determinants of 'bargaining power'.

Some of the propositions upon which the 'bargaining model' depends can exist logically only when there is monopoly in the supply of investment, and therefore effectively bilateral monopoly in government–investor relations. Most applications of the 'North American' model do not make this last requirement explicit, although there are praiseworthy exceptions (for example, Mikesell, Bartsch, and Bahrman, 1971).

The bargaining model was revealed in an exchange led by Vernon, reported in the proceedings of the 1967 annual meeting of the American Society of International Law (Vernon, 1967). It was developed and refined in North America, mainly by Harvard scholars and by Mikesell at Oregon. There have been three major attempts to define the model, by Mikesell, Bartsch, and Bahrman (1971), Smith and Wells (1975: chapter 1) and Moran (1975: chapter 6). The articulation of the bargaining model in several widely read and persuasive articles and books has caused it to have pervasive influence, and it has been applied in a variety of situations. It is reflected, for example, in works by Fritzsche and Stockmayer (1978), Waelde (1977), Zorn (1977), and Gillis *et al.* (1978).

The dichotomy between the two approaches is not always as clear as is suggested in this presentation. Most authors within the 'Commonwealth' tradition would recognize that ignorance of market conditions or incompetence on the part of governments can lead to the striking of bargains that leave resource rent with investors, and that this will often give rise to political pressures for renegotiation that lead either less or more revenue to be collected in total than would have been the case if the government had sought only to tax the resource rent. They would recognize as well that an investor's exercise of monopoly power can lead to many of the consequences predicted by the 'bargaining model'. The main practical differences derive from the view within the 'resource rent' approach that it is commonly the case that even the government of a small and weak country can often establish fiscal arrangements for mining that claim for the state as much revenue as it is ever wise for it to claim, and that if it acts wisely to this end there are good prospects for establishing relatively stable arrangements.

We might distinguish these two analytic approaches to the definition and distribution of resource rent from the prevailing ideologies that guided government–investor relations in developing countries before, and with declining influence after, the Second World War. There was a widespread belief that natural resources belonged to the corporation that found them. Governments were entitled to levy royalties that were limited in form and level in ways that effectively precluded collection of more than a small proportion of the economic rent, and the terms were laid out within 'concession agreements' that were negotiated at the beginning of investment and were typically current over very long periods. This attitude was supported ideologically by profession of belief in the sanctity of contracts between investors and governments; in practice it amounted to accepting the rule that finding was keeping. Most concession agreements related to colonial states, the metropolitan controllers of which accepted this general ideology and the sanctity of concession agreements. Independent states in developing countries were constrained in their management of relations with mineral investors by the possibility of diplomatic and military intervention by the investors' home governments.[1] These attitudes are now of little importance in the real world of government–investor relations, although they sometimes inhibit attempts by inexperienced administrations to increase charges for access to resources.[2]

The early work of Penrose on petroleum concessions in the Middle East seems to be within the modern 'Commonwealth' tradition, while acknowledging the presence of circumstances in which the main propositions of what became the 'North American' model would be valid.

Penrose (1959:245) noted that, if there were effective competition in the supply of investment, the share of profits from oil investments accruing to the government would be pushed 'up to the point where the company obtaining the concession retained only "normal" profit which, for our purposes, may be defined as the amount of anticipated profit that would just induce the company to enter or continue to invest in the oil industry of the country'.

Penrose noted that the supply of investment in Middle East oil in 1959 was characterized more by monopoly than by competition. Much of the valuable oil property had been allocated to companies by past concession agreements that yielded little of the resource rent to governments, and companies were able to some extent to resist demands for renegotiation. Political intervention by the Western governments strengthened the position of their national investors in negotiations of new concession agreements. Host government officials often had little information on the rent value of new concessions. Only a small number of corporations in the world possessed the financial, technological, and marketing strength to undertake large-scale petroleum production in the Middle East. Collusion among major petroleum companies further reduced competition in the supply of investment.

[1] The possibility of strong diplomatic or even military intervention by governments to secure access of its nationals on favourable terms to the resources of other countries probably still exists for those mineral resources that are seen to be of great national importance to major world powers—for example, Middle East oil to the USA. However, this does not appear to have been important in determining the terms of agreements between governments and investors in recent years.

[2] The attitudes survive in the occasional statements of beliefs by mining companies and commentators on the mining industry, to the effect that the proper distribution of resource rent between governments and investors depends on 'value judgements'.

The arrival on the scene of new investors tended to increase competition, but this remained peripheral in many areas of the Middle East. The presence of monopoly in the supply of investment meant that resource rent was in practice allocated through a process of bargaining within the context of bilateral monopoly. The bargains that were struck from time to time were affected considerably by the strength of domestic political pressures in host countries for increased rent shares.

Penrose (1959:249) observed that efficiency in petroleum production and distribution required large-scale operations. A large number of producers each operating on a smaller scale might possibly generate less resource rent, although the increased competition would increase the government's share of it.

Vernon's (1967) early articulation of the 'North American' model focused on the conditions upon which resource concessions were made available in developing countries. The developing country was distinguished from the advanced industrial country by the absence of general laws and policies, including laws and policies over taxation, within which the exploitation of natural resources could proceed. Thus a system of law was developed in the process of negotiating a 'concession agreement'.

The bargaining position of the investor was strong in the negotiation of the first concession agreements in a developing country. The host government was typically ignorant of the value of natural resources under its control. Besides, early investments were associated with such great uncertainty that the rent value of many concessions was *ex ante* very low. Hence early concession agreements typically tied up large resources for long periods with very small fiscal levies.

With the success of investments under the early concession agreements, the environment changed considerably. The host government began to develop general laws, policies, and administrative approaches for the natural resource sector. The risk associated with new investments was reduced considerably. New concession agreements were on fiscal terms very much more favourable to the government. The conclusion of the new agreements contributed to the formation of beliefs in the host countries that the early concessions had been 'unfair', and pressures developed for the renegotiation of concession agreements. Investors opposed changes in the terms of agreements with the exaggerated hostility of negotiators on the defensive in a weak position. A bilateral monopoly existed between the government and the established concessionnaires, and agreements were changed in favour of governments, usually in a series of steps.

The renegotiation of concession agreements was occasionally blocked by pressures from the home governments of investing companies. Although the basis of the resistance was weak in international law

(Brown, 1977), it was occasionally supported by diplomatic pressures of considerable importance (Lowenfield, 1967). However, in the post-war period the pressures became much less effective, and by the later 1960s governments' 'rights to renegotiate' had been established for all practical purposes.

In Vernon's exposition of the history of concession agreements, the bilateral monopoly was established by the granting of a concession with vested rights over resources. It did not depend on the presence of monopoly in the supply of new investment. But Vernon's analysis generated a substantial literature on the general 'bargaining model' of government–investor relations and this model came to be applied to new, as well as to established, concessions, and independently of the state of competition in the supply of investment.

In the 'bargaining model' articulated by Mikesell, Bartsch and Bahrman (1971), Moran (1975), and Smith and Wells (1975), the concession agreements upon which resource projects are launched are a matter of market place bargaining. There may be an element of monopoly in the supply of investment, but this is not thought to be essential to the analysis. The initial terms of the agreement are highly unfavourable to the government, because of the investor's aversion to risk and the high risk (including a large element of what is described in this book as sovereign risk) associated with the investment. When the investment has been proved successful, the terms of the original agreement appear to be unfair, and the host government seeks to renegotiate them. At issue is the distribution between the government and the investor of the whole of the surplus income above the current expenditures necessary to produce output. This surplus is referred to as 'rent' (Mikesell, Bartsch and Bahrman, 1971:33), although it includes quasi-rents of exploration and mine development. The company, in resisting renegotiation, asserts the sanctity of contract and perhaps seeks support from international law; but the most important constraints upon the host government's collection of the 'rent' are its supposed effect on the perceptions of foreign investors in future, the government's ignorance about how far it can push the bargain without jeopardizing current or future production, and for short periods the foreign corporation's use of its wealth to manipulate domestic politics. The main articulators of the 'North American' model believe, however, that political interference does not secure the investor's interests for long.

The bargaining model literature is characterized by the focus on the distribution of a short-term surplus, corresponding roughly to accounting profits, rather than to the collection of an economic rent. The domestic political process is important to the allocation of the surplus: the government's share will be larger if it is able to mobilize popular support around the issue, or if it is under challenge from radical nationalist or socialist pressure groups. The government's bargaining power tends to rise over time with the mobilization of political interest, increased knowledge, and improved negotiating and administrative skills, so that successive renegotiations increase the government's share of the surplus. The government is more likely to demand renegotiation and to make severe demands in renegotiation of concession terms when the realized profitability of investment is very high.

There is on public record advice that is strongly influenced by scholars of the Harvard Institute of International Development to the governments of Indonesia (Smith and Wells, 1969), Papua New Guinea (Wells, 1973), and Bolivia (Gillis *et al.*, 1978) on mineral rent policy. The advice is characterized by the emphasis on shares of profit rather than on extraction of resource rent; by the promulgation of the idea that the fiscal terms of resource concessions are fixed only for brief periods between the successive renegotiations of agreements; and by the idea that a government's 'bargaining power', and so the share of profits that it can obtain, is determined by the practice in other parts of the geographic region of which the mineral-owning country is part. The Harvard advice to the Indonesian and Papua New Guinea governments is discussed further in chapters 13 and 14.

The 'North American' model appeared to provide a reasonable description and explanation of the historical experience of concession agreements, in particular from the 1950s to the early 1970s. However, it was only of limited help to governments seeking to develop appropriate mineral rent taxation policies or to investors in assessing the fiscal terms that were likely to apply to resource developments. It emphasized rather the indeterminacy of the process through which such terms were set. It seemed to establish that investors' fears about instability in concession terms were fully justified and based on an inevitable but unpredictable historical process. Investors might yet judge that the commitment of funds to mineral projects was worthwhile, but such commitments should be made after generous allowance for sovereign risk.

Several authors who have accepted the 'North American' model as an apt description of historical experience have sought to draw normative conclusions about types of agreements that would be less vulnerable to instability in future, and these attempts have led them close to the ground of the 'Commonwealth' model. Uncertainty about fiscal terms was recognized as inhibiting investment. If periodic renegotiations were inevitable, then it would contribute to the reduction in business uncertainty if governments could be persuaded to seek changes in fiscal terms only in agreed circumstances, selected for their correspondence with periods during which strong pressures for renegotiation could be expected. As early as 1966, Johnson had suggested that it would be sensible to provide for the review or revision of concession agreements in one of three situations: after a fixed number of years; upon the investor's recovery of his outlays together with a 'reasonable' return on outstanding balances; or when certain specified circumstances changed. A similar yearning for stability appears in Oliver's comments on the paper by Vernon which launched the North American model:

> Ambassador Covey J. Oliver remarked that . . . we have noted the pull and haul between legitimate expectations that contracts be performed as agreed upon and tendencies offered to the contract obligation that develop late in the performance life of the arrangements. He found himself wondering why we do not do more regarding the pre-planning of the various phases of the life of a long-term contract. If the contract itself could be structured so as to provide for different degrees of participation at different stages of the relationship, this probably would reduce greatly the tendency to cast off contract obligations. . . . [*American Society of International Law – Proceedings* 1967:106]

Vernon rejected this suggestion (*American Society of International Law – Proceedings*, 1967:106), because the concessionnaire and the host government 'do not envision the day of the divorce at the outset'. Other Harvard proponents of the 'North American' model have adopted a similar position (for example, Gillis *et al.*, 1978:116).

Mikesell is an author whose early works lie within the 'North American' tradition, yet who in recent years has come to emphasize the cost of instability in fiscal terms, both to investors and to host governments, and has suggested the need for automatic review procedures (Mikesell, 1975:33). Mikesell's later work recognized that the circumstance that is most likely to prompt pressure for renegotiation is the emergence of high profitability from projects that *ex ante* appeared to be possibly marginal and that were therefore subject to light fiscal charges. His first remedy was for the parties to agree to the review of the terms every five to seven years after the full recoupment of the initial capital investment. If at the time of the review the past and projected net cash flows to equity-holders suggested an internal rate of return above some agreed level, the agreement should be renegotiated in favour of the government. Otherwise, the original terms should be maintained. Later he argued that the application of the resource rent tax as defined by the present authors could lead the required adjustments to be made within the terms of the initial contract, without introducing the uncertainties associated with renegotiation (Mikesell, 1981).

The recent proposals by Mikesell take the discussion within the bargaining model tradition into the mainstream 'resource rent' or 'Commonwealth' approach. Authors within the 'resource rent' tradition have sought solutions to the acknowledged practical problems associated with high uncertainty and government ignorance at the time concession agreements are negotiated, and with the subsequent instability of agreements when terms later seem to be exploitative, through the use of internationally funded public exploration in order to reduce uncertainty about geological aspects of the project (Faber and Brown, 1980), or through the use of fiscal arrangements that compress as much as possible the dispersion of the probability distribution of after-tax outcomes to investors (Herfindahl, 1974:242–6; Garnaut and Clunies Ross, 1975).

Herfindahl (1974:242–6) observed that the expectation that concession terms might be changed by the government raised the supply price of investment and reduced government revenue, whether or not the changes were implemented. The present authors' resource rent tax proposal (see chapter 7 below) was thought to reduce this problem to the greatest possible extent through the automatic increase in the government share of revenues when rates of return corresponding to the supply price of private investment had been realized: the arrangement might possibly appear to the investor as credibly stable; and the government did not leave the investor with a windfall through the failure to collect revenues that had not been anticipated by the investor.

The 'resource rent' approach depends on the presence of competition in the supply of investment at the time at which investment funds are committed. The presence of monopoly creates the indeterminacy in the allocation of resource rents that is analysed in the 'bargaining' model. However, in many of the cases for which the 'North American' approach was developed the monopolistic relationship had its origins in the law of the host country rather than in the investor's exclusive possession of technology or managerial or marketing skills that were essential to the exploitation of the deposit. It is only where the investor's bargaining power arises from the latter source that an unavoidable bilateral monopoly exists between the government and the investor.

Competition and Monopoly in the Supply of Investment

The owner of a resource (which is a government in most jurisdictions) is in a monopolistic position with respect to the sale of access to that resource. Where the owner is seeking to sell access to the resource in a competitive market, he is in a position to exact charges equal in value to the expected economic rent of the resource. However, where the owner faces monopoly or oligopoly in the supply of investment, a relationship is established that is analogous to bilateral monopoly, with rent charges determined through a bargaining process, and with the economic rent shared between the owner of the resource and the investor.

The distinction between 'market' and 'bargaining' contexts for the setting of rent charges is of considerable importance. It is worthwhile examining the conditions under which each would be relevant. For purposes of exposition, it is taken that monopoly in the supply of investment, rather than the presence of competition, needs to be explained.

When government officials are sitting opposite mining company officials at the negotiating table, and discussing the terms of an agreement, the situation certainly feels like one of bilateral monopoly. But whether or not a real situation of bilateral monopoly exists depends on whether, if the investor abandoned the negotiations, the government would be able to deal with other investors and receive benefits comparable to the best that the first company, negotiating uncertainties aside, would be prepared to offer.

Monopolistic elements might enter the supply of investment for one of two types of reason. The first relates to normal considerations of industrial organization: barriers against access to markets, technology, and resources; or one firm's ownership of the range of managerial qualities that are necessary to establish a mineral investment. The incidence of this source of monopolistic organization may vary between the various mining industries, between regions, and through time. The second relates to the investor's tenure in the relevant mineral properties and accepted morality within the mining industry about such property rights: if the investor has an exclusive right (through holding a mining lease or through other means) to develop some resource, the government may not be prepared to deal with other potential investors; or else they may not be prepared to deal with the government.

In some of the mineral industries, a small number of vertically integrated firms have dominated production and trade. Governments wishing to have deposits of the relevant minerals developed have had to deal with one or other of these companies. The tendency towards monopolistic organization has been increased by the regional concentration of the activities of some of the major firms.

Monopoly in some of the mineral industries sometimes occurs because one firm has exclusive access to new technology, when recent innovation has led to the development of previously uneconomic mineral deposits. Offshore natural gas and deep offshore petroleum provide recent examples, and deep sea mining for manganese nodules may soon provide another. In a few cases, monopoly in extraction arises from monopoly in processing and marketing. (Trade in bauxite and alumina provides a possible example.) Whenever there is monopolistic organization in mineral production and trade, it complicates the setting of charges to collect the rent.

Even when the investor has no monopolistic control of technology or access to markets, the bargaining model of government–investor relations may be applicable if the government is constrained for other reasons from dealing with other companies. This situation arises most commonly when a company has undertaken exploration expenditure and discovered a mineral deposit before the settlement of the terms and conditions of rent charges. Indeed, the company may actually have been granted a right to mine at the time of exploration and before agreement on fiscal arrangements. More commonly, it will merely rely on a 'prospecting authority', or some similar arrangement that legally carries no exclusive right to develop the property, but does carry an implication of tenure, or at least of some priority in negotiations. Any requirement that the government should reach agreement with a single company establishes a bilateral monopoly. Alternatively, if the government accepts the implication of priority but not of absolute tenure for the first investor, it must offer terms that, to investors as a group, appear to give a sufficiently large share of income to compensate for exploration expenditure. If the prospecting authority did not carry some implication of tenure, the expected private benefits of exploration would not justify spending upon it. The government is forced either to undertake exploration itself or to grant rights to mine with prospecting rights. The granting of absolute tenure with exploration rights would establish the bilateral monopoly for any future investment negotiations in which agreement was sought during or after the exploration phase. However, a competitive environment could be maintained over the fixing of fiscal terms and conditions if agreement on them were reached at or before the start of exploration.

A second situation in which bargaining of a bilateral monopoly kind is likely is one in which, because of inexperience, the government is unaware that there are other possible investors than the one with which it is currently negotiating, or in which the government does not trust its capacity to enlist them. Alternatively, the government may be unaware that the terms offered by one investor are out of line with international market conditions. The remedy for this situation is straightforward if the negotiators seek appropriate advice. This source of unnecessary bilateral monopoly has become much less important in developing countries since the early 1970s, when specialist technical assistance began to be provided by several of the international agencies.[3]

[3] First in the field was the Technical Assistance Group of the Commonwealth Secretariat, which continues to be a major source of advice on mineral tax regimes to Commonwealth developing country governments. The United Nations, through the Centre for Transnational Corporations and through the Natural Resources Division, has technical assistance capacity which it makes available to all developing countries. The various arms of the World Bank and especially the Energy Department now also provide some advisory services to developing country governments.

Ironically, the problem of ignorance about the competitive environment is probably now as great or greater within industrial country governments as in the developing countries, especially when mineral leasing powers lie with state or provincial governments.

Of course, there is always an element of bilateral monopoly when the government is dealing with the operator of an established mine, because of the dislocation of production and the shock to business confidence that would be associated with the departure of the investor. Thus, much of the analysis of the 'bargaining model' may seem to be relevant to the renegotiation of unsatisfactory arrangements for established mines. But so long as there is competition for new investment, the 'resource rent' model can provide more reliable guidance even in this situation. A government seeking to impose on an established investor terms that were similar to those that might be developed within the 'Commonwealth' model for application to new investors would incur much less cost in terms of increasing sovereign risk than if the new terms were chosen arbitrarily.

Early or Late Agreement

It can be argued persuasively that the amount of rent that the government is able to secure from a mineral deposit will be maximized by the longest possible delay in reaching agreement on fiscal charges. However, the government is likely to be forced into making early agreements if it wishes to secure private investment in exploration and feasibility studies, at the same time as maintaining competition for the supply of investment while negotiations are proceeding.

It is sometimes tempting for a government to delay reaching agreement with an investor because it recognizes that successful exploration represents sunk expenditure, so that recovery of the cost of exploration is not essential for inducing development investment in a particular project. However, if fiscal arrangements arising from late negotiations do not allow a return on exploration expenditures, no exploration expenditure will be forthcoming in future.

A stronger case for delay hinges on the company's risk aversion. The earlier an agreement is reached, the greater the variance that will have to be supposed in possible investment outcomes. Additional expenditure on exploration and feasibility studies reduces the perceived variance of outcomes, although it may raise, leave unchanged, or lower their expected value. The reduced variance of outcomes, however, lowers the supply price of investment and this tends on balance to raise the mineral rent available for collection by the government.

However, this statement of the benefits of delay ignores the costs of the prolongation of uncertainty about the terms of the agreement, namely its operation on the company's risk aversion to raise the supply price of investment and in particular to reduce the willingness to explore. This effect is compounded by the common assumption by companies that the government will take as much revenue as it possibly can from the project. The insecurity of tenure greatly increases uncertainty, unless the government virtually eliminates it by convincing the company that it will not allow other companies to develop the deposit, thus creating a bilateral monopoly.

The greater uncertainty over the terms of an agreement, in fact, seems likely to eliminate any possibility that delay will increase the amount of economic rent, unless the government either gives what almost amounts to security of tenure with an exploration lease, or else finances exploration expenditure itself. A government seeking to maximize receipts from charges on mineral rents must reach agreement before there are major private exploration expenditures if it is to work within a competitive framework in the setting of rent charges. Even if bilateral monopoly is inevitable, early agreement will still be desirable because otherwise investors may be less willing to take the first steps in the face of any given set of possible outcomes.

The cost of early agreement, in terms of forgone rent, will depend heavily on the form of rent charges. It will be less the more expected rent payments are concentrated on the higher end of the probability distribution of expected outcomes, and we believe that with an appropriate form of tax it can be virtually eliminated.

Stable or Flexible Rules

The 'North American' model supposes that there is an inexorable tendency towards renegotiation of agreements as events unfold. However, the later Mikesell version of the model admits that the likelihood of instability depends to come extent on the form of the fiscal arrangements. In addition, where stability of agreements can be shown to benefit governments, the absorption of this reality into government consciousness may lead to greater commitment to stability in agreements as a matter of policy.

It cannot be denied that political forces will sometimes force renegotiation of agreements, whatever the balance of advantages of stable as against flexible rules. Presumably, agreements that contain automatic adjustments of investor–government profit shares with rates of return on investment will be less vulnerable to irrational political pressures for change.

A company facing the possibility of instability in its fiscal terms will discount the possible outcomes that it perceives under proposed arrangements in order to take account of the various possible changes that might be made in future renegotiations. There will probably be some systematic downward over-adjustment to anticipated outcomes, since companies will fear the worst. It is quite rational for a company to make major downward adjustments in its assessments of the higher end of the probability distribution, since these are most likely to prompt renegotiation. Given that companies must almost inevitably discount such high returns in their investment decisions, there are good grounds for governments to make prior arrangements for collecting these high returns rather than pretending to make an agreement and then renegotiating it. The possibility of drastic adjustment to terms of agreements is likely to increase substantially the uncertainty about investment outcomes, and hence to raise the supply price of investment and to lower the expected rent available for collection.

It follows that governments have a strong interest in stable arrangements, which are perceived to be stable, and which include charges that operate most heavily on the higher end of the probability distribution of outcomes. Where inherited arrangements for major resource investments diverge markedly from normal competitive standards, it may be possible to renegotiate an agreement and yet maintain a fairly high perception of future stability, provided the renegotiation is conducted within principles that themselves appear likely to stand the political tests of time. Even if renegotiation is thought to be wise on economic grounds or necessary on political grounds, there is a strong case for announcing general rules directed at the appropriation of mineral rent and designed to be stable rules, and for bringing established mineral producers into these arrangements.

To cover all possibilities, we should mention one situation in which changes in the rules at a particular historic point might be profitable for some countries individually, although they would be damaging to resource-owning countries taken as a whole. When a government has formed the view that its major mineral deposits have already been discovered and brought into production, or that it is no longer obliged by its shortage of domestic skills and capacities to require direct foreign investment in the mineral sector, it may be able to increase its revenues by plundering quasi-rents of established producers without fear that this will lead to lower rent receipts in future. Recognition of this possibility presumably raises the supply price of investment through reducing the pay-back period of investments in all countries, and especially in countries whose governments might be expected to find themselves in this position.

General versus *ad hoc* Rules

Some countries apply general rules to all mineral investments within each of several categories. Others negotiate separate arrangements for each project. In practice, most governments apply some general and some special rules, with the relative importance of general and *ad hoc* rules varying from country to country.

The proponents of the 'North American' model have generally favoured the negotiation of *ad hoc* arrangements (Gillis *et al.*, 1978: 116), but there are good reasons why this may not be the most satisfactory approach. *Ad hoc* arrangements, in principle, have the advantage that they allow the collection by the government of a higher proportion of the rent from some projects without similarly burdening others that are not able to support such imposts. This possible advantage is reduced severely in practice by the large amount of information and analytical capacity that the government must possess if it is to extract the mineral rent project by project. Certainly the advantages of separate arrangements for each project are likely to be negligibly small if agreement is reached before there have been major exploration expenditures and feasibility studies.

One further advantage of general rules is that their application requires less detailed information on individual projects and less sophisticated negotiating capacity within the government. They also avoid the administrative costs and delays of project-by-project negotiations.

These advantages are additional to the major advantage already discussed above over the choice between fixed and flexible terms: that fixed rules increase rent by reducing the investor's uncertainty. With general rules, the investor knows, even before substantial outlays of executive time have been made, what the fiscal conditions are.

General rules must always be announced in advance of investment, while *ad hoc* rules can be negotiated either in advance or in the course of investment. The advantages of *ad hoc* agreement are more likely to be realized if negotiations are completed after exploration information is available. Thus *ad hoc* rules are likely to be associated with late negotiations, and by implication with conditions of bilateral monopoly. But, even where bilateral monopoly exists, it is at least possible that the government would be able to establish its strongest bargaining position by announcing general rules and refusing to negotiate any changes in them, so long as its position is credible. This approach can draw support from the literature that discusses bilateral monopoly in mining: various writers argue that the government's share of the 'surplus' is greater when it is clear to the investor that the government has no scope politically to make concessions (Penrose, 1959; Mikesell, Bartsch and Bahrman, 1971), and the government may be able to place itself in this position through the announcement of firm and unalterable policy.

It is possible for general rules to be applied, but to differ among the different parts of the mineral industry, in so far as risk and other factors affecting the supply price of investment are thought to be different. Another variant is for general rules to be set, but to be varied over time with factors that are thought to bear on the supply price of investment. In the latter case, the rules applicable to a particular investment might either remain those of the historical period of its establishment or move with the general rules from time to time. It is also possible for *ad hoc* negotiations to determine the details of fiscal arrangements, but for the negotiations to be conducted within a clear, pre-announced general framework.

Ad hoc rules are probably less likely to be stable rules, since any cases of especially favourable treatment to an investor may be difficult *ex post* to justify politically. Against this, general rules that fail to extract a large part of the rent from a very rich mineral deposit may appear suspect to investors, and renegotiation can not be excluded unless the form of the rules allows automatic adjustment of company and government share to profitability.

The Relative Merits of the Models

It is arguable that the 'bargaining' model's power is specific to the particular historical period for which it was developed: the immediate post-colonial period in which the governments were seeking to change unsatisfactory concession agreements related to established mines the owners of which held exclusive and hence monopolistic tenure over particular valuable resources. Host governments operate in a more competitive environment when they issue licences in new areas. And there is evidence that the industries that were characterized by monopolistic control of technology or markets in earlier periods are becoming more open, as seems to be the case with petroleum (Penrose, 1959) and copper (Moran, 1975).

The 'North American' approach is not very helpful in explaining or predicting any particular bargaining *outcome*. There are some situations in which a government would undoubtedly forgo mineral rent that it could otherwise collect if it looked to the prescriptions of the 'North American' model when it was, in fact, selling access to its resources in a competitive market. Moreover, as discussed in the preceding section, the government may establish the strongest possible bargaining position for itself in bilateral monopoly by acting as if it thought there were competition in the supply of investment and refusing to depart from general policies. It may be the case, however, that a monopolistic supplier of investment to some industry is in a strong position to hold out for special terms and conditions. Whether this situation exists is a question of fact to be resolved in the individual circumstances of each country and industry. When general rules have been promulgated, for example in relation to taxation, a government wishing to avoid the loss of useful investment in an industry characterized by monopoly might be moved to offer special arrangements in other areas, for example in relation to the terms of government investment.

While the existence of bilateral monopoly suggests the relevance in some circumstances of the 'North American' model, it leaves unanswered the question of whether there is any need for the bilateral monopoly to continue. There is no need for it to continue if it depends simply on ignorance, but where there is ignorance about the real nature of the investor's power, delay in negotiations may be appropriate until the government can inform itself better. There is also no need for it to continue if it exists only because of one company's formal or implied tenure in access to the resource at the time at which the fiscal charges are settled. In these cases, the government's revenue from its minerals can probably be increased by a revision of the whole system of mineral rights so as to remove the exploring company's tenure in access to the deposit before the terms have been settled. This revision would suggest others: the fixing of terms before there have been major exploration expenditures; hence the application of forms of rent charges that take a particularly large share of profits when profits are high and thus are suitable to very uncertain outcomes; and hence also commitments to more stable fiscal arrangements and perhaps to rules of general application. The argument of earlier sections of this chapter suggest that such revision of policy on mineral rights and taxation would be worthwhile in order to avoid leaving an unnecessarily large share of mineral rent with the investor and generally to avoid paying an avoidably high penalty to overcome the company's aversion to risk.

Summary

To summarize our own view on these matters, bilateral monopoly and competition are both models that might in certain conditions describe the position of government and investor in negotiation over the terms of access to minerals. At the present time, however, we believe that the situation is intrinsically one with considerable elements of competition, and that monopolistic elements, where they exist, are often introduced arbitarily by conventions that give the exploring firm a moral right to mine and by delaying agreement until much exploration investment has been done.

Accepting this view, we favour that we have called the 'Commonwealth' as against the 'North American' approach as a general guide on dealing with the fixing of terms and conditions.

Provided a certain condition can be fulfilled, we believe that, so long as there is a competitive element, government will achieve the best results for its own revenue by clarifying the terms early rather than late in the exploratory process, by making them fixed rather than subject to renegotiation, and by laying down rules that are general rather than devised separately for each project (though the general rules may vary from one industry to another and may contain within themselves automatic adjustments for changing circumstances). It is our view that the same approach will be most rewarding even in many situations in which there is monopoly in the supply of investment. The condition that must be fulfilled for these propositions to hold is that the arrangements themselves should render the amount of tax revenue highly sensitive to realized profitability.

If this condition can be observed, then, our strong tendency would be to favour a system in which rights to explore are auctioned among reputable applicants or, less desirably, given to the first reputable applicant, and in which those rights once granted imply a right to mine subject to fiscal provisions and other requirements that are fixed and known in advance of exploration. An alternative, perhaps even more satisfactory if it can be arranged, without loss of efficiency, is for the government itself to undertake exploration, and to auction mining rights with pre-fixed fiscal provisions when information is freely available. In either case, the fiscal provisions, we repeat, must be such so to ensure that a very great proportion of any exceptionally high returns can be collected in tax.

7

TYPES OF RENT TAX AND CRITERIA FOR EVALUATION

The extraction of mineral rent is generally recognized as requiring some form of tax, charge, or fee additional to ordinary business taxation on the project. The main forms of levy that have been, or might be, used for this purpose will now be listed and their characteristics considered. Tables 7.1–7.3 at the end of the chapter will schematically summarize their relative strong and weak points. These will be expounded in the following three chapters.

In the tables and in chapters 8, 9 and 10, a number of criteria will be used to judge the various tax systems. The criteria are interdependent, and we start by explaining them and their interrelationships.

The immediate obejctives that a government may reasonably have over its taxation are at least four, and we shall discuss them as if they were its actual objectives. First, the government will want the tax system to be neutral, that is, not without some definite compensatory purpose to affect the allocative decisions made by various agents in the economy. This will not preclude deliberate measures to compensate for what are called 'externalities', that is to say, cases in which the interest of private agents conflicts with social efficiency. But such correctives need to be deliberate and purpose-built; they correspond only by rare accident with the unintended allocative effects that many taxes have. Second, the government will want to maximize its revenue from the tax system. Third, it will want to reduce the variability and uncertainty of the revenue outcome for itself: in a broad sense, to minimize the risk. Fourth, and probably least important, it will have some preferences over timing of the revenue and may want to avoid undue delay in its receipt.

We proceed to analyse the second of these objectives further. What factors will affect a system's capacity to raise revenue? First, neutrality itself will be relevant. On the whole, a more efficient allocation of resources within the government's jurisdiction implies a higher real income in that economy at large, and this on the whole means more taxable capacity. A system that deters investments from overseas that would otherwise go forward clearly reduces taxable capacity. One that encourages over-investment in particular industries or areas at the cost of under-investment in others has the effect (by definition) of reducing real income, leaving less to be taxed overall as a consequence. It is not necessarily the case, however, that, where a tax gives incentives to over-investment *from abroad*, it decreases its own government's revenue base. Yet in general neutrality can be regarded as contributing to revenue in the sense of increasing the revenue base.

Three characteristics of a tax system besides its neturality may be important to its capacity for raising revenue. One is its tendency to reduce perceived risk to the investor. This in turn can be seen as a resultant of two further characteristics: the perceived stability of the tax system (the likelihood that it will persist unaltered), and the intrinsic tendency within the system to reduce risk. If there is aversion to risk on the part of investors, that is to say, if they need to be rewarded for taking risks, it follows that, the less risky any project is, the more of their profits investors will be prepared to surrender in the form of government revenue before they come to regard a project as ineligible. Thus, the more the tax system reduces risk, the more severe it can be made consistently with any given level of investment. Thus, either an increase in the perceived stability of the tax system or an increase in its tendency to compress the probability distribution of returns (or to reduce the probability of very low returns, if that is the relevant criterion for reducing risk) will increase the revenue-raising potential of the tax. There is, however, a second test with a bearing on the revenue-raising capacity of a tax system. This is the question of how much revenue can be raised on any given project when the marginal rate of taxation on income or cash flow is at the maximum consistent with incentives to efficiency (a question referred to here as the 'maximum-rate test'). Third, its adaptability to realized profit will have a considerable bearing on its capacity for raising revenue when applied in general form, that is, under uniform rules across an industry.

It is probably simplest to take each of these factors individually: neutrality, stability, intrinsic investor risk, the 'maximum-rate test', adaptability, government risk, and revenue delay. In addition, however, there are practical questions of implementation that deserve to be considered. One, which we call *imposition*, covers the difficulties in satisfactorily fixing the 'parameters' (such variables as rates or amounts of duty). Some systems can be made reasonably satisfactory but only if there is considerable knowledge and sophisticated estimation on the part of the authorities. Such knowledge may be costly or simply unavailable. A second we call *administration*. This term is taken to cover the cost and difficulty of making the tax enforceable. This is likely to be the greater if a radically new system has to be devised since experience may be needed before loopholes are recognized and closed; and it is greater with some forms of tax than with others.

The factors listed are to some extent interdependent. The following three points should be made.

First, there will in general be a tendency for the risk and delay borne by the government on particular taxable projects to vary across tax systems inversely with the risk to the investor. Keeping the government's risk and delay low and its revenue high would to that extent be conflicting objectives. On this assumption, Leland (1978), supposing governments and investors to be risk-averse or potentially so, treated the problem of getting the best tax from the government's viewpoint as one of appropriate risk-sharing between the two parties. The more risk-averse the government is in relation to the investor, the more its interest is to choose taxes that push the risk on to the investor with the loss of potential revenue to the government that that entails. The less risk-averse relatively the government is, the more its interest is to bear the risk itself, with a corresponding increase in its potential revenue. Risk in practice, however, is somewhat elusive as a variable to be measured. Government and investor may not see the risk associated with a project in the same way. One of the two may measure risk as the probability of returns below a certain level; the other may measure it simply by the dispersion of the probability distribution. Furthermore, investor risk is not the only factor affecting the capacity of a tax to raise revenue for the state. A form of tax that is more risky to the investor may conceivably perform better on the other main test relevant to its revenue-raising capacity.

Second, both neutrality and revenue realization depend on how easy it is to impose the tax in something approaching ideal form and how effectively it can be administered. Generally we shall discuss the neutrality and revenue realization questions as if imposition and administration were ideal, but we recognize of course that in practice the conclusions on the first and third questions should depend partly on those on the fifth and sixth.

Third, the concept of neutrality needs to be clarified in its bearing on risk. One view of appropriate policy towards the risk associated with projects and the risk aversion of investors is that they should be regarded jointly as a source of externality, that is to say, as a factor that tends to distort decisions from the socially optimal. On this understanding, it would be desirable for investors to take decisions as if all investments were riskless and to respond simply to the expected returns. This follows from the view that the risk of individual projects involves no cost to the public, which can expect to gain on the swings what it loses on the roundabouts. In general we favour this view, though we realize that there are some cases in which individual projects are so large that the public (and the state on its behalf) has to discount for the risk associated with them. On the other hand, it could be argued that when we talk about a tax system's being *neutral* among projects we should imply that it leaves the projects (risk and all) in the same order of eligibility that they would have in the absence of tax. To insist upon this interpretation, however, would make judgements about neutrality difficult and would reduce the relevance of neutrality as a criterion of evaluation. The effects of a tax on risk and on expected returns are two separate dimensions, and we know of no plausible assumptions that for our present purpose would enable us to suppose how investors' decisions balance one against the other. In addition, it is difficult to determine whether, and if so in what conditions, a particular tax is likely to alter the relative riskiness *among projects*. Hence, when we talk of a tax system's being neutral between projects, we imply simply that it will leave the order of their expected yields unaltered. We treat its effect on risk as a separate question and assume that risk reduction to the investor is in itself socially useful, not only because it tends to increase the revenue that can be collected from the industries concerned, but also because on general allocative grounds it tends to increase social product.

We are left then with nine criteria: (i) neutrality, (ii) stability, (iii) intrinsic investor risk, (iv) maximum-rate test, (v) adaptability, (vi) government-risk, (vii) delay, (viii) imposition, and (ix) administration. The second, third, fourth and fifth are important because of their bearing on the capacity of the taxes to raise revenue for the state. The first is independently important but also has a considerable bearing on the realization of government revenue. The sixth, seventh, and ninth criteria are less vital than the first five except in special circumstances: the country with small administrative resources may have difficulties about adequately framing and enforcing certain kinds of tax; and the country with one large mineral project may rightly be concerned about the risk or the delay to its revenue that certain forms of tax entail. Yet generally these problems are superable. Imposition dilemmas, the concern of the seventh criterion, are somewhat different. They may be critical in making a tax good or bad, efficient or inefficient. Yet again they are likely to be far more of a concern in countries with meagre administrative resources. Put simply, the questions to be asked of each are: how far it avoids biasing the choice between industries, projects, amounts extracted, and techniques; to what extent taxation arrangements in force at the time investments are made can be expected to remain unchanged for the life of the mining project; how far the system, if maintained, reduces risk to the investor; how much revenue the system can be expected to realize when the maximum marginal rate of tax on net receipts is as high as is consistent with maintaining incentives to efficiency; how closely tax revenue adjusts to realized profitability; how far, for a given expected revenue, it reduces the dispersion of possible outcomes and avoids delay in receipts from the viewpoint of the taxing authority; how easy it is to discover the tax rates appropriate for maximizing government revenue or social income; and how easy it is to administer effectively.

The premiss of the discussion is that some extra tax, above and beyond the ordinary profit tax, must be imposed so as to collect for the public some part at least of the rent or property value of a mineral resource that is handed over to a private extractor. Six main methods of attempting to do this (with three further variants) will be considered, with potentially three modes of imposition for each method. The three methods of imposition are: 'general' (that is, imposed on the same terms over all projects in the industry); 'negotiated' (the terms negotiated for each project separately); and 'auctioned' (with terms again determined for each project separately but with some important variable fixed by auction). It will be seen that one form of imposition may have a particular comparative advantage for certain types of tax.

Types of Rent Tax

1 *Fixed fee (FF)*

Mining rights may be sold by charging a fixed sum of money independently of the outcome of the investment or indeed of whether there is any investment at all. The cash need not all be collected at once; collection may be spread over time, but the principle is the same.

To impose a FF in general form is most appropriate when the government has little idea of the value of what it is selling or the investor of what he is buying, and where the actual proceeds will be hard to assess in any case. Thus it may be suitable for the sale of exploration, as distinct from extraction, rights and for small-scale unorganized mining activity such as occurred in the Californian and Victorian gold rushes.

Negotiating or auctioning the fixed fee is a means of relating it to the actual value of the resource, at least as it appears to the investor. Setting of fixed fees determined by auction ('cash-bonus bidding') is used in the US federal jurisdiction as a means of distributing the rights to extract offshore oil and gas, and also onshore oil and gas from established fields (Dam, 1976:147). It is commonly regarded as a successful method off the US mainland (Mead, 1977). Some of the evidence is given below in chapter 18, (pp. 307–13). Doubts, however, have been raised about its effectiveness in collecting rent from the larger Alaskan oil fields (Gaffney, 1977) because the size of the sum required to reflect the value of the resource would in such cases limit the competition. Auctioning in principle leads the investor to set his own value on the resource, and it charges him an amount equal to that value.

Under conditions approaching certainty over the productive power of the asset, and over stability in government policy and a high degree of competition among investors, the auctioned FF would be the ideal method, completely neutral and collecting as much as the investor is willing to pay. It is less obviously useful if these conditions do not operate. Knowledge is in fact deficient in all the mineral industries. In mining for metals, exploration rights are normally given at a point at which there is very little knowledge of the extent of deposits and hence when those rights would not command a price in any way comparable with the true rent of a profitable project. So there is little value in auctioning exploration rights as a principal method of collecting rents. With metals, however, exploration is normally taken to give a company first claim over actual mining in the area. If this were not so, exploration would not be worth undertaking. So auctioning of mining rights *after* exploration is not acceptable. Furthermore, the organization of some of the metal mining industries means that the number of companies that would consider bidding for any lease would be small.

Why, then, does the auctioned FF seem to have been reasonably successful in collecting rents from US offshore oil and gas, at least over the 1950s and 1960s. If the evidence on this is rightly interpreted, it may perhaps be explained by the comparatively large number of companies interested in petroleum extraction in the USA (a circumstance not repeated with other industries, and perhaps not with the oil industry in all other countries), by the low level of 'political' or 'sovereign' risk there, and by the possibility that, in spite of uncertainty over any individual oil lease, large producers, with numerous diverse blocks, may treat the overall risk of their extractive operations as comparatively low. Statistically, a high proportion of blocks is unsuccessful; but US outer continental shelf (OCS) blocks individually are small, so that a number of tickets can be taken in the lottery; and presumably dry holes do not ruin oil executives' careers. Furthermore, in these apparently successful applications, the auctioned fixed fee has been combined with at least one of the conditional taxes described below.

The FF is of course very easy to administer, but imposing it in 'general' form would raise difficult questions of the appropriate compromise between too high and too low a level, while on the other hand attempting to negotiate an appropriate level for each project also faces special difficulties requiring detailed knowledge of the individual project and its prospects.

2 *Specific or ad valorem duty (SAVD)*

Specific or *ad valorem* duties (commonly called 'royalties') on the amount or value of the product are among the most popular additional mining taxes. They are commonly included as one element in the fiscal regime in most petroleum and mining industries, for example in Australia, Papua New Guinea, Botswana, Tanzania, the USA, and the UK from among the countries featured in the case studies of later chapters. In general form they have less deterrent effect to projects than a general FF of the same expected revenue on projects actually undertaken; or, to put it another way, they are capable of raising more revenue on projects actually undertaken without being any more deterrent in their effect on marginal projects. While a 'general' SAVD does not have the same deterrent effect on marginal projects as the general FF with the same expected revenue, the SAVD has the additional distorting effect that, in raising unit costs of extraction, it tends to reduce the pace and extent of extraction (the 'high-grading' effect), with the result that deposits are left in the ground even though their price exceeds the total social unit cost of extracting them.[1,2] This is socially wasteful; it means that the capital investments and the disturbances to land or water are made without full use of their productive capacity, and hence that such investments and disturbances tend to be excessive.

The attraction of these taxes is their extreme simplicity in administration. In the case of the specific duty, the base of the tax is the amount produced, which is easy to check. In the case of *ad valorem* duty, some formula can be prescribed for determining value for tax purposes from quantity produced in case there should be any room for doubt as to what is the true market price. However, there must be difficulty in knowing *at what rate* to impose the tax if it is applied in general form.

Potentially the SAVD too could be negotiated or auctioned. Again, auctioning could potentially reduce the distorting effect of the tax in deterring projects, but it would leave other distorting effects (the high-grading effects) more or less intact. Auctioning would in principle have some of the same advantages in ideal circumstances as for the FF, but the same practical problems would arise, and, since the general form of the SAVD has more capacity for raising revenue than the general form of the FF, the case for having it in auctioned form if at all is not quite as compelling.

[1] This effect was considered above in the discussion of figure 3.1.

[2] A royalty of this form on oil may none the less actually increase the total amount extracted because a slower initial rate of extraction, such as is favoured by the royalty, may permit a higher pressure to persist within the oil-bearing structure after a given quantity has been extracted. There is no presumption, however, that in such a case the slower extraction is more socially efficient—unless the investor is discounting time at a higher rate than is socially appropriate, that is to say, unless the investor is in too much of a hurry.

The auctioning of the SAVD (or, for that matter, of the HRIT, PPT, or RRT discussed below) would not, moreover, avoid the possibility of the tax's deterring projects where there was imperfect knowledge at the time of the auction and where more information became available before the decision to establish a project was finally taken. The difference between the FF and the other taxes arises from the fact that the amount bid successfully when the FF is auctioned is paid in full and cannot be avoided by subsequent development decisions, whereas the amount bid for SAVD need never be paid if the project is not undertaken. The investor, in other words, may 'walk away'. High bids for SAVD could be based on the understanding that only projects that were expected to be exceptionally profitable in the absence of tax at the time of investment would actually be undertaken. At worst, a company might bid high with the aim of excluding others and with little commitment to mining on its own part. The tendency for firms to bid too high and then not to proceed might be called the 'self-deterrent effect'.

One method of attempting to avoid the distorting effects of an *ad valorem* or specific royalty is to apply it on a sliding scale. In an extreme version the royalty might disappear after a certain number of years. This is one of the experimental methods tried recently in the USA, for which purpose the sliding scale follows a logarithmic formula. If the timing of the sliding-scale royalty is exactly right, it will avoid much of the SAVD's typical distorting effect on recurrent extraction decisions, but it may conceivably lead to delay in extraction as a means of tax avoidance. If applied in general form, it could furthermore still deter some projects that were eligible before tax. And, if the scale were linked to time, no general form could be exactly right for all projects. So there are potentially problems of imposition, and the tax, even if these could be overcome, would not even be ideal for application in general form. Yet if for administrative reasons a specific or *ad valorem* duty were adopted, use of a sliding scale is certainly to be recommended; the authorities might either tailor it to each project or class of projects or else construct the best approximation available to an appropriate general form for the industry.

3 *Higher rate of income tax (HRIT)*

One way of imposing an additional tax on mining is simply to tax profits (as normally assessed for tax purposes) at a higher-than-normal rate. This is the procedure adopted in the Indonesian petroleum industry, where the 'production-sharing' arrangements are designed to have a fiscal effect identical to that of a high rate of income tax. This has the advantage, in cases where an income tax is already established, of simplicity in the introduction of the additional levy, and the administration of the tax is as strong or as weak as that of income tax itself. Income tax does of course entail greater problems of administration than such simple levies as the FF or SAVD. An income tax, however, if properly administered, has a far greater adaptability to the circumstances of the project. Hence the HRIT, or any profit-related tax, could be expected to be imposed in general form with much higher revenue-raising prospects over a range of projects than the SAVD or FF in general form. The imposition of the HRIT in general form does, however, still raise problems of judgement. If the additional rate were too high, too many projects would be deterred; if it were too low, revenue would be unnecessarily forgone; and, as with the FF and SAVD, it would be difficult to be sure of making the best compromise.

The 'higher rate' could in principle be negotiated or determined by auction. Auction has the same kind of potential value and the same practical problems as for the FF or SAVD, together with the 'self-deterrent' effect (or possibility of walking away) mentioned above, though this last would be less serious than with the SAVD. Because of the much higher adaptability of a general HRIT than of a general FF or SAVD to a project's circumstances, the advantage of a project-by-project approach is much less marked.

See tables 7.4 and 7.8 for a numerical example of how the HRIT could work.

4 *Progressive profit tax (PPT)*

Rather than simply charging an additional rate of income tax no matter whether the project is highly or barely profitable, the authorities could borrow the principle of the progressive individual income tax, and tax the more profitable project at a higher rate than the less profitable.

One way of doing this that is closely analogous to the method of the progressive individual income tax is to tax, at an additional rate, additional profit that raises total annual profit above a certain ratio to the value of the accumulated investment. This ratio will be called the 'threshold' rate. Thus, if the accumulated investment amounted to £100 million and the threshold for additional tax were 15 per cent, then taxable profit up to £15 million in any year would be taxed at the normal rate, say 50 per cent, but profit in excess of £15 million would be taxed at a higher rate, say 75 per cent, so that the additional tax would be the extra 25 per cent on annual profit in excess of £15 million. (A further higher rate of tax, say an extra 10 per cent, making 85 per cent in all, could be imposed on the excess of annual profits over a further, higher, threshold, for example over the amount that would bear a ratio of 25 per cent to the investment.)

The existing rules for determining income for tax purposes could be used. The important new entity that would have to be defined by legislation would be the amount of the investment outlay. Whether the depreciation that is allowed on capital expenditure in the particular income tax system was to be used to reduce the value of the investment for the purpose of deciding at what level of profit the higher rate of tax comes into play would be a matter for decision. We believe that the logic behind such a tax argues that it should be so used. Deducting depreciation from the capital investment for purposes of assessing the extra tax would increase the tax collected, other things being equal. Once all investment had been fully depreciated, the higher rate of tax would in that case be collected on all subsequent taxable profit. There is logic in this, but at any given tax rate it would make the PPT, with any given threshold and rate of tax, into a much stiffer imposition than if, with the same threshold and rate of tax, investment were taken as the undepreciated cumulative value of investments simply added together. This latter form, 'PPT Mark I', is the one used in three actual systems discussed in later chapters of this book. It is in this latter form that we shall mainly consider the PPT here, but with reference too to what we consider its more satisfactory variant. In the alternative form, 'PPT Mark II', with a threshold based on depreciated value of investment, it becomes rather more like the tax that we shall consider next, the resource rent tax, particularly if the depreciation used for tax purposes approximates to the 'true' depreciation of the value of the project.

The PPT has its burden more clearly adapted to profitability than the HRIT, or *a fortiori* than the FF and SAVD. It thus entails lower risk to the investor than the HRIT. This means that, if applied in general form, it can be allowed to collect a bigger expected revenue from projects actually undertaken, without deterring any more of the marginal projects, than the HRIT.

In certain ideal conditions, namely if the positive cash flows from the project were to continue unchanged for ever and there were no depreciation allowances against taxable income (except for actual maintenance and replacement expenses) even the PPT Mark I, with a threshold equal to the investors' discount rate, would represent a tax entirely on rent, and if in addition cash flows above the threshold level in each year were certain in advance, the PPT with this threshold would be neutral. However, as will be shown in chapter 8, the PPT in actual conditions will not be entirely neutral with such a threshold—and probably not with any one particular threshold. In principle, however, with true depreciation deducted, PPT Mark II, applied to a project that generates sufficient positive cash flow over a finite period after an initial investment, could be framed in such a way as to be, on fairly plausible assumptions, a tax on true economic rent. (This is explained in the Appendix to chapter 8.) Apart from this presumed advantage in principle of the Mark II form, there is the practical one that the method enables a given revenue to be raised with a given threshold at a lower rate of tax. The advantage of the lower rate in reducing the hazard of incentives to avoidance through over-investment or under-investment will be explained in chapter 8.

See tables 7.5 and 7.8 for numerical examples of how the PPT in both forms might work.

The PPT has the administrative disadvantage as against the FF and SAVD of being an income tax with all that that entails in problems of definition and assessment; its administrative advantage over some other profit-based taxes is that it can use existing tax legislation to a large extent, but there is the additional large question to be answered of the definition of capital investment.

The principle of the PPT has been applied recently to the copper, gold, and silver project of Bougainville Copper Limited in Papua New Guinea (since 1974), and also in highly diluted form (see chapter 17) to oil investments in the UK sector of the North Sea (since February 1975), for which the additional profit tax is referred to as the 'petroleum revenue tax'. It has been incorporated in a modified form into Indonesia's 'third generation' mining agreements. An older example is the special tax applied to income from the base metal mines at Broken Hill in Australia.

The adaptability of the PPT means that decisions on appropriate rates are rather easier than in the cases of the three taxes previously mentioned. The conflict between too high and too low is less acute; the stakes as it were are not so high.

5 *Resource rent tax (RRT)*

The PPT will often tax two projects with the same net present value differently if they have different time patterns of cash flows. This is largely avoided with what we have described (in Garnaut and Clunies Ross, 1975) as the 'resource rent tax' (RRT). This aims to tax positive net present value (NPV) as it is realized, assessing it at some discount rate. If the discount rate corresponds to the investor's discount rate, then the RRT will (except in one case that may not be very common) tax entirely and only the positive NPV as the investor assesses it. The exception is the case in which the NPV attained by one year is reduced thereafter by subsequent net outflows of cash which are not again compensated by further inflows.

The working of the tax is most easily explained if it is supposed to be assessed on a project rather than on a company. In this case all cash inflows and outflows generated by the project are summed in each year to give a net inflow figure. The net cash flow figures are accumulated year by year at an interest rate (the 'threshold') until a cumulative positive value is attained. That value is then taxed at the tax rate. So long as subsequent annual net cash flows are positive from then on, they are simply taxed at that rate. If in any later year the net cash flow is negative, then from that year onward the net inflows are accumulated once again at the threshold interest rate until a cumulative positive value is attained; this value is then taxed, as is any subsequent stream of positive annual inflows; and so on.

If the RRT is imposed after a company profit tax or any other tax, the payment of such a tax is treated as a cash outflow for purposes of assessing the RRT. Alternatively, the RRT may be collected first and treated as a cost in the assessment of the profit tax.

Various matters for decision in the application of the RRT are discussed in chapter 10. Table 7.6 gives an example of how the RRT would be calculated at a tax rate of 60 per cent, with a threshold rate of 10 per cent, on an imaginary project. The positive and negative cash flows taken into account are supposed to be summed in each year to a positive or negative figure for 'net assessable receipts', and this is accumulated each year with 10 per cent added to the cumulative value of the previous year. Any payments of taxes assessed 'before' the RRT can be taken as included as negative items in 'net accessable receipts'. If company profit tax is collected after RRT, RRT can be treated as a deduction for company profit tax.

See tables 7.6 and 7.8 for a numerical example of how the RRT might work.

The RRT, in tying tax more closely than the PPT to actual NPV, reduces further the risk to the investor. Except in the case in which the project ends with losses after taxable profits have been earned, the RRT cannot convert a pre-tax actual positive NPV to a post-tax actual negative NPV if the investor's discount rate (or one higher than that) is used as threshold, though, as shown in chapter 8, it may convert a pre-tax positive *expected* NPV to a post-tax negative one. Again, with the correct discount rate (or a lower one), it cannot convert a pre-tax negative actual NPV into a post-tax positive one.

The obverse side of reduction of risk to the investor tends to be an increase of risk to government revenue and delay in its receipt, but, as suggested at the end of chapter 9, some taxes may provide less conflict than others, and timing of flows for the RRT and the BT (considered below) appears likely to reduce the conflict for those two taxes, because of possible differences between governments' and investors' discount rates.

Administratively, the introduction of the RRT would require definitions of the assessable receipts and deductible payments: the positive and negative cash flows that are netted and accumulated to form the base of the tax. This is not particularly complex, except (as with income tax) where it is complicated by intra-firm transfer payments. Since a mining company often has diverse interests, some definition is needed of the receipts and expenses of the project or activities to be taxed. Broadly, all receipts and payments concerned with the project are included except those that are capital, dividend, or interest transactions with equity-holders, creditors, or debtors. The fact that the tax is based on cash flows, rather than income, means that there is no need to have rules for depreciation or for valuation of stock, two of the main bugbears of income tax legislation. There is also no need (by contrast with the PPT) to define capital investment.

Because of the adaptability of the tax, the dilemma of imposition is if anything even less than with the PPT. Its structure will make it easy to ensure that it gets high revenues from profitable projects and low revenues or none from unprofitable ones. There is, however, a real difficulty over fixing a correct discount rate that will encourage neither over- nor under-investment.

The RRT as described here has been embodied in legislation in Papua New Guinea as an additional tax applying to future projects for extracting metals, petroleum, and gas. The legislation is fairly simple, and, though there is no proper experience yet of its administration, it applies to one very large project in process of development, in which it was acceptable to the major investor involved. The general form, at the parameters chosen by the Papua New Guinea government, has been consistent with the launching of several new exploration projects. The RRT has recently been incorporated into major new mining agreements in Tanzania and several other developing countries, and its application in general form to the petroleum and mining industries is under discussion in Australia.

6 *Brown tax (BT)*

We give the term 'Brown tax' to the form of tax proposed by E. Cary Brown (1948). He suggested it as a substitute for standard profit tax, but it could be applied as a special tax on mining income. This carries even further the business of devising a fully 'adaptable', because fully neutral, profit tax in which the risk to the investor, whether measured by dispersal of the probability distribution of outcomes, or even by the probability of negative discounted value for the project, is reduced by the tax.

Under the Brown tax applied to a project, the authorities would tax all cash flows generated by the project at a constant proportional rate, collecting positive taxes in years in which there are positive net cash flows and paying subsidies (negative taxes) in years in which net cash flows are negative.

See tables 7.7 and 7.8 for a numerical example of how the BT might work.

Apart from its effect in reducing risk, this tax would be completely neutral and would tax economic rent, no more and no less. Since no accumulation for time is needed, there is no 'threshold' rate, and hence it would be unnecessary for the authority to estimate the investor's discount rate; this is an important advantage over the PPT and RRT. Because of the complete neturality of the tax, the rate could be set very high without deterring any projects; indeed, the only desirable limit would be set by the rate at which incentives to economize start to weaken significantly. That being the case, there might in practice be no case for a negotiated or auctioned rate rather than a general rate. Risk to the investor would be reduced to a greater extent for a given rate of tax than under the RRT (or *a fortiori* any of the other four taxes so far discussed) since this tax alone reduces negative as well as positive NPV outcomes.

A disadvantage of the BT is that, of the taxes so far considered, it entails the greatest risk to the government. On a very large project this risk might be unacceptable. The BT is the only one of the forms of tax in which the state lays itself open to actual financial loss. Subsidizing a project for making losses might also be difficult to 'sell' politically, even though the subsidies would not in principle convert the losses into gains for the investor. This pill might be sweetened by giving the government the same rights to participate in control of the enterprise that it would have had if its initial capital contribution (through the negative tax) had been rewarded by equity in the same proportion as the private contribution. In this case a 40 per cent BT would make the government legally holders of 40 per cent of the equity, but with a right to receive 40 per cent of any net positive cash flows rather than any dividends that the company might choose to pay. Here taxation shades off into equity participation, paid for, moreover, at full price. It may well be that, if a government were prepared to accept the risk entailed by the BT, it would choose rather to be a holder of conventional equity shares.

A final possible disadvantage is on grounds of stability of the fiscal regime, as seen by the investor. It may be difficult for investors to be completely confident that subsidies to future capital outlays will continue to be paid at some very high rate. Thus investors may just possibly react to a BT system as one involving greater risk or a higher expected net tax burden than its formal character justifies.

Imposition of the BT is as easy as it could be. The only matter that would have to be decided is the rate of tax at which incentives to economize would begin to weaken. Administration of the BT, however, has much the same problems as that of the RRT. It is a new tax, for which new legislation, albeit rather simple, would be needed, and time might be required to discover the loopholes that needed to be blocked. Thus, set against the neutrality and extreme simplicity of the BT are disadvantages, mainly cosmetic but some possibly substantial.

7 *Product-sharing (PS)*

Product-sharing between host government and investor has a certain intuitive appeal, and some governments have adopted it. There is a variety of possible forms and twists that can be given to this device. In the simple form in which it consists of paying for the resource with a certain fixed proportional share of the output, which the government can then sell, PS has virtually exactly the same effect as an *ad valorem* duty. The caution that has to be sounded is that very high effective rates of duty may be disguised under product-sharing. For a product sold on a competitive world market, 50 per cent product-sharing would be the equivalent of a 50 per cent *ad valorem* duty (at whatever point it is collected). This would have very large distorting effects on extraction in the direction of high-grading unless the authority were in effect controlling the extent and speed of extraction regardless of the private investor's interest. In a more complex form, in which full deductions are allowed for all costs, including depreciation, PS has similar financial effect to HRIT. This is a form approximated in Indonesia. For our purposes, PS can be treated simply as a variant of the SAVD or of the HRIT, and its properties need not be considered separately.

8 *Resource rent tax with equity (RRTE)*

Whenever a government insists on the acquisition of equity in a project without paying what could be considered to be a market price, it is imposing a cost on the investor that is similar in its fiscal effect to some additional taxation. Equity to the host government could be substituted for tax rights. Equity-holding has an appeal for some governments because of the impression it gives of ownership and control. Equity-holding (even majority holding) is in fact neither a necessary nor a sufficient condition for effective control: not necessary because the required legislation to establish control can be passed *ad hoc*, and not sufficient because effective control demands appropriate expertise at the disposal of the government as well as formal authority. Nevertheless, equity-holding has been popular, and there is a case for having clear rules on equity acquisition that fit with the logic of the tax system.

One sensible rule over government participation is that, when a government contributes to investment, it should be rewarded with equity in no less than the same proportion as the private investor is rewarded for his own investment outlay. There may indeed often be economic value in the government's acquisition of equity in mining projects by paying for it. The circumstances in which this is so are discussed in chapter 5.

The BT puts the government in a rather similar position to one that acquires equity by paying for it; and, as suggested above, this position might be formalized by giving it legally the equity that it has paid for during the investment phase.

A similar result to the RRT could be approximated by giving the authority gratis say a 50 per cent equity in the project at the point at which it would otherwise become due for 50 per cent RRT, that is when the original investment has been 'repaid' at the threshold rate.

The effect need not be exactly the same as with the RRT since there need be no obligatory dividend payment corresponding to the tax payment, and, unless it were precluded under the arrangements, the government might inherit a share of outstanding debt so that debt servicing would claim a share of future cash flows.

If this device were to be used, some limitations would probably have to be set on the share structure of the firm. Whatever the proportion of equity that the government was to receive, there would need to be a decision on whether it was or was not intended to give the government formal control, and the share structure would need to be arranged accordingly. If the state were likely to acquire control, the legislation or agreement with the firm might need to limit the state's extraction of dividends in some way, for example to those that would be payable to it under the corresponding rate of RRT. Alternatively, if the state did not have formal control, it might be guaranteed access at its discretion to dividends up to this amount. With such limitations, the RRTE would be close enough to the RRT in its implications for neutrality, for expected revenue, and for public revenue risk, and, in its imposition and administrative problems, to be considered a variant of the RRT.

9 *Carried interest (CI)*

This term is used for arrangements in which the state uses its revenue to acquire (compulsorily) equity in the project. The revenue may be converted into equity as it accrues, or (as in Zambia under the copper nationalization of 1970, in Papua New Guinea under various petroleum agreements, and in a recent uranium mining agreement in Tanzania) equity may be acquired in advance through a loan from the company which the profits accruing to the state are then used to repay. Various forms may be devised for achieving the object of the exercise, which is to use the revenue that would otherwise accrue in order to purchase equity, presumably for the sake of closer control of the enterprise or for some symbolic reason.

In so far as carried interest is written into the terms of a company's permit to operate, it has implications for the timing of the company's after-tax cash flows. If a certain part of what would otherwise be taxed does not actually have to be paid out to government but is used instead to allow government to acquire equity, the risk to the firm of liquidity crises and of the consequent costs is reduced. Apart from this effect in reducing risk, however, it is probably best to regard the effects of CI as depending on the kind of tax on which the finance for acquiring the equity is based. The common case of the government's repaying, out of its 'share' of the profits, a sum of money representing its 'share' of capital expenditure at an agreed interest rate (as in recent Papua New Guinea petroleum and Tanzanian uranium agreements) is similar in its fiscal effect to a tier of RRT with a tax rate equal to the government's share of the equity and a threshold rate equal to the agreed interest rate.

Combining Rent Taxes with Other Taxes

The discussion in this chapter has proceeded as if the choice of one form of rent tax precluded the use of any other, and has ignored the presence in all economies of general taxes that affect mining along with all other industries. It is necessary for us now to address the question of whether special taxes on mining should replace or complement taxes of general application such as corporate income tax and import duty.

Taxes of general application generally raise some revenue from mining enterprises relatively early in the life of a project. For example, during the mine development phase, and probably even during the exploration phase, investment in mineral development will generally make some contribution to the revenue through import duties and through income tax on the personnel or suppliers to investors in the mineral sector. This contribution to the revenue, coming early and with a high degree of certainty, weakens the case for seeking large amounts of earlier and more certain revenue through the FF or SAVD.

But, like other sources of early and more certain revenue, these taxes tend to reduce the present value of economic rent generated in mining. They may also deter investments that would otherwise be commercially viable. The question arises of whether mining activities should be exempted from them, and subjected only to special taxes on mineral rent.

Such exemption would usually be administratively inconvenient, and it is ruled out in some countries on these grounds. More commonly, it is ruled out on grounds of perceived equity in taxation. It is impossible to make general judgements about whether exemptions would promote or detract from the neutrality of the whole taxation system: the distortions created by the more favourable treatment of mining may be less than, equal to, or greater than the distortions removed by the abolition of taxes that are in themselves distortionary. It depends on the circumstances. The case for exemption is strongest when the general taxes are most distorting, and, from a national viewpoint, when a higher proportion of investment is drawn from other countries.

These same issues arise around the question of whether standard corporate income tax should be replaced by mineral rent taxes. Corporate income (or profit) tax exists in most countries. In its ideal form, in which there is proportional tax on profits after depreciation, and depreciation is equal in each year to the fall over that year in the present value of the cash flows expected from the project over the remainder of its life, the tax is neutral as between projects and techniques. However, it is not neutral between saving and consumption as uses of income, and in principle it biases decisions in favour of present, as against future, consumption (Swan, 1976). In fact, its neutrality between projects and techniques depends on conditions that can never strictly be fulfilled. To base depreciation for tax purposes on the real reduction in present value of the project would require accurate foreknowledge of the future on the part of the tax authorities and the calculation of depreciation *ad hoc* on the basis of clairvoyance for each project in each year. In practice, depreciation has to be calculated on some arbitrary rule, and in those circumstances the tax may discriminate between different time patterns of receipts and hence between different techniques, products, and industries. The effect of corporate income tax is to reduce post-tax returns, and to reduce the post-tax supply price of investment below its pre-tax level, at least in the corporate sector. To remit this tax altogether on mining projects and to replace it by a rent tax extracting all returns above a pre-tax supply price of investment might, on the face of it, divert capital into mining that would be more productive in other industries. Alternatively, the post-company-tax supply price of investment could be used for the purpose of determining rent from mining projects, and in this case the remission of corporate tax on mining might be more satisfactory. A further possibility is to collect both company profit tax and tax on the remaining rent, with either treated as an expense for the purpose of assessing the other. If the rent tax is applied first, the rent is assessed as the surplus over what is necessary to pay the pre-company-tax supply price of investment; whereas, if the rent tax is applied after the company tax, the rent is assessed as the surplus over the post company-tax supply price of investment. There is no reason why the consistent application of a rent tax in this way on top of company income tax should introduce any distortion of productive decisions over and above what may have been already inherent in the company income tax regime. The prevalence and general acceptance of company income tax, and the fact that certain international tax provisions treat it more favourably than certain other kinds of tax, may well suggest the desirability of retaining it for mineral projects and applying any specifically rent-collecting tax as an 'additional tax'.

There is, however, a set of circumstances that might dictate a different approach. Many countries will derive their mining investment, and some also a large part of their modern sector investment generally, from overseas. In these circumstances it may not be important to avoid differences in basic tax treatment within the country between mining and other industries, since there may well be no private investors or financiers actually deciding between investment in the country's mining industry and, say, in its agriculture. We shall refer to countries that approach this position as 'case 1' countries, and those that rely to a considerable extent for mineral investment on a domestic capital market as 'case 2' countries. Indonesia, Tanzania, Zambia, Papua New Guinea and Botswana are examples of case 1, and the USA of case 2. The UK and Australia, though more dependent on investment from abroad than is the USA, probably should also be classified as case 2. Regardless of whether company income tax is applied in non-mineral industries, there may in case 1 countries be no significant distortion introduced into the allocation of resources by applying simply a rent tax to the mining sector. For reasons developed in chapter 10, there is some presumption that, with a given revenue expectation for a project, this will do less to deter marginal projects than the combination of company income tax and rent tax. However, the institutional reasons for maintaining company income tax in addition to any rent tax on mining projects may still be strong.

Combining Rent Taxes with One Another

The fact that different forms of rent tax have different advantages and disadvantages raises the possibility that some combination of taxes might be devised to deliver the positive, without the negative, features of both. Governments often do impose a variety of special taxes. The USA, in its federal offshore oil leasing, has maintained a combination of an auctioned element and one that is either 'general' in the sense used here or else determined before the lease is put on offer. This principle, continued for a number of years through an auctioned fixed fee and a general *ad valorem* royalty, has more recently been carried into an experimental phase in which the auctioned and pre-determined elements may each take a number of forms. The auctioned part is considered necessary as a means of efficient and impartial allocation of leases.

We shall suggest in later chapters that, provided the auctioned fixed fee is not ruled out because of private mineral rights, the best result all round may often be achieved by combining an auctioned fixed fee with one of the profitability-related taxes. This combination is advocated not only because it makes possible an appropriate sharing of risk between investor and government when both are risk-averse, enabling the government to take the best position in its trade-off between expected revenue and risk, as Lloyd (1981) following Leland's analysis (1978) has shown, but also because the two taxes may complement each other's contributions to revenue maximization since one adapts to actual profit while the other avoids the need for any maximum-rate limit, and because they allow an appropriate choice by government in its trade-off between the desired timing of revenue and its expected amount. The adaptation to actual profit prevents sovereign risk (and any lack of competition that might arise from some companies' financial incapacity to make economic bids) from corroding the revenue through their effects on the amount bid as a fixed fee.

Tables 7.1–7.3 show very tentatively the relative advantages and disadvantages of the six main forms of tax (1–6). On certain of the main points an attempt will be made to justify and to qualify these comparisons in the course of chapters 8–10. In the comparisons made over neutrality and revenue realization, it is assumed that any dilemmas of imposition and problems of administration and legislation are solved. Where the rates are fixed by negotiation or auction it is assumed, for purposes of comparing neutrality and revenue realization, that the negotiation or auctioning reaches the result desired by the authorities. In practice, of course, this is by no means necessarily true. Negotiation is difficult and chancy, and in many circumstances auctioning may be out of the question.

It is also to be noted that the '+' (for 'good') and '—' (for 'bad') scores are purely relative ranks on one criterion: '—' on neutrality is better than '——' on neutrality, but it is not necessarily worse than '+' on the components of revenue realization.

Summary

In general, the taxes to the right of tables 7.1–7.3 perform better on neutrality, stability, investor risk, and (except in case of auctioning) imposition dilemmas than those to the left of the tables. The taxes to the right, however, tend to perform relatively poorly on the maximum-rate test, government revenue risk and delay, and ease of administration. The tests that favour the taxes on the right of the tables are generally far more important than those that favour those on the left. Certain combinations of taxes may reap a mixture of advantages.

The judgements presented in tables 7.1–7.3 are crude and require qualification. This task is undertaken in the three following chapters, which concentrate most on the four taxes to the right of the tables. Chapters 8–10 suggest that the relative merits among HRIT, PPT, RRT, and BT depend to some extent on the circumstances in which the tax is to be applied, but they do provide a strong presumption in favour of the RRT and, if it is politically acceptable, the BT. They also provide a case for combining one of these taxes with an auctioned fixed fee where circumstances appropriate for its use exist or can be created.

Appendix
Notes on Tables 7.3–7.8

The two projects supposed for these tables are purely hypothetical and are presented for illustrative purposes only. They may well not be typical of any kind of resource project. One has markedly higher and less regular returns than the other.

Income tax is assumed to operate in a simplified form in which depreciation is deducted on a straight-line basis over ten years, and in which there is no carrying-forward of losses whatever. The income tax rate is supposed to be 50 per cent. The HRIT and PPT are supposed to operate on income as defined for income tax purposes. Initially there are supposed to be no interest payments (or, alternatively, interest is not deductible for purposes of income tax). In table 7.8(c) this assumption is removed and figures in brackets give the alterations that would be made if interest, on a hypothetical schedule oulined below, were deductible for purposes of income tax, HRIT, and PPT.

All the rates of taxes supposed in tables 7.4–7.7 are set so that, combined with income tax at 50 per cent, they would give a top marginal rate of tax on cash flow of 80 per cent. The eight systems represented in tables 7.8(a) and 7.8(c) all have top marginal rates of 80 per cent. Thresholds for PPT and RRT are all 10 per cent, except where otherwise indicated in table 7.8(c).

PPT at 60 per cent before 50 per cent income tax gives the same result (at least, in the simplified rules) as PPT at 30 per cent after 50 per cent income tax, provided income tax is based on income net of PPT in the former case, and PPT is based on the appropriate slice of income gross of income tax in the latter case.

Where, as in the two examples, there is no final period of negative cash flows after RRT has been collected, RRT and BT alone at the same rate will give the same present value of revenue when discounting is done at the RRT threshold rate. This is exemplified in Table 7.8(c).

Table 7.1

Comparison on various criteria of the six major methods of rent taxation on minerals: general form

	(1) Flat fee	(2) Specific or *ad valorem* duty or simple production sharing	(3) Higher rate of proportional income tax	(4) Progressive profit tax (Mark I)[11]	(5) Resource rent tax	(6) Brown tax
Neutrality[1]	——[7]	——[7]	—	+	++	+++
Government revenue maximization						
Stability[2]	———	—	+	+++	+++	++[8]
Intrinsic investor risk (low = +)[3]	——	—	+	++	+++	++++
Maximum rate test[4]	+++[10]	+++[10]	++[9]	——[9]	—[9]	—[9]
Adaptability[5]	——	—	+	++	+++	++++
Government revenue risk and delay (low = +)	++++	+++	++	+	—	——
Imposition dilemmas[6]	——	—	+	++	+++	++++
Administration, legislation	++	+	—	——	———	———

Notes:

For purposes of all criteria except 'imposition' and 'administration', imposition dilemmas and administrative–legislative problems are assumed to be resolved as far as is in principle possible. The rent taxes are assumed to be applied alone, that is without income tax. For HRIT and PPT, this means that the 'extra' element in the tax is applied alone. The same applies in tables 7.2 and 7.3.

[1] Neutrality of each tax is considered when it is assumed to be imposed in such a way as to extract the maximum revenue of which it is capable (in this table, over projects as a whole; in tables 7.2 and 7.3, over each individual project), provided that the 'threshold rate' in taxes (4) and (5) is set so as to approximate most closely to neutrality.

[2] 'Stability indicates that perceived political risk is low.

[3] 'Intrinsic investor risk' is the risk there would be to the investor if there were no political risk.

[4] This refers to the relative capacity of taxes (3)–(6) for raising revenue on a particular, typical, and still eligible project when the rate of tax that will give the maximum revenue consistent with maintaining incentives to cost minimization is applied. Thresholds of taxes (4) and (5) may be regarded as being set so as to approximate to the investors' discount rate.

[5] 'Adaptability' refers to the capacity of the tax, applied in general form (i.e., with uniform parameters) across an industry, to adjust revenue receipts to realized profitability of projects.

[6] This refers to the ease with which correct tax rates and thresholds for revenue maximization and for approaching neutrality can be determined.

[7] No straight comparison between these two can be made, since (1) distorts the choice between the affected industry and others, while (2) (imposed so as to maximize tax revenue) will distort that choice less but will also distort choice over the timing and extent of extraction.

[8] It is possible, but by no means certain, that investors might perceive tax (6) as more unstable than (4) or (5).

[9] The appropriate order among these four on this criterion is not entirely clear and would depend on the details of the tax legislation, on the scale and timing of returns, and on the extent of loan finance. Combining with income tax tends to favour revenue from the PPT forms to a greater extent than this ordering suggests. See tables 7.8(a)–(c).

[10] Since there is no limit on the maximum rate set by the need to avoid removing incentives to cost minimization, it seems appropriate to regard taxes (1) and (2) as best on this test, though it is difficult to compare them with the others on this criterion in an entirely meaningful way.

[11] The 'Mark II' form of this tax would tend to perform better on 'neutrality' and the 'maximum-rate test' than the 'Mark I' form. Examples in tables 7.8(a) and 7.8(c) suggest that the PPT Mark II may generally perform better on the maximum-rate test than the RRT and BT at the same rate, even when all are applied alone and without income tax.

Table 7.2

Comparison on various criteria of the six major methods of rent taxation on minerals: negotiated form

	(1) Flat fee	(2) Specific or *ad valorem* duty or simple production sharing	(3) Higher rate of proportional income tax	(4) Progressive profit tax (Mark I)[10]	(5) Resource rent tax	(6) Brown tax
Neutrality[1]	++++	−	+	++	+++	++++
Government revenue maximization						
Stability[2]	− −	−	+	+++	+++	++[7]
Intrinsic investor risk (low = +)[3]	− −	−	+	++	+++	++++
Maximum-rate test[4]	+++[9]	+++[9]	++[8]	− −[8]	−[8]	+[8]
Adaptability[5]	n.a.	n.a.	n.a.	n.a.	n.a.	n.a.
Government revenue risk and delay (low = +)	++++	+++	++	+	−	− −
Imposition dilemmas[6]	− −	−	+	++	+++	++++
Administration, legislation	++	+	−	− −	− − −	− − −

Notes:

For purposes of all criteria except 'imposition' and 'administration', imposition dilemmas and administrative–legislative problems are assumed to be resolved as far as is in principle possible. Subject to the conditions that the 'threshold' in taxes (4) and (5) should be such as to approximate most closely to neutrality, the negotiation process in each case is supposed to be conducted with complete success on the government's part so that it extracts the maximum expected present value in the form of revenue that the investor would be prepared to concede beforehand.

1, 2, 3, 4, 5, 6 See notes 1–6 respectively of table 7.1.

Table 7.3

Comparison on various criteria of the six major methods of rent taxation on minerals: auctioned form

	(1) Flat fee	(2) Specific or *ad valorem* duty or simple production sharing	(3) Higher rate of proportional income tax	(4) Progressive profit tax (Mark I)[11]	(5) Resource rent tax	(6) Brown tax
Neutrality[1]	++++	—	+	++	+++	++++
Government revenue maximization						
Stability[2]	——	—	+	+++	+++	++[7]
Intrinsic investor-risk (low = +)[3]	——	—	+	++	+++	++++
Maximum-rate test[4]	+++[9]	+++[9]	++[8]	——[8]	—[8]	+[8]
Adaptability	n.a.	n.a.	n.a.	n.a.	n.a.	n.a.
Government revenue risk and delay (low = +)	++++	+++	++	+	—	——
Imposition dilemmas[6]	++++	—[10]	+[10]	+[10]	+[10]	+[10]
Administration, legislation	++	+	—	——	———	———

For notes see page 112

Notes to table 7.3:

For purposes of all criteria except 'imposition' and 'administration', imposition dilemmas and administrative–legislative problems are assumed to be resolved as far as is in principle possible. Only rates of tax are assumed to be auctioned. 'Threshold rates' in taxes (4) and (5) are assumed to be fixed so as to approximate as closely as possible to neutrality. Subject to this condition and to any limit to bids set by the authorities on taxes (2)–(6) in order to avoid excessive distortion and to prevent abandonment of cost minimization, the auctioning process is assumed to be conducted with complete success from the government's point of view so that the winning bidder offers the maximum expected present value that anyone in the market would be prepared to concede beforehand.

[1,2,3,4,5,6] See notes 1–6 respectively of table 7.1.

[7,8,9] See notes 8–10 respectively of table 7.1.

[10] Uninhibited auctioning of the rates of these taxes might remove the incentive to economize or, especially with tax (2), otherwise distort production decisions unduly. The need to limit bids provides a 'dilemma' for taxes (2)–(6) that is not present in tax (1).

[11] See note 11 of table 7.1.

Table 7.4

Example of 'Higher Rate of Income Tax' (HRIT)

(1) Year	(2) Net cash flow	(3) Depreciation (straight line over 10 years)	(4) Income for tax purposes (= (2) − (3) if positive)	(5) Income tax at 50% (= 50% of (4))	(6) HRIT additional tax at 30% (= 30% of (4))	(7) Total tax (= (5) + (6))
0	−200 capital					
1	−200 capital	20				
2	+ 50 recurrent	40	10	5	3	8
3	+100	40	60	30	18	48
4	+100	40	60	30	18	48
5	+100	40	60	30	18	48
6	+100	40	60	30	18	48
7	+100	40	60	30	18	48
8	+100	40	60	30	18	48
9	+100	40	60	30	18	48
10	+160	40	120	60	36	96
11	−200 capital					
	+160 recurrent					
	=− 40 net	20	140	70	42	112
12	+180 recurrent	20	160	80	48	128
13	+180	20	160	80	48	128
14	+180	20	160	80	48	128
15	+150	20	130	65	39	104
16	+120	20	100	50	30	80
17	+ 90	20	70	35	21	56
18	+ 60	20	40	20	12	32
19	+ 30	20	10	5	3	5
20	+ 20	20	0	0	0	0
				760		1216

Note: See Appendix.

Table 7.5

Example of 'Progressive Profit Tax' (PPT)

(1) Year	(2) Net cash flow	(3) Depreciation (straight line over 10 years)	(4) Income for tax purposes (= (2) − (3) if positive)	(5) Capital investment for purposes of PPT Mark I	(6) Income exempt from PPT Mark II (= 10% of (5))	(7) PPT at 60% Mark I (= 60% of [(4) − (6)] if positive	(8) Capital investment for purposes of PPT Mark II	(9) Income exempt from PPT Mark II (= 10% of (8))	(10) PPT at 60% Mark II (= 60% of [(4) − (9)] if positive
0	−200 capital								
1	−200 capital	20					200	20	
2	+ 50 recurrent	40	10	400	40	0	380	38	0
3	+100	40	60	400	40	12	340	34	15.6
4	+100	40	60	400	40	12	300	30	18.0
5	+100	40	60	400	40	12	260	26	20.4
6	+100	40	60	400	40	12	220	22	22.8
7	+100	40	60	400	40	12	180	18	25.2
8	+100	40	60	400	40	12	140	14	27.6
9	+100	40	60	400	40	12	100	10	30.0
10	+160	40	120	400	40	48	60	6	68.4
11	−200 capital								
	+160 recurrent								
	=− 40 net	20	140	400	40	60	20	2	82.8
12	+180 recurrent	20	160	600	60	60	200	20	84.0
13	+180	20	160	600	60	60	180	18	85.2
14	+180	20	160	600	60	60	160	16	86.4

15	+150	20	130	600	60	42	140	14	69.6
16	+120	20	100	600	60	24	120	12	52.8
17	+ 90	20	70	600	60	6	100	10	36.0
18	+ 60	20	40	600	60	0	80	8	19.2
19	+ 30	20	10	600	60	0	60	6	2.4
20	+ 20	20	0	600	60	0	40	4	0

Notes: See Appendix. The 'threshold' assumed is 10%.

Table 7.6

Example of 'Resource Rent Tax' (RRT)

(1) Year	(2) Net cash flow (net assessable receipts)	(3) Accumulated (at 10%) net assessable receipts where previously untaxed; otherwise current net assessable receipts	(4) RRT at 60% (= 60% of (3) if positive)
0	−200	−200	0
1	−200	−420	0
2	+ 50	−412	0
3	+100	−353	0
4	+100	−288	0
5	+100	−217	0
6	+100	−139	0
7	+100	− 53	0
8	+100	+ 42	25
9	+100	+100	60
10	+160	+160	96
11	− 40	− 40	0
12	+180	+136	82
13	+180	+180	108
14	+180	+180	108
15	+150	+150	90
16	+120	+120	72
17	+ 90	+ 90	54
18	+ 60	+ 60	36
19	+ 30	+ 30	18
20	+ 20	+ 20	12
			761

Notes: See Appendix. The 'threshold' assumed is 10%.

Table 7.8(c) is designed to show how the eight systems, with the same top marginal rate, would compare in revenue on what is called, in chapters 7 and 9, the 'maximum-rate test'. This of course does not show, or even strongly suggest, which is best at revenue-raising overall since the hypothetical projects are not designed in any way to be typical, and since there are a number of other considerations beside the maximum-rate test that are relevant to the capacity of a system to raise revenue. Nevertheless, tables 7.8(a) and 7.8(c) do point to various considerations that may affect performance on the maximum-rate test (that is, on the important implied assumption that a project would still be eligible under all systems). The other three perform better in relation to the HRIT the higher the realized returns (and vice versa). The PPT Mark II performs consistently better than the PPT Mark I (for obvious reasons). If the government discounts the

Table 7.7

Example of 'Brown Tax' (BT)

(1) Year	(2) Net cash flow	(3) BT at 60% (= 60% of (2))
0	–200	–120
1	–200	–120
2	+ 50	+ 30
3	+100	+ 60
4	+100	+ 60
5	+100	+ 60
6	+100	+ 60
7	+100	+ 60
8	+100	+ 60
9	+100	+ 60
10	+160	+ 96
11	– 40	– 24
12	+180	+108
13	+180	+108
14	+180	+108
15	+150	+ 90
16	+120	+ 72
17	+ 90	+ 54
18	+ 60	+ 36
19	+ 30	+ 18
20	+ 20	+ 12

Notes: See Appendix.

future at less than the threshold rate, the BT alone is likely to perform better from the government's viewpoint than the RRT alone at the same rate of tax.

Interest

Interest payments are normally deducted for income tax purposes, and it is assumed that they would be deducted also for purposes of HRIT and PPT. To give a possible example of the effect of this on tax revenue, we suppose that in both the hypothetical projects the 200 investment of year 1 (half the initial investment) is financed by a loan at 6 per cent interest, the principal to be paid back in six equal instalments beginning in year 3. Then there will be interest of the following amounts to be offset against income.

Year	Interest	Year	Interest
2	12	6	6
3	12	7	4
4	10	8	2
5	8		

Table 7.8(a)

Examples of taxes of Tables 7.4–7.7 combined with income tax or otherwise raised to give a top marginal rate of 80 per cent on the same project

(1) Year	HRIT with income tax				PPT Mark I with income tax				PPT Mark II with income tax			
	(2)	(3)	(4)	(5)	(6)	(7)	(8)	(9)	(10)	(11)	(12)	(13)
	Base (Y) of income tax (see table 7.4)	Income tax (T) at 50% (= 50% of (2))	HRIT at 30% (= 30% of (2))	Total tax (= (3) + (4))	PPT Mark I at 60% (see table 7.5)	Taxable income after PPT (= (2) − (6))	Income tax at 50% (= 50% of (7))	Total tax (= (6) + (8))	PPT Mark II at 60% (see table 7.5)	Taxable income after PPT (= (2) − (10))	Income tax at 50% (= 50% of (11))	Total tax (= (10) + (12))
0	0	0	0	0	0	0	0	0	0	0	0	0
1	0	0	0	0	0	0	0	0	0	0	0	0
2	10	5	3	8	0	10	5	5	0	10.0	5.0	5.0
3	60	30	18	48	12	48	24	36	15.6	44.4	22.2	37.8
4	60	30	18	48	12	48	24	36	18.0	42.0	21.0	39.0
5	60	30	18	48	12	48	24	36	20.4	39.6	19.8	40.2
6	60	30	18	48	12	48	24	36	22.8	37.2	18.6	41.4
7	60	30	18	48	12	48	24	36	25.2	34.8	17.4	42.6
8	60	30	18	48	12	48	24	36	27.6	32.4	16.2	43.8
9	60	30	18	48	12	48	24	36	30.0	30.0	15.0	45.0
10	120	60	36	96	48	72	36	84	68.4	51.6	25.8	94.2
11	140	70	42	112	60	80	40	100	82.8	57.2	28.6	111.4
12	160	80	48	128	60	100	50	110	84.0	76.0	38.0	122.0
13	160	80	48	128	60	100	50	110	85.2	74.8	37.4	122.6
14	160	80	48	128	60	100	50	110	86.4	73.6	36.8	123.2

15	130	65	39	104	42	88	44	86	69.6	60.4	30.2	99.8
16	100	50	30	80	24	76	38	62	52.8	47.2	23.6	76.4
17	70	35	21	56	6	64	32	38	36.0	34.0	17.0	53.0
18	40	20	12	32	0	40	20	20	19.2	20.8	10.4	29.6
19	10	5	3	8	0	10	5	5	2.4	7.6	3.8	6.2
20	0	0	0	0	0	0	0	0	0	0	0	0
Undiscounted total		760	456	1216	444		538	982	746		387	1133
Total discounted at 10%		280	168	448	157		201	358	257		151	408
Total discounted at 6%		405	243	648	232		289	521	383		213	596

(1) Year		RRT before income tax			RRT after income tax				RRT alone at 80%
	(14)	(15)	(16)	(17)	(18)	(19)	(20)	(21)	(22)
	RRT at 60% (see table 7.6)	Taxable income after RRT (= (2) − (14))	Income tax at 50% (= 50% of (15))	Total tax (= (14) + (16))	Net cash flow after income tax (from (3) and table 7.6)	Base of RRT (as (3) in table 7.6)	RRT at 60% (= 60% of (19) if positive)	Total tax (= (3) + (20))	RRT at 80% (= 80% of col. (3) of table 7.6 if positive)
0	0	0	0	0	−200	−200	0	0	0
1	0	0	0	0	−200	−200	0	0	0
2	0	10	5	5	45	−417	0	5	0
3	0	60	30	30	70	−389	0	30	0
4	0	60	30	30	70	−358	0	30	0
5	0	60	30	30	70	−324	0	30	0
6	0	60	30	30	70	−286	0	30	0

Table (*cont.*)

(1) Year	(14)	RRT before income tax (15)	(16)	(17)	(18)	RRT after income tax (19)	(20)	(21)	RRT alone at 80% (22)
	RRT at 60% (see table 7.6)	Taxable income after RRT (= (2) − (14))	Income tax at 50% (= 50% of (15))	Total tax (= (14) + (16))	Net cash flow after income tax (from (3) and table 7.6)	Base of RRT (as (3) in table 7.6)	RRT at 60% (= 60% of (19) if positive)	Total tax (= (3) + (20))	RRT at 80% (= 80% of col. (3) of table 7.6 if positive)
7	0	60	30	30	70	−245	0	30	0
8	25	35	17.5	42.5	70	−200	0	30	34
9	60	0	0	60	70	−150	0	30	80
10	96	24	12	108	100	− 65	0	60	128
11	0	140	70	70	−110	−182	0	70	0
12	82	78	39	121	100	−100	0	80	109
13	108	52	26	134	100	− 10	0	80	144
14	108	52	26	134	100	89	53.4	133.4	144
15	90	40	20	110	85	85	51.0	116.0	120
16	72	28	14	86	70	70	42.0	92.0	96
17	54	16	8	62	55	55	33.0	68.0	72
18	36	4	2	38	40	40	24.0	44.0	48
19	18	0	0	18	25	25	15.0	20.0	24
20	12	0	0	12	20	20	12.0	12.0	16
	761		390	1151			230	990	1015
	219		172	391			50	330	292
	353		231	584			91	496	470

(1) Year	BT before income tax				BT alone at 80%
	(23) BT at 60% (see table 7.7)	(24) Taxable income after BT (= (2) − (23) if positive; zero if above negative)	(25) Income tax at 50% (= 50% of (24))	(26) Total tax (= (23) + (25))	(27) BT at 80% (= 80% of col. (2) of table 7.7)
0	−120	0	0	−120	−160
1	−120	0	0	−120	−160
2	30	0	0	30	40
3	60	0	0	60	80
4	60	0	0	60	80
5	60	0	0	60	80
6	60	0	0	60	80
7	60	0	0	60	80
8	60	0	0	60	80
9	60	0	0	60	80
10	96	24	12	108	128
11	− 24	140*	70	46	− 32
12	108	52	26	134	144
13	108	52	26	134	144
14	108	52	26	134	144
15	90	40	20	110	120
16	72	28	14	86	96
17	54	16	8	62	72
18	36	4	2	38	48
19	18	0	0	18	24
20	12	0	0	12	16
	888		204	1092	1188
	219		62	281	292
	393		97	490	524

*The negative Brown tax of this year is assumed to be treated, for income tax purposes, as a contribution to capital investment rather than as an addition to income.

Table 7.8(b)

Cash flows of two hypothetical projects used in Table 7.8(c)

Year	Project 1 (as in Tables 7.4–7.8(a))		Project 2	
0	–200 capital	= –200	–200 capital	= –200
1	–200 capital	= –200	–200 capital	= –200
2		50		100
3		100		0
4		100		400
5		100		200
6		100		0
7		100		400
8		100		200
9		100		0
10		160		320
11	–200 capital + 160 recurrent	= – 40	–200 capital + 320 recurrent	= 120
12		180		0
13		180		720
14		180		360
15		150		0
16		120		540
17		90		180
18		60		120
19		30		60
20		20		40

The rate of 6 per cent is chosen on the supposition that prices are stable and that all rates are indexed. A similar assumption lies behind the choice of the 10 per cent threshold and of the discounting, in table 7.8(c), of tax revenue at 6 per cent as a possible approximation to the government discount rate. The rates actually likely to operate if the loans were unindexed under an expectation of 10 per cent or so annual inflation would do more to reduce revenue in real terms from income-tax-based systems.

Carrying forward of losses

If losses were carried forward for income tax (and HRIT and PPT) purposes, this would tend to reduce the revenue from those systems including income tax and even more from those based entirely on income tax, though in the hypothetical examples the effect would be very small.

Relationship with tables 7.1–7.3

Apparent differences between the results in tables 7.8 and those predicted in tables 7.1–7.3, with respect to the 'maximum-rate test', and particularly to the relative performance of the PPT on it, are explained by the fact that tables 7.1–7.3 do not include PPT Mark II and suppose that comparison is between the 'extra' or 'additional' elements in the four taxes alone without standard income tax. On this basis, as Table 7.8(a) shows, the PPT Mark I gives markedly less revenue on Project 1 at both discount rates used than the RRT or BT at the same rate. On Project 2 (the irregular returns project), however, the performance of PPT Mark I in relation to RRT varies with threshold chosen and discount rate applied for comparison.

Thresholds and approaches to neutrality

The introduction of income tax necessarily interferes with any approach to neutrality that the PPT, RRT, or BT on its own might display. When the RRT is collected after income tax, as in column (5) of tables 7.8(a) and 7.8(c), then (with threshold equal to investor's discount rate) it will not turn an actual negative after-income-tax NPV into a positive one or vice versa. (For an analogous statement to be made about PPT Mark II, imposed after income tax, the PPT would need to be based on income net of income tax. This, with the same threshold and a 60 per cent rate of PPT, would give somewhat less revenue in total than the system in column (3).) Where the extra tax is imposed before income tax, as in columns (2), (3), and (4) there seems to be no satisfactory statement about neutrality that can be made. It is probable that, in columns (2), (3), and (4), of Table 7.8(c), the threshold of 20 per cent, used in the second (asterisked) set of figures, is the most appropriate one where 10 per cent is the investor's discount rate and a 50 per cent tax is to follow the PPT or RRT. Thus the second set of totals in those columns probably comprises the ones that should be compared with those in columns (5)–(8), since they would be appropriate for the same investor discount rate.

Table 7.8(c)

Total revenue from various profit-based tax systems with top marginal rate of 80 per cent on two hypothetical projects

		HRIT at 30% additional to income tax at 50%	PPT Mark I at 60% before, or 30% after, income tax, with income tax at 50%	PPT Mark II at 60% before, or 30% after, income tax, with income tax at 50%	RRT at 60% before income tax, with income tax at 50%	RRT at 60% after income tax, with income tax at 50%	RRT at 80% alone	BT at 60% before income tax with income tax at 50%	BT at 80% alone
Project 1: moderate, smooth cash flow	Undiscounted	1216 (−42)	982 (−39) *829 (−26)	1133 (−39) *1056 (−32)	1151 (−26) *917 (−26)	990 (+9)	1015	1092	1184
	Discounted to year 0 at 10%	448 (−28)	358 (−26) *302 (−18)	408 (−26) *373 (−21)	391 (−18) *312 (−18)	330 (−8)	292	281	292
	Discounted to year 0 at 6%	648 (−33)	521 (−30) *440 (−20)	596 (−30) *550 (−25)	584 (−20) *465 (−20)	496 (−4)	470	490	524

Table (*cont.*)

		HRIT at 30% additional to income tax at 50%	PPT Mark I at 60% before, or 30% after, income tax, with income tax at 50%	PPT Mark II at 60% before, or 30% after, income tax, with income tax at 50%	RRT at 60% before income tax, with income tax at 50%	RRT at 60% after income tax, with income tax at 50%	RRT at 80% alone	BT at 60% before income tax with income tax at 50%	BT at 80% alone
Project 2: high, irregular cash flow	Undiscounted	2764 (−29)	2672 (−29) *2504 (−25)	2802 (−29) *2757 (−26)	2747 (−18) *2684 (−18)	2662 (−4)	2574	2680 (−18)	2688
	Discounted to year 0 at 10%	1052 (−20)	1009 (−20) *944 (−17)	1048 (−20) *1029 (−18)	989 (−12) *949 (−12)	927 (−5)	845	877 (−12)	845
	Discounted to year 0 at 6%	1497 (−23)	1442 (−22) *1347 (−20)	1502 (−22) *1477 (−21)	1441 (−14) *1393 (−14)	1368 (−5)	1282	1343 (−14)	1321

Notes: See Appendix. Figures in brackets are the reductions from totals for interest where interest, on the terms supposed in the Appendix, is deducted from income from income for income tax, PPT, and HRIT purposes. All unmarked PPT and RRT figures are based on a 10% threshold, as in tables 7.5, 7.6, and 7.8(a).

*Indicates 20% threshold rate for PPT and RRT purposes.

8

THE NEUTRALITY OF RENT TAXES

In this chapter we examine more closely some of the propositions put forward in chapter 7 about the neutrality of various taxes on mineral income, and especially of the three types of profit-based tax devised for collecting the economic rent: the progressive profit tax (PPT), the resource rent tax (RRT), and the Brown tax (BT).

In demanding neutrality of a tax applied to a particular industry and not to others, we are asking that it should not have the effect of diverting investment to or from that industry. We are also asking that it should not alter decisions on *which* projects within the industry are undertaken and which are not, or on the techniques, the extent, or the pace of production.

Broad generalizations were made in chapter 7 about the degrees to which various rent taxes departed from the ideal of neutrality in the exploration, mine development, and production phases of the exploitation of mineral deposits. We shall return to the issue of neutrality in these three phases in the concluding pages of this chapter. In the meantime, we find it convenient to make a different distinction: that between decisions on outlays of a *recurrent* kind, which have an immediate impact on the level of production, and those in the nature of *investment*, which have a delayed effect. In general, recurrent expenditures are important in the mineral recovery phase and investment expenditures in the exploration and mine development phases, as well as sometimes in the recovery phase.

Neutrality and Recurrent Expenditure

It was observed in chapter 3 that any tax that reduces the marginal revenue to producers below the sale value of minerals is likely to cause lower-grade but economic parts of deposits to be bypassed. It is in the nature of mining business that ore that is economic if mined with higher-grade adjacent minerals may be uneconomical to mine separately at a later date. Thus, when taxes that are non-neutral with respect to decisions on current outlays cause lower-quality ores to be bypassed, that may lead potentially economic production, and potential mineral rent, to be lost for ever.

Specific and *ad valorem* royalties are non-neutral with respect to recurrent expenditure decisions. The non-neutrality is least severe for *ad valorem* royalties, greater for royalties that are specific to the mineral content of ore, and greatest for royalties that are specific to the volume of ore independently of its mineral content. This follows from the latter royalties' greater effect (if all are levied to yield similar amounts of revenue from some specified intra-marginal ores) in reducing marginal receipts to producers below marginal sales revenues.

The other mineral taxes discussed in chapter 7 are for most purposes nearly neutral with respect to recurrent expenditures. The fixed fee, resource rent tax, and Brown tax all avoid reducing marginal receipts to producers below the recurrent marginal costs of extraction.

The qualification to the neutrality of some of the taxes listed in the previous paragraph with respect to recurrent expenditures derives from their possible non-neutrality in the choice of techniques. To the extent that these taxes discourage some investment, they will tend to bias recovery techniques in a capital-sparing direction, and this may affect the extent of mineral recovery. These aspects of non-neutrality with respect to recurrent expenditures are best considered as deriving from non-neutrality in investment, as is discussed in the following section. It is initially assumed in the discussion that, when a particular tax is imposed, investors may as a result alter the discount rate that they apply (for example, because of the effect of the tax on risk), but that, both before and after such a change, they will not apply different discount rates to different projects, or to different investments on the same project that are subject to the same tax regime. This may not of course be strictly true, and the assumption will later be relaxed as it relates to different investments on the same project.

Neutrality over Investment Decisions and Techniques

There are two questions about each tax system. First, in what circumstances if any could an ideally negotiated or auctioned, 'tailor-made' tax of this type be neutral with respect to decisions about the technique, pace, and extent of production? Second, in what circumstances if any could a general tax of this type applied to one industry be neutral among industries and projects? The answer in brief is that any of the six taxes discussed can in principle be tailor-made to be neutral among projects but, for all but the fixed fee and the Brown tax, distortions might remain as to *how* the project was undertaken. With a general tax, all but the Brown tax might introduce distortions, even in the choice of projects, but the importance of these distortions would vary greatly from one tax to another given some ground of comparison such as equal expected revenue over a range of hypothetical projects that would be undertaken under all types of regime.

Here as before competition in the supply of investment is assumed. This means that any project or subsidiary investment that achieves a positive expected net present value (ENPV) at the ruling discount rate will be undertaken. A tax is not neutral with respect to decisions on investment if it changes the ENPV of a project or subsidiary investment from positive to negative or from negative to positive. We describe the former effect as an incentive to under-investment and the latter as an incentive to over-investment.

The more crude forms of mineral taxation, including FF, SAVD, and to a less extent HRIT, if fixed in 'general' form across projects, do not perform well on this criterion. The object of attempting to confine a tax to the collection of economic rent is to avoid such effects. How well can the PPT, RRT and BT meet this requirement? To answer this it is necessary to consider first their operation in the conditions of a simplified model.

Model 1: Riskless Investments with Regular Yields

Imagine first a world in which it is possible to think of every investment as entailing a negative cash flow (K) on a particular date which will be followed thereafter by an annual stream of net positive cash flows known with certainty and continuing at the same rate for ever, each year's positive cash flow (P) meeting this description after expenditures necessary to maintain the cash flows for the future have been netted out. The ratio that the annual net positive cash flow bears to the original negative cash flow P/K is the internal rate of return (IRR) on the investment. If the investor's discount rate (DR), represented by D^*, is lower than the IRR, the investment will have a positive NPV and will be undertaken. If the DR (D^*) is lower than the IRR, the investment will have a negative NPV and will not be undertaken.

The progressive profit tax (PPT)

Suppose first that the PPT with a threshold equal to the DR were applied at rate T $(1 > T > 0)$ to investments in this kind of world. The tax (R_1) collected each year from the annual proceeds, P_1, of an investment K_1 would be given by

$$
\begin{aligned}
R_1 &= T\left(\frac{P_1}{K_1} - D^*\right)K_1, & \frac{P_1}{K_1} &> D^* \\
R_1 &= 0, & \frac{P_1}{K_1} &\leq D^* \qquad (8.1)
\end{aligned}
$$

The tax, in other words, is positive if $P_1/K_1 > D^*$ but it does not reduce the IRR to D^*, and is zero if $P_1/K_1 < D^*$.

The tax applied to the proceeds of a single investment will thus leave a pre-tax positive NPV still positive and a pre-tax negative NPV still negative. This result continues to hold if there is a second potential investment, K_2, which is part of the same project and hence has its proceeds (a stream of P_2 per year) taxed with the proceeds of K_1.

The tax collected (R_{1+2}) would be given by

$$R_{1+2} = T\left(\frac{P_1 + P_2}{K_1 + K_2} - D^*\right)(K_1 + K_2), \quad \left(\frac{P_1 + P_2}{K_1 + K_2} - D^* > 0\right)$$

$$R_{1+2} = 0, \qquad \left(\frac{P_1 + P_2}{K_1 + K_2} - D^* \leq 0\right). \tag{8.2}$$

The former of these expressions (8.2) can be written as

$$R_{1+2} = T\,[P_1 + P_2 - D^*\,(K_1 + K_2)] \tag{8.3}$$

and from (8.1),

$$R_1 = T\,(P_1 - D^*\;K_1). \tag{8.4}$$

Subtracting (8.4) from (8.3), we get

$$R_{1+2} - R_1 = T\;(P_2 - D^*\;K_2). \tag{8.5}$$

This is exactly the same as the expression for R_2, the tax due per year if K_2 had been undertaken independently (and $P_2/K_2 > D^*$).

Thus, if tax would be due on the proceeds of the second investment taken alone, the same amount of tax is due on the proceeds of the second investment when it is taxed jointly with the first. Hence in this case the second or subsidiary investment, if it gives a positive NPV before tax, will give a positive NPV after tax, whether the subsidiary investment is taxed alone or with another taxed undertaking.

Suppose that $P_2/K_2 < D^*$ but that $P_1/K_1 > D^*$ and that $(P_1 + P_2)/(K_1 + K_2) > D^*$. Then, from the same reasoning given above, the addition of the second investment increases the tax due by $T\,(P_2 - D^*\,K_2)$, which is negative in this case, or in other words reduces it by $T\,(D^*K_2 - P_2)$.

However, the amount per year before tax that is needed to turn a negative NPV from the second investment to a positive NPV is given by

$$(D^*K_2 - P_2).$$

But this is greater than the extra tax saved by the second investment. Thus, the extra tax saved will not turn the subsidiary investment from one with negative to one with positive NPV.

Thus the tax in these circumstances is neutral.

Suppose, however, the threshold rate for tax is below the discount rate, i.e., is equal to D_1, such that $D_1 < D^*$, and that the pre-tax P/K for a particular investment is higher than D^* so that there is a positive pre-tax NPV determined by the flow of $(P/K - D^*)K$ per year, which is the amount by which the annual flow exceeds that required to make the IRR equal to DR.

Tax per year on the proceeds of the investment will then be $T(P/K - D_1)K$. But

$$(P/K - D_1) > (P/K - D^*)$$

in this case. It is possible then, if T is high enough and $(D^* - D_1)$ high enough, that the extra tax, $T(P/K - D_1)K$, may be greater than the annual pre-tax cash flow that provides the positive NPV, namely $(P/K - D^*)K$, which would mean that the tax would turn a positive NPV into a negative one. This will happen if

$$T > \frac{P - D^*K}{P - D_1 K}.$$

Hence it is in general more likely to happen the higher is the tax rate (T). It is also more likely to happen the larger the gap between the threshold (D_1) and the discount rate (D^*). And it is more likely to happen the lower is the annual stream of profit (P) in relation to the threshold level of annual profit (D^*K).

If, on the other hand, $D_1 > D^*$, we have the possibility of the tax's turning a negative into a positive NPV. In this case the proceeds of a second investment K_2, with pre-tax negative NPV, following on an investment, K_1, with positive NPV, would incur no tax in itself but would reduce the tax per year payable on the proceeds of K_1 by

$$T(P_1 - D_1 K_1) - T[P_1 + P_2 - D_1 (K_1 + K_2)]$$

i.e., by

$$-T(P_2 - D_1 K_2)$$

i.e., by

$$T\left(D_1 - \frac{P_2}{K_2}\right)K_2.$$

But the amount per year by which the pre-tax proceeds of K_2 fall short of the annual flow required to give zero NPV is

$$\left(D^* - \frac{P_2}{K_2}\right)K_2.$$

Thus, if

$$T\left(D_1 - \frac{P_2}{K_2}\right)K_2 > \left(D^* - \frac{P_2}{K_2}\right)K_2$$

i.e., if

$$T > \frac{D^* K_2 - P_2}{D_1 K_2 - P_2} \tag{8.7}$$

(since $D_1 K_2 - P_2 > O$), the pre-tax negative NPV resulting from K_2 will be converted into a post-tax positive NPV. It follows from the condition (8.7) that this distorting in the direction of over-investment is more likely to occur the higher the rate of tax (T), the closer the annual pre-tax profit (P_2) is to the threshold profit level ($D^* K_2$), and the higher the gap between the threshold (D_1) and the discount rate (D^*).

Thus a threshold below the investor's DR will produce an incentive to under-investment, and one above the investor's DR will produce an incentive to over-investment. The incentive will in either case be more likely to be effective the higher the tax rate and the bigger the gap between tax threshold and discount rate.

This over-investment effect is essentially the same as the one described by Averch and Johnson (1962) for public utility price regulation. The case they supposed was one in which there was in effect a 100 per cent tax on proceeds above a certain rate of return. Their conclusion has been criticized on the ground that public utility regulation does not work exactly like this. A PPT, however, would work exactly in the way they suppose, except that the rate of tax would be less than 100 per cent.

Averch and Johnson, with a 100 per cent effective tax rate, concluded that, if the threshold (the regulated rate) were exactly equal to the discount rate, the level of investment would be indeterminate: the investor makes no positive NPV whatever decision he takes. Here, with a tax rate below 100 per cent, the PPT, with threshold equal to the discount rate, is neutral.

The incentive to over-investment will be effective only if there is a prior investment already subject to tax at the threshold. The higher the threshold is, on the other hand, the less likely it is that a prior investment will be taxable. This is presumably the reason why the Averch–Johnson literature supposes that the higher the allowed rate of return, the less the over-investment will be.

It must be stressed that these conclusions apply rigorously only in the model world we have supposed.

The resource rent tax (RRT)

Now let us consider the neutrality of the RRT in this same model world.

The RRT would collect no tax from the annual stream of net proceeds (P per year) of an investment (K) until the NPV of the project discounted at the threshold rate had just exceeded zero, that is, until the NPV of the stream of *P* values discounted at the threshold rate to the original investment date just exceeded the value of *K*. From that point, however, the RRT would take a fraction of the whole positive cash flow, not simply a fraction of the excess above a certain minimum, which is what the PPT would collect.

Suppose two identical projects (of the regular, everlasting kind we have supposed) were subjected respectively to the PPT and RRT with the same threshold rate (D_1) and tax rate (T). Each would reduce by a fraction (T) the NPV of the project derived from the cash flows over its whole life discounted at D_1. The PV of revenue extracted (discounted at rate D_1) would thus be the same in both cases, and hence the NPV after tax (discounted at D_1) would also be the same in both cases.

It follows that, if D_1 (the threshold rate) is identical with D^* (the investors' discount rate), the post-tax NPV to the investor is the same under the RRT as under the PPT. From this we can conclude that in this model the RRT, like the PPT, is neutral provided its threshold rate is set at the investors' discount rate.

If, however, the threshold is above or below the discount rate, there is the same tendency to distortion under the RRT as under the PPT. This can be shown by similar reasoning, except that, instead of looking at a single year's proceeds and the effect of the tax upon them (which will be the same for all years in the case of the PPT in this model, in which annual profits are assumed constant), we look at the NPV over the whole life of the enterprise. The RRT aims to take a part of this NPV if it is positive, but leaves it alone if it is negative. For this purpose, however, the cash flows are discounted by the threshold rate to give the NPV.

Suppose the threshold (D_1) is below the discount rate (D^*). Then a project that has a NPV, assessed at the investor's own discount rate, with a positive value of A^*, will be taxed under the RRT on a value of A_1 (where $A_1 > A^*$), so that the tax collected is higher than if the threshold had been D^*.

A project that in the absence of tax would have a NPV (at the investor's own discount rate), of A^*, just above zero, will have to pay tax of TA_1, which may well be higher than A^*, so that the NPV after tax is negative. Again, (i) the lower the NPV in the absence of tax, (ii) the higher the tax rate, and (iii) the bigger the gap between the threshold and the discount rate, the more likely it is that the distorting effect will operate. If we suppose a threshold, D_1, above the discount rate, then the reverse effect—a tendency to over-investment—may occur.

Any investment that can be conceived of as following on a previous investment whose proceeds are taxable and are taxed with its own will be able to offset, against any loss (negative NPV) that it would make in the absence of tax, the reduction in tax that it introduces to the proceeds of the previous investment.

We have seen that, if $D_1 = D^*$, the tax is neutral. This means that an additional investment, K_2, whose NPV (at the investors' discount rate), $A_2{}^*$, is lower than zero before tax, will have its proceeds taxed with those of the earlier investment, K_1, (whose NPV discounted at the threshold rate, namely $A_{1.1}$, is assumed to be above zero so that the proceeds would have been taxable) and the tax will be applied to a value

$$A_{1.1} + A_{2.1}$$

where

$$A_{2.1} < A_2{}^* < \text{O}.$$

Thus, the addition of K_2 reduces the tax payable by $-TA_{2.1}$.

Now the loss (difference between O and NPV) on the investment K_2 in the absence of tax would be

$$-A_2{}^*.$$

Thus a gain of $-TA_{2.1}$ will be offset in the investor's calculation by a loss of $-A_2{}^*$. Since

$$A_{2.1} < A_2{}^*$$

i.e.,

$$|A_{2.1}| > |A_2{}^*|$$

it is possible that the magnitude of the gain, $|TA_{2.1}|$, exceeds that of the loss, $|A_2{}^*|$, so that the investment, K_2, may be rendered eligible by the tax in spite of being ineligible in its absence.

Again, (i) the higher the NPV in the absence of tax, (ii) the higher the tax rate, and (iii) the greater the gap between the threshold rate and the discount rate, the more likely it is that the distorting effect will operate.

The Brown tax (BT)

This tax does not require any discounting for time. In the conditions of this model, the outcome under the BT of any project that would be undertaken in the absence of tax must have positive NPV. The Brown tax will simply reduce any positive or negative cash flows by a proportion, T. Thus, whatever the discount rate applied by the investor, a positive outcome will remain positive after tax since each flow is reduced as it occurs by a constant proportion; and similarly, a negative outcome will remain negative after tax.

Model 2: Riskless Investments with Irregular Yields

We now suppose that, though there is still no uncertainty, so that the investor knows exactly what the cash flows from his investment will be, the positive flows do not continue from year 1 at the same annual rate for ever but continue only for a finite time or in some other respect are irregular from year to year. Three stylized forms of irregularity can be considered: (i) where the annual cash flows are positive and equal to one another in every year from 1 to n but zero thereafter; (ii) as in (i), except that after year n there is a certain number of years of negative cash flow and then after that no flows at all; (iii) where the flows continue from year 1 for ever, are equal over each series of alternate years, but are higher in years 1, 3, 5, etc. than in years 2, 4, 6, etc., when they are below the investors' discount rate times the value of the investment. The taxes will be considered in the reverse order to that used above.

The Brown tax

The Brown tax is still neutral.

The resource rent tax

The resource rent tax, with its threshold at the investors' discount rate, is in these new circumstances neutral except in case (ii), where a negative cash flow follows a positive one. In that special case the investor will have been taxed on NPV at the appropriate rate up to year n, but the negative cash flows thereafter reduce the NPV, and no tax is rebated to adjust the total tax collection to the appropriate fraction of the NPV. Thus the base of the tax will be greater than the actual pre-tax NPV, and it is conceivable that the amount of tax will exceed the pre-tax NPV and so shift a pre-tax positive to a post-tax negative NPV. In the other two cases the base of the tax will be the NPV, and the tax will be neutral.

The progressive profit tax

The PPT, with its threshold at the investors' discount rate, differs from the RRT in that its incidence on projects with equal pre-tax NPV is typically affected by the time pattern of cash flows.

In case (i), with the cash flows positive, continuous, and regular up to a certain date and thereafter zero, the annual positive cash flow in each year from 1 to n would need to exceed D^*K in order to attain even a zero NPV. In other words, it would be possible for the positive cash flow in each of those years to exceed the level at which PPT becomes due even though the pre-tax NPV would be zero or negative. It follows that a positive pre-tax NPV might be turned by the tax into a negative post-tax NPV. The same applies *a fortiori* to case (ii), where there are some years of negative cash flow after year n.

To understand the effect of PPT (with the threshold D^*) in irregularity of the type in case (iii), suppose that an original annual pre-tax cash flow of P is altered by being raised in years 1, 3, etc., to $(P + y)$ and lowered in years 2, 4, etc., to $(P - y)(1 + D^*)$ where this expression is lower than D^*K. Then the pre-tax NPV over the whole life of the project is the same as if the annual cash flow were P each year. But in years 1, 3, 5, etc., there will be extra tax of Ty (over and above the tax payable in the case of constant cash flow), while in years 2, 4, 6, etc., there will be a reduction in tax; but, since $(P - y)(1 + D^*)$ is by assumption less than D^*K, the reduction in tax will be less than $Ty(1 + D^*)$. Thus there will be an increase of tax in each of years 1, 3, 5, etc., that has a present value greater than the reduction in tax in the following year in each case. So the PV of the tax bill will be increased by the discounted value of an amount that can be regarded as occurring in alternate years and that is higher (i) the higher the rate of tax, and (ii) the greater the difference between cash flows in 'high' and 'low' years. Since the latter of these variables can in principle be extended without limit and is certainly in no way limited by the post-tax difference between the annual positive cash flow in the equivalent continuous-flow case and D^*K, it is clear that this kind of irregularity in cash flow under the PPT (with threshold equal to the investor's discount rate) will allow a pre-tax positive to be changed to a post-tax negative NPV.

Thus the BT remains neutral in these circumstances; the RRT with threshold equal to the investor's discount rate is neutral except in a case in which sufficient negative cash flows follow the last positive ones from any project; the PPT with the same threshold will tend to have a bias to under-investment under any condition of irregularity that lowers the cash flow in some years below the value of the threshold.

Model 3: Investments with Uncertain Irregular Yields

We now suppose realistically that there is uncertainty about outcomes, so that what the investor is comparing with some standard is an *expected* cash flow; he will invest if the expected NPV (ENPV) is positive at a certain discount rate (which we suppose is raised on account of uncertainty). The expected value is taken in the usual sense to be the mean of the perceived probability distribution of values.

The progressive profit tax

The effect of uncertainty on the neutrality of the PPT is an exaggeration of the effect of irregularity. Thus, with a threshold equal to the investor's *DR* (D^*), the bias towards under-investment will be increased by uncertainty. This is because the expected tax on a probability distribution of values, some of which fall below the tax threshold, will be greater than the tax that would be due on the expected value. And this happens because (most markedly when the pre-tax present value is close to zero) there are extra taxes on outcomes above the expected value not fully balanced by reductions in taxes on outcomes below the expected value.

The resource rent tax

The RRT (with threshold equal to D^*) has a bias towards under-investment under conditions of uncertainty for essentially the same reason as that given above for the PPT. The expected tax on a probability distribution of outcomes will be greater than the tax appropriate to the expected outcome.

The Brown tax

The Brown tax will remain neutral, even under conditions of uncertainty, and will actually reduce the risk of the investment (by most measures of risk) since positive and negative cash flows are treated symmetrically for tax purposes and there is no threshold at which tax becomes due.

Variant: Depreciation and the Progressive Profit Tax

Depreciation allowances have been ignored in the foregoing demonstrations by supposing that all allowances are for actual expenditures to maintain the productive capacity of the mine for the period over which it is to produce. This is appropriate for the RRT and BT, but if the PPT follows the definitions of profit that are characteristic in income tax, it will make allowances for depreciation. The tax rules for depreciation are not uniform or necessarily very logical, and it is therefore not possible to say in general terms what their impact will be. In the previous chapter we differentiated PPT Mark II, which applies depreciation to the treatment of capital investment value (as well as to the assessment of annual income), from PPT Mark I, which does not. See table 7.4 for examples of both forms.

To deal with depreciation, we can suppose either that it follows some ideal pattern or that it conforms to a simple rule such as straight-line reduction. The model in the Appendix to this chapter supposes an ideal or true depreciation which reflects the reduction each year in the remaining present value of the project as it moves through its life. To allow for economic rent as distinct from the returns 'repaying' the cost of the investment, the initial value of the investment is taken to be the present value of a stream of cash flows that, discounted at the investor's discount rate, would equal the cost of the original investment. It is this amount that is depreciated as the hypothetical stream of 'normal' returns is realized. Net receipts in excess of the amounts in this hypothetical stream are regarded as constituting the economic rent, the return in excess of normal. The Appendix to this chapter does not suppose that the actual stream of cash flows is infinitely prolonged or that it is regular from year to year, but it does assume that in each year it is no less than the hypothetical 'normal' cash flow of that year.

Of course, true depreciation in the sense of this model can not be precisely applied in a real tax system. To the extent that actual depreciation allowances may bear some approximation to the true depreciation of this model, it may be of interest that, as shown in the Appendix, PPT Mark II with a threshold at the investor's discount rate can be applied in such a way that it takes a share of the economic rent only. This means that, provided the investor is completely confident that in each period he will receive income no less than the 'normal' return supposed for that year, the tax is neutral.

However, for reasons given above in this chapter under 'Model 2' and 'Model 3', if the investor is not completely convinced that actual returns will not fall below these 'normal' levels in any year, the PPT Mark II with a threshold at the investor's discount rate will have a bias to under-investment.

The effect of the methods of depreciation actually used will depend on their form and the time over which the depreciation is applied. Price inflation, which is normally ignored in actual depreciation for tax purposes, will reduce the real value of depreciation allowances, and, if the capital value is not indexed for purposes of applying the PPT threshold, there is a further burden on the investor, and inflation will exacerbate any bias towards under investment. (It is not out of the question that indexation should be applied for *both* purposes, but this would be a further departure of the PPT from income tax rules.) If depreciation is allowed over a period shorter than the ideal, this will reduce the present value of income for tax purposes but will at the same time reduce in some years the capital values to which the threshold is applied. These two effects will work in opposite directions in their impact on the severity of the PPT Mark II.

When depreciation is included in determination of the PPT Mark I, that tax becomes less severe than the PPT Mark II for any given combination of threshold rate and tax rate. Thus, inclusion of depreciation allowances appears to remove or reduce any presumption that the tax, with its threshold at the investor's discount rate, will have an under-investment bias. This is true at any rate of its effect on potential investments that are subject to depreciation for tax purposes.

It is clear that the use of depreciation allowances embodied in actual income tax legislation renders it more difficult to make statements about the conditions of neutrality for the PPT, but, with the PPT Mark II at least, the presumption is still that the use of the investor's discount rate as threshold will introduce some bias to under-investment.

Thus the BT is neutral, but the RRT and PPT Mark II have a bias towards under-investment when the threshold is at the discount rate. This tendency would appear to be more marked and consistent under the PPT Mark II than under the RRT for any given tax rate. These distortions apply, in the case of the *general* forms of the taxes, to the choice among projects, but, in the case of ideally tailor-made taxes, only to 'subsequent' decisions.

It would seem that the best approximation to neutrality could be obtained by setting the thresholds for the PPT Mark II and RRT close to, but somewhat above, the discount rate. If depreciation allowances are applied as for income tax, the verdict on the PPT Mark I is doubtful. If the investor could be taken to base his decisions on the assumption that he would certainly achieve a positive or zero NPV both over the whole project life and over the last x years of the project life taken together, where x is any number less than the total number of years in the project's life, his discount rate would provide a neutral threshold rate under RRT. This condition is not sufficient, however, for the investor's discount rate to provide a neutral threshold rate under PPT in the absence of depreciation allowances, or under PPT Mark II with depreciation allowances approaching true depreciation; the corresponding sufficient condition here is that the investor should be certain of achieving an income not lower than the value of the threshold (D^*K) in each year. There would be no entirely satisfactory threshold covering all cases unless all possible investments and increments to investment had the same degrees of risk and (in the case of the PPT) the same typical time patterns of yields. But we may for the purpose of discussion talk about a threshold rate that best approximates to neutrality with a certain tax for a certain class of investment. This might be called a 'quasi-neutral threshold' or a 'threshold approaching neutrality'. It will tend to be higher for the PPT than for the RRT with the same tax rate for a given class of investments.

The discussion of the neutrality of these three taxes should be taken by comparison with the main alternatives: the fixed fee (FF), the specific or *ad valorem* duty (SAVD), and the higher rate of ordinary income tax (HRIT). The FF, though neutral between techniques and over all decisions subsequent to the choice of the project, will, unless determined ideally by negotiation or auction, have a bias to under-investment, in the sense of knocking out marginal projects. The SAVD, unless ideally negotiated, will also have a clearly deterrent effect on marginal projects beside its effect already mentioned on the extent of production. The HRIT in 'general' form will, even under ideal conditions of regular cash flows and certainty, require higher pre-tax returns of a project that would be needed in its absence (and hence higher than in industries not subject to the HRIT), so that it would have a bias to under-investment even in circumstances in which the PPT could approach complete neutrality. With the HRIT, by contrast with the PPT, this disincentive effect could not in principle be removed by raising the threshold (since there is no threshold to raise) but only by reducing the rate of the extra tax towards zero and so removing the additional revenue from highly profitable, as well as from marginally profitable, projects. *A fortiori*, the HRIT is inferior on these grounds also to the RRT and the BT.

This account abstracts from the practical problems of fixing appropriate thresholds in the case of the PPT and RRT. Dowell (1980) suggests, by the use of simulation procedures, that serious distortions from neutrality may occur through fairly small errors in the fixing of the RRT threshold, and that these errors may seriously reduce the revenue yield of the tax. He argues that, if the tax is applied in general form, the threshold is likely to be wrong for at least some projects since investors will apply different discount rates to different projects. He proposes instead the use of a combination of auctioned FF and an *ad valorem* tax on output that is arranged to disappear when a certain appropriate point in the life of each project has been reached so as to avoid the high-grading effect. (Such a disappearing royalty, but with its disappearance discretionary, is planned to operate over UK oil and gas, and royalties sliding downwards according to a formula have recently been used experimentally on some federal oil and gas leases in the USA.) Whatever the merits of this proposal, it should be pointed out that the BT at least avoids the distortions of which Dowell complains as allegedly arising from the RRT. It is, however, difficult to see a disappearing royalty constructed in such a way that, applied in 'general' form, it can be a fine instrument for tapping residual rent. The onus of proof is on the proposer to outline a practically applicable method of doing so. If not in general form, the royalty would need to be fixed arbitrarily by project in advance of bidding on the fixed fee. This is because auctioning two tax parameters is out of the question, as presumably is negotiating one parameter with the investor after he has won his mining rights by bidding on another. Though, as we have shown, the PPT and RRT may have distorting effects if their thresholds are *wrongly* fixed, Dowell's analysis does not seem to justify the strong practical implications that he seems to derive from this danger.

Importance of Neutrality

What practical conclusions are to be drawn from these observations? In chapter 7 (p. 105), the distinction was made between 'case 1' and 'case 2' countries, and the general presumption was advanced that neutrality, of obvious value in case 2, is also generally to be sought in case 1. The incentive to under-investment in mining means in the case 2 country that too few of its resources are directed to the mining industry; in the case 1 country, that productive investment that could be attracted to the country is being excluded; and in both, that the techniques in mining may tend to be unduly capital-sparing. The incentive to over-investment means in the case 2 country that too many of its resources are directed to mining; in the case 1 country, that it is making certain investments possible by subsidizing them through remission of taxes on other investments; and in both, that techniques in mining are encouraged to be unduly capital-using. It is not inconceivable that there are positive externalities from directing capital to mining that justify the subsidy, or negative externalities that justify the deterrent, but the case needs to be fairly clearly made in either case and not just taken for granted; and there may well be more appropriate ways of allowing for such externalities if they exist.

Other Decision Rules Affecting the Supply Price of Investment

The above discussion of neutrality with respect to investment is premissed on the view that investment proceeds when the discounting of expected net cash flows at the investor's discount rate yields positive expected net present value. As discussed in chapter 4, we believe that this is the most important single criterion applied by investors in large-scale mineral production.

But it was also observed in chapter 4 that investors commonly require possible mineral investments to pass other tests as well. Perhaps the most important additional tests are the bankers' requirement that loan-funded expenditures (which are typically the major capital expenditures in large investments) be repaid with interest in a relatively short period of time with very small probability of default; and the managers' requirement that there should be very small probability of absolute loss or of low profitability. To the extent that these considerations are important, the conclusions of the previous discussion of non-neutrality with respect to investment decisions may be modified.

When investment decisions are made according to the 'bankers' criterion' of short pay-back periods with low probability of default, taxation systems are more nearly neutral the more their impact on producers' earnings is concentrated later in time, and in particular after the initial investment has been recouped with interest. The FF fares poorly by this criterion if the amount is set in general for all projects. (However, if the FF is auctioned, its effect in raising the supply price of investment is reflected, if at all, in low revenue yield rather than in non-neutrality.) By the bankers' criterion alone, HRIT and PPT introduce substantial non-neutrality into investment decisions by delaying pay-back. The RRT, however, with its threshold equal to the bankers' discount rate, is neutral in the sense that it does not delay pay-back. The BT, since it simply reduces the scale of all inflows and outflows and does not affect the pay-back period whatever discount rate is applied, is also neutral provided the 'sovereign' risk (that the rules may be altered during the life of the project) is discounted.

When investment decisions are made according to the 'managers' criterion' (that there must be a very low probability of low realized returns), taxation systems are more nearly neutral the more their impact on producers' earnings is concentrated on income that would represent a rate of return on investment in excess of the minimum acceptable (see figures 4.1 and 4.2). For a given expected revenue on each project, again the FF (in general form) fares worst, and specific duties, *ad valorem* duties, HRIT, and PPT, in that order, progressively less badly. If neutrality by this criterion means simply that there is no increase or decrease in the probability that the investor's NPV would actually fall below zero, the RRT is neutral provided that its threshold is precisely at the investors' discount rate and that the investor ignores the possibility that a period of negative net receipts may close the life of a project and follow a period in which positive NPV has been realized over the project's life to that time. By the same criterion, the BT is neutral if sovereign risk can be ignored.

Neutrality and Sovereign Risk

It was observed in chapter 4 that it is investors' subjective expectations about after-tax cash flows that are relevant to investment decisions. Whatever the state of law or policy on the taxation of mineral income at the time of investment, the investor, if he expects that the law or policy may change in future, will take that prospective change into account.

Thus, in addition to raising the supply price of investment, as observed in chapter 4, sovereign risk affects expectations of after-tax income in the direction of leading the investor to expect tax demands higher than those due under current law or policy. (Strictly, since sovereign risk would exist in the absence of any tax, it could be argued that neutrality would involve retaining it rather than removing it. For convenience of exposition, however, and in order to enable 'neutrality' to be associated more closely with desirable allocation, we shall adopt the fiction that sovereign risk is introduced by a system of taxation, so that maintaining it will be potentially non-neutral.) Both the raising of the supply price of investment and the raising of expectations about tax payments may possibly deter investments that would be undertaken in the absence of sovereign risk. Sovereign risk may thus be regarded as a source of non-neutrality, so that there is a tendency for tax systems that seem more likely to remain unchanged over long periods of time to be more nearly neutral.

There is no simple relationship between the form and the stability of a tax system. However, as discussed in chapter 4, there is general presumption that it is more likely to be stable the more it has the effect of compressing the probability distribution of after-tax investment outcomes. The FF and SAVD are strongly non-neutral by this test since they do not reduce the dispersion of the probability distribution of after-tax outcomes. The HRIT may have some tendency to reduce dispersion, but this is not very strong. The PPT, however, and especially the RRT and BT, do compress the distribution of outcomes as measured by NPV. Nevertheless, the BT may have another reason for incurring sovereign risk in the fact that it subsidizes investments and losses. Politically this may sit rather uneasily. There may thus, on balance, be some advantage for the RRT and PPT over the BT. Although expectations of instability in law and policy can not be entirely eliminated for the PPT and RRT, they are likely to be less important with these two than with other forms of taxation.

Differing Neutrality Considerations with Respect to Exploration, Mine Development, and Mineral Production

The analysis in this chapter suggests that it is likely that all forms of mineral taxation may affect decisions on investment or production in some circumstances. Where these effects represent simply risk reduction, they are probably benign. If sovereign risk can be ignored, this seems to be the case with the only changes brought about by the Brown tax. The other taxes, even if they are considered 'stable' by the investors, raise greater or lesser possibilities of undesirable distortion. The non-neutralities of the various forms of tax, however, are of differing relative importance for the three phases of decisions that have to be made: namely, on exploration, on mine development, and on actual mineral production.

The SAVD and the FF, set at the same rate for all mineral industries and projects, are strongly distorting at the stage at which the choice of project is made, and the SAVD is distorting at all stages of investment and production, and especially in the recovery of minerals from established mines.

For those same decisions on recovery from established mines, the FF and BT are for practical purposes neutral. The RRT, with its threshold set at the investors' discount rate, is also close to being strictly neutral in this phase: the tendency, discussed above, for it to discourage some investment decisions is virtually eliminated when outlays in any year are less than gross revenues and the tax collection has already begun; at such a stage the cost of investment is effectively reduced at the tax rate, so that the RRT is then very like the BT. The PPT and HRIT are successively farther from strict neutrality in this production phase, but their divergence from it would still seem to be of minor importance.

Neutrality with respect to exploration and mine development decisions seems to dictate the use of BT, RRT, or PPT in preference to any of the alternatives apart from the auctioned form of the FF. If we abstract from the effects on sovereign risk and ignore the risk-reducing effect of the BT, the auctioned FF and the BT, and they alone, are strictly neutral. The practical superiority of BT, RRT, and PPT over the auctioned FF has to depend on the latter's lack of immunity from perceptions of sovereign risk and its high risk in general, hence on what may in many circumstances be its lower capacity for raising revenue, and the related question of whether, in particular industries under the usual conventions about rights, an effective auction can really be held. The last two questions are discussed in chapters 9 and 10.

It has been shown, however, that complete neutrality cannot be secured in practice by the RRT or PPT except under rather restricted conditions involving certainties on the investor's part. If tax regimes could be set project-by-project it would be possible to avoid discrimination among projects by setting the rate of additional tax as high as it can be set without removing the incentive to economizing, and fixing by negotiation or auction the lowest threshold that the investor would accept. This, if it were practicable, would be apparently a way of maximizing revenue without deterring marginal projects. It would be strictly neutral, however, only if the project came as a single immutable package with fixed techniques. In so far as subsequent investments or any variation in techniques were possible, the threshold and rate of tax arrived at by the negotiation (or auctioning) might have over-investment or under-investment effects.

In addition to these disadvantages,

(i) auctioning may not be practicable with a number of the mineral products and is of questionable use with the taxes other than the FF;

(ii) a government will almost certainly not have the information to achieve the 'right' threshold by negotiation;
(iii) project-by-project fixing of tax regimes may discourage exploration because of the uncertainty involved.

For these reasons (explained at greater length in chapter 6), we believe that in general the wisest policy is to have a pre-determined tax regime for each mining industry, a regime that may be changed over time but so that the change applies only to newcomers to the industry concerned, provided of course that the initial regime established is one fairly rationally designed to pursue national objectives.

If this is done, the threshold and tax rate adopted for an industry with either PPT or RRT will inevitably have some potentially distorting effects. There will be some conceivable socially acceptable investments that it will deter, or others that it will inappropriately subsidize, or both. The threshold chosen can only be a compromise. The general line of the argument in this chapter would support the view that, for any industry, the threshold appropriate for the PPT without depreciation allowance or for the PPT Mark II would be higher than that for the RRT with the same tax rate, and that either should be somewhat higher than a rate that could be taken as the investor's discount rate; and moreover that the possibility of distorting effects is more serious with the PPT than with the RRT. Furthermore, the more project-based, rather than company-based, the additional tax regime is, the less are the incentives to over-investment (and the erosion of the tax base that arises from them) likely to operate, and therefore the higher the threshold may reasonably be set.

To the extent that eligibility is determined partly by the bankers' criterion or the managers' criterion, the RRT and the BT will have all the more marked advantages in terms of neutrality over against the other taxes. These special influences on the supply price of investment are more important in the mine development than in the exploration phase. Sovereign risk, however, may cast some shadow over the BT's otherwise great advantages on these criteria and may tell against it, particularly for the mine development phase when very large expenditures are incurred. By contrast, since the PPT and RRT are potentially more distorting in relation to investments with a more dispersed, than to those with a less dispersed, probability distribution of outcomes, their distorting effect is probably more important to exploration than to mine development decisions. Thus the BT's advantages in neutrality over the PPT and the RRT are likely to be particularly important in exploration decisions. Hence the RRT may have the advantage over the BT in the mine development phase and the BT over the RRT in the exploration phase.

Though the choice of tax system must take account of characteristics other than neutrality, as discussed in the next two chapters, the considerations just mentioned here might suggest the use of a hybrid system of rent charges in which the BT principle was applied to exploration expenditures (which would mean that they were subsidized, say at the mineral tax rate) and RRT to mine development and production (which would mean that from the beginning of mine development capital expenditures should be immediately deductible and losses should be carried forward at an interest rate equal to the investor's discount rate). This combination would make it reasonable to have a lower threshold rate for the RRT element than would approach neutrality best on balance if an exclusively RRT scheme were adopted. This is because exploration is a more risky investment than development and production. Since the lower discount rate might well be closer to those actually applied by investors over development and production decisions, the neutrality of the tax over such decisions might be thereby enhanced. However, as suggested at the end of chapter 6, it may be better still for the government to carry the principle of subsidizing exploration even further and to undertake exploration itself as far as feasible, auctioning mining rights in that case when their value is more fully known.

To apply the auctioned fixed fee ('cash bonus bid'), in circumstances where it can raise substantial amounts of revenue, *with* a PPT, RRT, or BT either increases revenue or makes it reasonable to reduce the rates of those taxes and so to reduce both any over-investment or under-investment bias owing to imperfect choice of threshold in the case of the first two and any tendency to gold-plating in all three. The addition of the auctioned fixed fee may thus be said to enhance the approach to neutrality of these profitability-based taxes or to increase total revenue, or both.

Summary

The Brown tax and the auctioned form of the fixed fee are neutral in relation to all production decisions if we ignore the fact that the Brown tax reduces risk, and also ignore the possibility that investors may expect the rules of the game to be changed. All the other taxes considered allow the possibility of non-neutrality in relation to choice of investment projects and to the techniques and pace of production, but, provided the taxes are imposed with sensible parameters, the distortions generally diminish in importance for any given expected revenue as one moves from 'general' fixed fee to specific or *ad valorem* duty to higher rate of income tax to progressive profit tax to resource rent tax, all imposed in general form. Any of these taxes ideally negotiated project by project might avoid non-neutrality over the choice of projects, but this is a reservation of no great practical importance. With the progressive profit tax and the resource rent tax, the best approach to neutrality is probably made by setting the threshold rate close to, but if anything somewhat above, the investors' discount rate and somewhat higher for the former tax than for the latter with any given tax rate. The specific or *ad valorem* duty is in addition non-neutral in relation to the extent of production.

The need to take account of the bankers' or managers' requirements in their implications for neutrality enhances the advantage of the Brown tax and resource rent tax. If sovereign risk is taken to impair neutrality, taking it into account reduces the advantage of the auctioned fixed fee or the specific or *ad valorem* duty and, perhaps but to a less extent, of the Brown tax and the higher rate of income tax, since they are the ones for which the rules seem most likely to be subject to alteration on account of political pressure, a risk from which the progressive profit tax and resource rent tax are likely to be largely immune.

Promoting neutrality may well be served best in practice either by the BT alone or by subsidizing exploration (or providing it by government enterprise, and auctioning leases for a fixed fee) and taxing mine development and mineral recovery by the resource rent tax with a threshold slightly higher than the investors' presumed discount rate.

Where PPT, RRT, or BT is used, the addition of an auctioned FF (cash bonus bid) will in appropriate circumstances increase total revenue or else allow their approach to neutrality to be enhanced, or both.

Appendix
Neutrality of the Progressive Profit Tax (PPT) Mark II

Suppose a project with a negative cash-flow, K_0 in year 0, and positive cash flows, $Y_1, Y_2 \ldots Y_n$ in years 1 to n, with no cash flows thereafter.

Let r be the investor's rate of discount for time.

Treat the excess of the actual present value of the project in year 0 over K as a windfall profit or economic rent, and suppose that the capital value of the investment is depreciated for tax purposes to represent what the true depreciation of the value of the project would be if the actual positive cash flows represented a schedule of what might be called 'normal-return' cash flows, $Y'_1, Y'_2 \ldots Y'_n$ in years 1 to n respectively such that $Y_1 \geqslant Y'_1$, $Y_2 \geqslant Y'_2$, etc., and the present value in year 0 of the stream of normal-return cash flows Y'_1, Y'_2, etc., were exactly equal to K_0.

Let K_1, K_2, etc., be the depreciated values of the capital investment in years 1, 2, etc., and let D_1, D_2, etc., be the depreciation deducted for tax purposes in years 1, 2, etc., such that $K_1 = K_0 - D_1$; $K_2 = K_1 - D_2$, etc.

Let Q_1, Q_2, etc., be income for tax purposes in years 1, 2, etc., such that $Q_1 = Y_1 - D_1$, etc.

True depreciated value in year 1 is taken to be the present value in year 1 of the remaining 'normal-return' cash flows for the following years. Thus,

$$K_1 = \left[K_0 - \frac{Y_1'}{(1+r)}\right](1+r) \tag{A8.1}$$

since the normal-return cash flow Y_1' has to be deducted, but all remaining normal-return cash flows are one year closer in time.

Since $D_1 = K_0 - K_1$, then from (A8.1),

$$\begin{aligned} D_1 &= K_0 - K_0 - rK_0 + Y_1' \\ &= Y_1' - rK_0 \\ \therefore Q_1' &= Y_1' - (Y_1' - rK_0) \\ &= rK_0 \end{aligned}$$

where Q_1' is the income that would be used as base for tax purposes in year 1 if actual positive cash-flow, $Y_1 = Y_1'$.

Similarly it can be shown that

$$\begin{aligned} Q_2' &= rK_1 \\ Q_3' &= rK_2, \text{ etc.} \end{aligned}$$

It follows that, if the project is exactly on the margin of profitability so that the present value in year 0 of the actual positive cash flows is exactly equal to K_0, the initial investment, a PPT (progressive profit tax) that takes only a proportion of the excess of income above rK_0 in year 1, above rK_1 in year 2, above rK_2 in year 3, etc., will collect no tax.

Indeed, provided the other conditions of the model as expounded above are fulfilled, such a tax will tax only the windfall profit or economic rent.

But note that those conditions include a positive cash flow in every project year after year 0, and moreover one not lower than the 'normal-return' positive cash flow used for purposes of calculating true depreciation. (Note also that true depreciation would need to be negative in any year m in which $Y_m' < rK_{m-1}$.)

In these restricted conditions, and provided all investors are certain that they will be fulfilled, the PPT Mark II, having its threshold rate equal to the investor's discount rate for time, and applying the threshold to the depreciated capital value of the previous period, would fall on economic rent or windfall profit only and would therefore presumably be neutral.

The conditions for neutrality of the RRT are less restrictive, and those of the BT, less again.

But it does not appear possible to state similar neutrality conditions for the PPT Mark I without adding assumptions much further from real-world possibilities, for example, that the project's positive cash flow continues at a constant rate for ever, so that there is no true depreciation.

9

GOVERNMENT REVENUE MAXIMIZATION AND THE DEGREE OF REVENUE UNCERTAINTY

This chapter examines the six basic types of tax introduced in chapter 7 and asks how they are likely to compare with one another in the amount of government revenue that can be raised and in the risk or uncertainty and delay of the government revenue flow from them. The greater part of the chapter analyses the main determinants of revenue-raising capacity, at first when the taxes are set at negotiated or auctioned rates and then when they are set at general rates announced in advance. The concluding pages of the chapter discuss uncertainty and delay in government revenue yields from various forms of taxation.

Revenue Maximization

There are four considerations in judging between the various taxes on capacity for raising revenue. First is their neutrality, discussed in the previous chapter. Second, the greater the propensity of the tax to reduce risk to the investor as he perceives it, the lower the investor's minimum requirement in expected returns and the higher the rent available to be collected. Third, the lower the top marginal rate of tax on profits or on net cash flows that can be associated with a given total revenue from projects actually undertaken, the more revenue can be collected from profitable projects without reaching rates of tax at which incentives to economize markedly diminish. Fourth, the more closely the tax burden is related to profitability as the investor perceives it, the higher the revenue that can be collected from profitable projects without deterring the marginal ones. We later examine how these last three considerations combine to influence the respective revenue-yielding capacities of the six taxes when neutrality *within* projects is preserved as far as possible, in particular if each tax is applied uniformly for all projects in an industry. The second and fourth of these criteria, investor risk and adaptability, rank the six taxes similarly if not identically.

Since risk to the investors, and therefore their discount rate, is reduced by credibly excluding renegotiation of the fiscal terms of a project, and *a fortiori* by fixing terms generally for the industry rather than leaving them to be negotiated, such arrangements for stabilizing and generalizing the rules will in themselves increase the rent to be extracted (see the argument in chapter 6). Admittedly, if knowledge were perfect and administrative costs nil, negotiation project by project would have to yield as much or more revenue overall. But in practice, quite apart from the losses in rent owing to the higher risk to the investor of project-by-project negotiation, there are administrative and other difficulties associated with this procedure. These difficulties are discussed in detail in chapter 10. There is thus a strong case for the establishment of general rules fixed in advance across each of the mineral industries. Some taxes can, and some cannot, be imposed so that, without variations in the rules to deal with the variety of projects or revision to meet changing circumstances, they can collect little or no revenue from unprofitable projects and high revenue from profitable projects. Some, in other words, are much better adapted than others for collecting the economic rents solely and thoroughly. It is those taxes most closely adjusted to collecting actual rents that are also the ones that can be imposed in general form at a standard rate with least sacrifice of marginal projects or of revenue from intra-marginal projects. Taxes that meet this requirement can also credibly be presented to the public as not subject to alteration. On this ground the FF and SAVD in general form are clearly unsuitable, and suitability generally increases as we move to the right of tables 7.1–7.3.

The advantages of general rules by themselves tell strongly in favour of the kinds of taxation favoured by the second and fourth of the four general considerations described above as determining the relative capacity of various taxes to raise revenue. This is just because these are the kinds that can reasonably be imposed in general form. On the question of the capacity of various taxes to yield revenue, there may be some conflict between the maximum-rate limit and the other three general considerations, but the closer examination below shows that there is also some consistency between their rankings.

In order to sort out what practical conclusions can be drawn about capacity to raise revenue, the following analysis at first makes some simplifying assumptions. These are later removed in the process of suggesting some realistic conclusions.

1 *Investor risk*

If we ignore for the present the limitation imposed on top marginal rates of income tax by the need to preserve the incentive to economize, and if we assume that thresholds for the PPT and RRT are applied either at the investors' discount rate or at some rate above it which is best calculated overall to promote neutrality, then, for any given ENPV of tax revenue (assessed at the investors' pre-tax discount rate), taxes further to the right in tables 7.1–7.3 will entail less risk to the investor than those further to the left. This is true whether risk is measured by the variance of the expected NPV outcomes or by the probability that NPV will fall below some specified level. At one extreme, the FF collects exactly the same amount from profitable and unprofitable projects and simply shifts the probability distribution of outcomes to the left without diminishing its dispersion. At the other extreme, the BT reduces positive and negative deviations from zero NPV in equal proportion; and each of the four taxes on the right, since its tax burden is related to profit, will presumably show less dispersion of after-tax outcomes than the SAVD and FF. It seems that the *ad valorem* duty will have its burden more closely related to profit, and hence will reduce dispersion more, than the specific duty, and the specific duty more than the FF.

The above paragraph has compared the risks for one project under the taxes on the assumption of equal ENPV of tax revenue. Suppose then that the tax systems are applied to six identical projects to give equal ENPV from each and that this level of tax leaves all six eligible. Then suppose that the tax rate is raised on all six to keep the ENPV of revenue still equal on all. As this process continues, the projects will one by one become ineligible, generally in order starting from those subject to taxes at the left of the table, which leave the risk highest. Thus generally each tax will be capable of raising more revenue on any project than all those further to the left of the table.

So much at least follows if the investors' discount rate is used to discount the revenue stream.

If the government's discount rate, supposed lower than the investors', is used instead to define what is meant by equal ENPV of revenue, this will increase the weighting given to revenue received later rather than earlier and diminish the weight given to early negative cash flows. It will therefore tend from the investor's viewpoint to diminish the relative severity of those taxes in which the revenue tends to be received later. The use of this lower discount rate on the revenue streams from the six systems will thus favour even more those further to the right, specifically telling in favour of all the others in relation to the FF, and of the RRT and BT in relation to the other four, and the conclusion already drawn is reinforced.

2 *Maximum-rate limit*

We consider this question first by discounting all cash flows at the investors' discount rate to find present values and by supposing (i) that the taxes can be applied with rates fixed individually for all projects, (ii) that the threshold rate is set in the case of the PPT and RRT at the investors' discount rate, (iii) that there are no interest payments to be met by the project, and (iv) that, in the cases of the HRIT and PPT, depreciation allowances either do not exist or represent only actual maintenance expenses required to keep the project producing and not otherwise deducted. Identical projects are supposed, with constant, certain, positive cash flows continuing from year 1 for ever. The four profit taxes (HRIT, PPT, RRT, and BT) are applied, one to each project, and at rates such that the same NPV of revenue as assessed at the investors' discount rate is obtained from each tax.

Then the tax rates of RRT and PPT will be the same. This can be shown as follows. If the RRT is applied at a rate of T%, it reduces the NPV by T%, since in the circumstances specified it is a tax on the NPV. In the case of the PPT the positive amount in each year from year 1 received in excess of an amount D^*K (where D^* is the investors' discount rate and K is the original cash flow in year 0) will, when discounted for time, sum to the NPV of the project. Reduction of each of these annual amounts by T% as it is received will reduce the NPV of the project by T%. Hence equal rates of the two taxes applied to equal pre-tax NPVs will reduce these NPVs in the same proportion and hence yield the same NPV of revenue accumulated at the investors' discount rate.

To yield equal revenue, the BT will also be applied at the same rate as the RRT. This can be seen from the following reasoning. The RRT revenue will begin to be collected just after a year m such that positive cash flows in years 1 to m, when discounted at D^* to year 0, exactly equal in magnitude the negative cash flow, K, in year 0. The BT, having applied the same rate of subsidy to the negative cash flow in year 0 and of tax to the positive cash flows in years 1 to m, will have yielded a zero NPV of revenue by year m. Thereafter, to raise the same revenue as the RRT, which begins to collect it from that time, it must have the same tax rate. The HRIT, however, will have a lower top rate of tax than the PPT to raise the same revenue. This is because the HRIT collects the additional tax on the whole of taxable profit, not just on the part above D^*K in each year. Thus under these conditions

$$T'_{\text{HRIT}} < T'_{\text{PPT}} = T'_{\text{RRT}} = T'_{\text{BT}}$$

where T' for each tax is the rate required to yield a certain given NPV of revenue under the conditions stated.

In the case given above, the depreciation allowances (other than actual maintenance expenses) of zero represent 'true' depreciation, since a project generating a constant rate of cash flow from year 1 for ever is not undergoing any loss of value from one year to the next.

Now, however, we alter the assumptions to suppose that the annual positive cash flows, while still regular, end after year n such that a pre-tax positive NPV has been achieved before year n ($n > m$). We also suppose that, in the case of the HRIT and PPT, straight-line depreciation allowances are applied to depreciate any investment completely over its productive life.

It will be clear that the equality of revenue between RRT and BT with a given tax rate is unaltered by these changes of assumptions. Let us consider, however, the effect of the changes on the relationship in revenue-raising capacity at a given tax rate between the PPT and these two taxes, taking first the PPT Mark I (PPT$'$). For this purpose, let the magnitude of the negative cash flow in year 0 be K, and the positive cash flow in each year from 1 to n be P. Then the depreciation allowance in each year will be K/n, 'profits' of the kind taxed under income tax will be $P - K/n$, and the part additionally taxable under PPT will be $P - K/n - D^*K$.

Suppose a given rate of tax, T_0, applying to both PPT and RRT. The present value of the revenue stream for the RRT will be given by:

$$PV_0\ (tax_{\text{RRT}}) = T_0 \left\{ \sum_{i=1}^{n} \left[\frac{P}{(1+D^*)^i} \right] - K \right\}. \tag{9.1}$$

Then the present value of the revenue stream for the PPT$'$ will be given by:

$$PV_0\ (tax_{\text{PPT}'}) = T_0 \left\{ \sum_{i=1}^{n} \left[\frac{P - K/n - D^*K}{(1+D^*)^i} \right] \right\}$$

$$= T_0 \left\{ \sum_{i=1}^{n} \left[\frac{P}{(1+D^*)^i} - \sum_{i=1}^{n} \frac{K/n + D^*K}{(1+D^*)^i} \right] \right\}. \tag{9.2}$$

The question is whether the second term in the braces in (9.2) above, with its sign changed, is greater than K. We call this term with its sign changed X. Then, if $X > K$, the PPT$'$ in these conditions raises less revenue than the RRT. As defined,

$$X = \sum_{i=1}^{n} \left[\frac{K/n + D^*K}{(1+D^*)^i} \right].$$

From summing this series,

$$X = \frac{K/n + D^*K}{(1+D^*)} \cdot \frac{1 - \dfrac{1}{(1+D^*)^n}}{1 - \dfrac{1}{(1+D^*)}}$$

$$= \frac{K/n + D^*K}{D^*} \cdot \frac{(1 + D^*)^n - 1}{(1+D^*)^n}$$

$$= \left(\frac{K}{nD^*} + K\right) \left[\frac{(1+D^*)^n - 1}{(1+D^*)^n}\right]$$

$$= K\left(1 + \frac{1}{nD^*}\right) \left[\frac{(1+D^*)^n - 1}{(1+D^*)^n}\right]$$

$$= K\left(\frac{nD^* + 1}{nD^*}\right) \left[\frac{(1+D^*)^n - 1}{(1+D^*)^n}\right]$$

If we substitute Y for the factors giving X/K, we see that

$$Y > 1 \text{ so long as } (1 + D^*)^n > nD^* + 1.$$

But, from the binominal expansion,

$$(1 + D^*)^n > nD^* + 1 \text{ provided } n > 1 \text{ (and provided, as assumed here, } D^* > 0).$$

Thus, so long as $n > 1$,

$$Y > 1,$$

and therefore $X > K$.

Thus, so long as the life of the investment is greater than one year (one tax period), the RRT in the circumstances stated will raise more revenue than the PPT Mark I. (If the life is one year, the straight-line depreciation, depreciating the whole in one year, will represent true depreciation, and RRT and PPT′ will then yield the same revenue for a given tax rate.) Clearly, if the straight-line depreciation takes place over less than the whole life of the project, the PV of revenue from the PPT′ will be less again. And, if the depreciation takes place at a constant proportional rate by the reducing-balance method, so as to reduce the undepreciated part to a negligible amount by the end of the investment's life or before, the PPT′ will yield a smaller value of revenue than by the straight-line method, since the reducing-balance method will mean larger depreciation in the early stages.

An unambiguous demonstration, even under these somewhat limited conditions, is not so easy in comparison between the RRT and the PPT Mark II (PPT″). The PPT″ will of course raise more revenue at a given tax rate (and threshold) than the PPT′ in the conditions listed and generally in all circumstances in which depreciation is involved. If it is used with true depreciation, as defined in chapter 8, and cash flow in each year is above the 'expected' cash flow on which the true depreciation is based, PPT″, with threshold rate at the investor's discount rate, takes a proportion (equal to the tax rate) of the economic rent (see the Appendix to chapter 8). The RRT in similar conditions (with the same threshold and tax rate) will take the same proportion of the economic rent. Therefore in these conditions their revenue-yielding properties will be the same. If, instead, straight-line or reducing-balance depreciation is used, the outcome will depend on whether this yields more or less revenue than 'true' depreciation. It can be shown from the equations in the Appendix to chapter 8 that, if the positive cash flows on which the value of the project for purposes of determining true depreciation is based happen to decline in proportion to the actual value of capital used for PPT purposes, then the method of depreciation used is true depreciation. That the PPT Mark II should give much the same revenue from a given project as an RRT with the same rate and the same threshold equal to the investor's discount rate (provided annual profits are always above the PPT threshold) is not unlikely, if there are not interest payments allowable for tax under the PPT. If, however, interest exists and is deductible under the PPT, or if the government's discount rate is below that of the investor, the RRT can be expected to yield more tax than the PPT Mark II; great irregularity of receipts, on the other hand, relatively favours revenue-raising under the PPT Mark II. (But see tables 7.8(a), (b), and (c).)

Thus, where T″ values are the required rates to give equal revenue under any of these conditions (and the single-year investment life is ignored),

$$T''_{\text{HRIT}} < T''_{\text{PPT}'} > T''_{\text{RRT}} = T''_{\text{BT}}. \tag{9.3}$$

Next we vary the assumptions by supposing uncertainty. Since the question we are posing is independent of expectations of investors, this does not affect the result. (PPT from here to p. 158 means PPT′.)

Now, however, we may suppose that the threshold is set somewhat higher than D^* (at $D^{*\prime}$) for the RRT, and somewhat higher again (at $D^{*\prime\prime}$) for the PPT, these rates being supposed to provide the best approach to neutrality (see chapter 8). This means that a somewhat higher rate than T''_{RRT} will have to be used in the RRT to raise the same revenue, and a rate even further above T''_{PPT} will have to be used for the PPT. If we denote these new required rates for the PPT and RRT by T''', then

$$T''_{\text{HRIT}} < T'''_{\text{PPT}} > T'''_{\text{RRT}} > T''_{\text{BT}}. \tag{9.4}$$

A further step is to suppose that, instead of being accumulated for social purposes at the investors' discount rate, D^*, the NPV of revenue is accumulated at a government discount rate, D^{G}, assumed to be lower than D^*. This will tend to increase the present value of revenue received late in relation to the present value of that received early. Under the BT, the original investment will then be 'repaid' by year k (earlier than year m). The positive receipts from year $k+1$ to year m will then have their taxes contribute to the NPV of revenue, whereas the RRT with the threshold at D^* will not raise revenue until year $m+1$. Thus, if accumulation is done at D^G, positive NPV of revenue is attained earlier under the BT than under the RRT with the same tax rate, whether its threshold is set at D^* or, *a fortiori*, at $D^{*\prime}$. In either case the tax rate under the RRT must be higher than that under the BT to produce the same NPV of revenue.

If the tax rates required for equality of NPV of revenue accumulated at the government's discount rate, D^G, under the various taxes, are given by $T''_{(G)RRT'}$ etc, then

$$T''_{(G)RRT} > T''_{(G)BT} \tag{9.5}$$

and

$$T'''_{(G)RRT} > T''_{(G)BT}. \tag{9.6}$$

The PPT will tend to receive its revenue earlier than the RRT, and hence lowering the discount rate will tend to increase the PV of the RRT revenue more than that of the PPT, so that the RRT may lower its tax rate somewhat to raise the same revenue (accumulated at D^G) as the PPT. This, combined with (9.3) and (9.4), gives:

$$T''_{(G)PPT} > T''_{(G)RRT}$$

and

$$T'''_{(G)PPT} > T'''_{(G)RRT}.$$

The relationship between T''_{HRIT} and T''_{PPT} is unchanged by the change in discount rate for accumulation since, in each year, the base of the HRIT is greater than that of the PPT (except when both are zero). Hence

$$T''_{(G)HRIT} < T''_{(G)PPT} > T''_{(G)RRT} > T''_{(G)BT} \tag{9.7}$$

and

$$T''_{(G)HRIT} < T'''_{(G)PPT} > T'''_{(G)RRT} > T''_{(G)BT}. \tag{9.8}$$

At the same time, the HRIT (with normal systems of depreciation which do not deduct past capital expenditures with interest, or else with depreciation simply for maintenance) raises revenue before year m or k (when the BT has reached a positive NPV of revenue on discount rate D^* or D^G respectively), and (unless depreciation allowances continue after year k or year m respectively) continues to yield revenue thereafter on the same base in each year as the BT. More precisely, if under HRIT there is straight-line depreciation over k years or m years, the depreciation will, in some years at least, be less than the positive cash flow, so that there will be some 'income' for HRIT purposes.

After year k or year m the HRIT will be levied on the same base in each year as the BT. Faster depreciation than this must still leave some tax to be collected *before* year k or m and the same *after* year k or m as under the BT. Slower depreciation than in the case given will give a higher NPV of tax revenue than in that case. Changing the method of depreciation will not overthrow this result. Hence the HRIT will require a lower rate to yield the same revenue as the BT; i.e.,

$$T''_{\text{HRIT}} < T''_{\text{BT}} \tag{9.9}$$

$$T''_{\text{(G)HRIT}} < T''_{\text{(G)BT}} \tag{9.10}$$

It follows from (9.6) and (9.12) that

$$T''_{\text{HRIT}} < T''_{\text{RRT}}$$

and from (9.7) and (9.12) that

$$T''_{\text{HRIT}} < T'''_{\text{RRT}}$$

and from (9.8) and (9.13) that

$$T''_{\text{(G)HRIT}} < T''_{\text{(G)RRT}}$$

and from (9.9) and (9.13) that

$$T''_{\text{(G)HRIT}} < T'''_{\text{(G)RRT}} .$$

The next special assumption to be removed is that there are no interest payments. Now if the HRIT and PPT follow the usual definitions of income for tax purposes by giving full deduction for interest payments, while the RRT and BT do not, this will tend to require the rates of HRIT and PPT to be higher than they would be if (as assumed to this point) there were no interest to be paid. It would be possible to impose the HRIT or PPT while giving no deductions for interest payments, but the particular attraction of these taxes is that they make it possible to follow existing tax law with few variations; so that it is reasonable to treat them as making the interest deductions. That being the case, it will still be true that, for equal revenue (assessed at government's discount rate), the rate of PPT must be higher than that of HRIT and RRT and the rate of RRT higher than that of BT, but we can no longer say anything confidently about the relationship of the rates of HRIT and BT or of HRIT and RRT. The presumption that PPT must have higher rates than RRT and BT is also greatly strengthened.

So far all considerations have suggested that at a given tax rate the BT will raise more revenue than the RRT and the RRT than the PPT. Two further relaxations of assumptions, however, have a tendency to weaken respectively one and both of these conclusions.

First, we remove the assumption of regularity of cash flow. As explained in Chapter 7, irregularity has the tendency to raise tax receipts under PPT to a higher level than with a regular cash flow of equal pre-tax NPV. There is no such tendency under RRT (unless, as discussed separately below, there is a final period or 'tail end' of negative cash flow) or under BT. This tends to raise the amount of revenue that can be expected from a given tax rate under PPT in relation to that under RRT or BT.

Second, we remove the assumption that each investment necessarily achieves a positive pre-tax NPV. If this is not the case, the BT will show negative net flow of revenue; the RRT will give zero revenue (unless a positive NPV has been realized at some time in the investment's life, in which case revenue may be positive); while the PPT and HRIT may readily give positive revenue even if positive NPV has never been realized. If the probability of negative NPV over the life of the investment is considered sufficiently great, this might conceivably cancel the weight of considerations suggesting a higher revenue for a given rate under BT than under RRT and under RRT than under PPT. The probability of a negative NPV outcome, however, must presumably be heavily outweighed in the assessments made by investors by the probability of a positive outcome. However, the presence of some probability of a negative NPV outcome, for example in relation to some exploration expenditures, removes the clear presumption that the BT at a given rate will raise more revenue than the RRT.

Given the deduction for interest, we consider it unlikely that these last two considerations would generally raise the revenue to be expected on a project from a given tax rate under PPT to or above that under RRT or BT. It is easier to believe that the second might raise the revenue to be expected under RRT to or above that under BT.

The effect of the maximum-rate limit is thus somewhat ambiguous in ranking the four profit-based taxes for revenue-raising. In itself, it sets the HRIT above the PPT, and the RRT and BT also very probably above the PPT. It sets the BT above the RRT in circumstances where the probability of negative outcomes is very low, but not otherwise. The FF and SAVD have the advantage that, not being based on profit, they are subject to no corresponding limit on this account.

3 *Combined effects of risk reduction and maximum-rate limit: negotiation project by project*

To summarize the considerations under 'Risk reduction' and 'Maximum-rate limit' for tax regimes set for each project separately, we can say that for the BT the rate can simply be set at the maximum rate T_{max}, consistent with economizing firm behaviour. This ought not to deter any investment that would be eligible before tax, unless the peculiarities of the BT cause the investor to fear that its terms may be changed.

The RRT, with its threshold set (for the sake of neutrality over techniques and supplementary investments) no lower than the investors' discount rate, will at tax rate T_{max} if anything collect less NPV of revenue (accumulated at the government's discount rate) project for project than the BT, and the rate will have to be lowered for marginally profitable projects to avoid deterring them. This will be true unless there is a tail end of negative cash flow, which may lead the RRT to collect more than the realized present value over the whole life of the project, or unless the project has some significant probability of realizing negative present value before tax. Thus, unless the neutrality requirements for supplementary investments and techniques are disregarded for the more prospectively profitable projects and the discount rate is lowered, or the probability of unsuccessful investment or of a tail end of negative cash flow is important, the RRT is capable of raising less revenue, project by project, than the BT.

There is also a fairly strong presumption that the PPT Mark I (with its threshold set to maintain neutrality as far as possible) will raise less revenue than either RRT or BT. Irregularity, a depreciation period longer than the life of the investment, or a high probability of unsuccessful investment might conceivably alter this relationship, but, if interest is at all important and is deductible for purposes of PPT but not of the others, this seems improbable. As we pass from more to less profitable projects moreover, the rate will have to fall to keep the projects eligible, and the fall will have to be greater than under the RRT. There is no guarantee, but it seems a fair presumption, that the PPT Mark I will raise less revenue than the BT or the RRT. With the PPT Mark II, which can collect more revenue at any given rate and threshold than the Mark I form, the comparison with the RRT and BT is much less certain.

There is more ambiguity again over the HRIT. On profitable projects it will compare favourably with the PPT because it will raise more revenue at tax-rate T_{max}. On barely profitable projects it will compare unfavourably since the greater risk must be compensated by lower rates to keep such projects eligible. No clear judgement can be given about its comparison with RRT and BT, but our guess is that deduction of interest from the tax base would tip the scales against it as a revenue-earner.

4 *Suitability for 'general' application*

As argued above and in chapter 6, there are good grounds for having common fiscal arrangements throughout an industry, and indeed for administrative reasons this may be virtually a necessity. (The important possible exception occurs where the fixed fee is determined by auction before much exploration expenditure has been incurred.) This being the case, the important question is not which tax performs best when negotiated project by project but which is best if a single rate (and, where relevant, threshold) has to be chosen for each tax.

To answer this question we may start by supposing the BT applied at T_{max}, the highest rate consistent with economizing behaviour. We have seen that the BT, even at this rate, will both reduce the scale of investment at risk to the investor and, even without the fall in the supply price of investment that this would entail, leave eligible any project that would have been eligible in the absence of tax since it will not reduce any pre-tax-positive ENPV to zero or below it. The RRT, applied at the same rate and with its threshold at the investors' discount rate, D^*, will deter certain pre-tax-eligible projects. The optimum rate for revenue from such an RRT may be at T_{max}. If so, it seems that the RRT will tend to give less revenue than the BT applied at that rate because certain projects will then not be undertaken (quite apart from the fact that, project by project, as we have seen, the RRT will tend to raise less revenue if PV of revenue is assessed at the *government's* discount rate, D^G, lower than the investor's). If the revenue-maximizing rate is below T_{max}, then it seems again that the RRT must also yield less revenue than the BT applied at T_{max}. Similarly, raising the threshold rate above D^* in order to avoid deterring marginal projects and outlays in spite of uncertainty will further reduce the RRT revenue on each project undertaken. To the extent that considerations of sovereign risk are important, however, the presumption of superiority of BT over RRT may be reduced, or even reversed, by doubts over the BT's stability, and hence possibly by a somewhat less favourable effect than the RRT's on the supply price of investment and a more severe impact on marginal investments.

Comparison with the PPT can be pursued by supposing, first, that there are no interest payments. The PPT applied at T_{max} will even then (for a given threshold) tend to deter more projects than the RRT applied at the same rate since it introduces greater risk and, even if the investors' discount rate is applied as threshold, will in some cases reduce pre-tax positive to post-tax negative ENPV purely on account of irregularity in receipts. Thus, to keep the same spectrum of projects eligible, the PPT must have a lower tax rate than the RRT. As we have seen, it is not clear whether, in the absence of interest payments, a given rate of tax and threshold will on average produce more or less revenue from a particular project under PPT than under RRT. Thus it is not unambiguously clear that the lower tax rate on the PPT will produce less revenue overall. If, however, we introduce interest payments and assume that deductions are made for them under the PPT, the PPT (whether Mark I or Mark II) seems very likely to yield less revenue for a given tax rate on a given project than the RRT. The deductibility of interest will thus greatly strengthen the conclusion that RRT raises more revenue than PPT in both the latter's forms. Thus there are strong grounds for treating the RRT as a better revenue-raiser than the PPT (Mark I and possibly Mark II as well) under the rules likely to be applied to the latter.

The comparison of PPT with BT is if anything more clearly unfavourable to PPT. If rates on both PPT and BT are set at T_{max} and the PPT threshold approaches neutrality, it seems very likely, as we have seen, that the BT will raise more revenue discounted at the government's rate on each project undertaken. The PPT will also deter some low-yielding projects, as the BT will not. It may be that the PPT can raise more revenue overall with a rate below T_{max} than with T_{max}. But if that is true, there will be an even stronger case for supposing that the maximum revenue obtainable from it on average will be lower than that from the BT. Thus there is a strong presumption that the BT is capable of raising more revenue overall than the PPT.

The comparison between HRIT and BT is more uncertain. HRIT at a given rate will certainly deter more projects. Thus it may well raise more revenue if applied at a rate below T_{max}. But if it *is* applied at T_{max}, it *may* (as tables 7.8(a,c) suggest) raise more revenue on the projects that are still undertaken than the BT. Thus the conclusion must be uncertain, though we believe it should favour the BT, especially if interest payments are deductible under the HRIT.

Confinement of consideration to uniform or general tax systems will greatly strengthen the revenue-raising power of the PPT in relation to that of the HRIT.

This argument, based on the application of the single most common investment criterion and ignoring sovereign risk, leads therefore to a strong presumption that, applied at a single optimum rate to a diverse range of projects and potential projects, the BT can probably raise more revenue than the RRT, and can almost certainly raise more than the PPT, and that the RRT can raise more than the PPT. It is less certain, though we believe it generally to be true, that the BT and RRT can be expected to raise more revenue than the HRIT, and less certain again that the PPT can raise more revenue than the HRIT. Where sovereign risk is important to the investor, the BT may not necessarily be capable of raising more revenue than the RRT, and there is a strengthened presumption that the RRT can raise more revenue than the HRIT; this is because both BT and HRIT seem likely to leave highly profitable investments more politically vulnerable than does RRT. Application of the 'bankers' criterion' (demand for quick pay-back and a ceiling for risk) and the 'managers' criterion' (demand for something approaching a floor to NPV) strengthen the presumption that BT and RRT can raise more revenue (by rendering more projects eligible for a comparable ENPV of revenue per project) than PPT and HRIT.

We also take it to be true, and scarcely needing demonstration, that these four taxes, related closely to actual profit, can raise more revenue when applied 'generally' in these circumstances than the SAVD or FF.

Revenue Uncertainty and Delay

The question of the uncertainty of the revenue flow is unimportant to a country with a large number of projects of comparable dimensions, but it may be important to a country that has only one large project or a very small number. Governments should not, and probably do not as a rule, apply anything equivalent to what we have called a 'managerial safety margin', but risk considerations cannot be entirely excluded.

'Risk' to the government associated with particular taxes, as applied to any project, is in general inversely related to risk to the investor. The only reservation to be made to this rule is due to the fact that government and investor may be applying different discount rates for time, so that relatively speaking the government may give more weight to later than earlier uncertainty, with the result that both parties may gain if the government shifts its share of the risk-bearing from the later to the earlier part of the lives of projects. The RRT, and even more the BT, would appear to score on this ground.

We can say that among our six taxes those providing less risk to the investor provide more to the government. Thus it would seem that any one tax entails greater risk to the revenue than any tax to its left in the tables in chapter 7.[1]

Delay in receipt of revenue tends clearly to increase as we move from left to right across the types of tax in tables 7.1–7.3. If this matters to a government, then the authorities might reasonably prefer

[1] There must be some uncertainty, however, as to whether RRT should be regarded as more risky to the investor and less to the government than BT. To be comparable on ENPV of revenue (assessed at the *government's* discount rate) on any project, the RRT (with its threshold at the *investors'* discount rate or higher) must have a somewhat higher tax rate than the corresponding BT even if the investors' discount rates applied to projects under the two tax systems were the same. This would mean that, for positive outcomes, the RRT would give greater risk to the government but less to the investor. For negative outcomes, however, which the BT subsidizes and the RRT does not, the RRT would give less risk to the government and more to the investor. If the government were more concerned about the negative outcomes and the investor about the positive ones, this could work to advantage of the RRT on both scores. Whether that premiss is true or not and whether or not the risk to the government should generally be accounted important, it may be a special drawback of the BT that it leaves the government open to actual financial loss.

the PPT to the RRT or BT. Alternatively, whenever one of the profitability-related taxes is used, concern over delay, like concern over uncertainty, favours the combination of such a tax with an auctioned FF if the conditions for making the latter an effective collector of revenue exist or can be created. The combination also has the advantage that, in shifting the burden of revenue-raising from either component alone, it increases the likelihood that the effective neutrality of either may be preserved: it favours the neutrality of the auctioned FF by reducing the amount of initial capital to be put up, and it favours the neutrality of the profitability-based tax by reducing its rate and so limiting the hazard of incentives to over- or under-investment.

Summary

Apart from neutrality, three main considerations determine the relative capacities of the various taxes to raise revenue: (i) the comparative risk that they impose on the investor, which affects the range of projects attractive to investors at any given pattern of expected tax revenues from projects; (ii) their comparative capacities for raising revenue from a particular project at the highest marginal rate of tax on profits consistent with economizing behaviour; and (iii) their relative responsiveness to profitability of various projects with any given set of tax rates and thresholds, which affects their suitability for application in general form.

If negotiation of tax regimes by project is allowed and supposed to be done with perfect success on the government's part, there is a general but very rough presumption that the revenue-raising power of those further to the right in tables 7.1–7.3 is higher than that of those to their left, except that the HRIT probably leads the PPT, and the advantage of BT over RRT and PPT is uncertain if sovereign risk is important.

If tax regimes are to be established uniformly across projects, as we believe is generally desirable except perhaps where conditions make the auctioned fixed fee useful, the relative advantages in revenue-raising of those taxes to the right in tables 7.1–7.3 are greatly increased, the PPT gaining in relation to the HRIT and probably surpassing it, and the RRT and BT gaining in relation to the PPT and HRIT.

Smaller risk and greater certainty in government revenue, on the other hand, are generally favoured by taxes further to the left on the tables. So is early receipt of revenue.

If, as we believe is commonly the case, uncertainty and timing of government revenue on any one project are relatively unimportant, the Brown tax and resource rent tax emerge from this chapter as the taxes of choice. The more the investor or his bankers are thought to be worried about possible changes in the rules under the Brown tax after investment has begun, and the more a government would be worried by making actual financial losses on a project, the more the resource rent tax (in one or other of its forms) will emerge as the tax of choice. On the other hand, the more variation there is in investors' discount rates and the more uncertainty on the part of the government about them, the more eligible will be the Brown tax.

Where there is a very small number of projects, however, so that uncertainty and delay in revenue loom large, the higher-rate income tax and the progressive profit tax gain considerably in eligibility by comparison with the resource rent tax and Brown tax.

Alternatively, where circumstances are such that the auctioned fixed fee is a potential revenue-raiser, its combination with one of the profitability-based taxes not only offers a government the choice of a compromise between high expected revenue on the one hand and early and certain revenue on the other, but also tends to limit the departures from neutrality that either component might incur if reliance were placed upon it to collect the mineral rent alone.

10

IMPOSITION AND ADMINISTRATION

This chapter discusses some of the problems that arise in the imposition and administration of mineral rent taxes and suggests what might be useful approaches to these problems. It considers mainly four forms of rent tax that are based on realized profit and that in a 'general' form are likely to perform relatively well for revenue maximization and neutrality: the HRIT, PPT, RRT, and BT.

The problems that arise in the imposition of taxes in general or negotiated form, that is, in fixing appropriately such variables as tax rates, are generally greater for the cruder forms of tax than for the taxes that perform better on neutrality and revenue maximization, those on the right of tables 7.1–7.3. On the other hand, purely administrative problems associated with the assessment of taxation liabilities once the system has been defined are generally *less* serious for the cruder forms of tax.

Auctioned, Negotiated, and 'General' Tax Parameters

It was observed in chapter 7 that each type of mineral rent tax could be applied with auctioned, negotiated, or general tax parameters (rates of tax and, where appropriate, threshold rates). There are characteristic problems of imposition and administration of each of these forms of tax, which cause one or other of them to be unsuitable in some circumstances.

Several considerations combine to make the auctioned form of taxation unsuitable unless *either* a great deal is known about the mineral deposit at the time of auction, *or* the individual bid is for one of a large number of small shares in a lottery of which the probability of success is thought to be known with reasonable confidence. This form is the more likely to be unsatisfactory if the scale of possible cash flows is fairly large in relation to the operations of mining companies. When much risky exploration expenditure is required to determine the potential of a mineral licence, it is only the fixed fee payable in advance of exploration (the US 'cash bonus bid'), or the fee payable later but guaranteed by a firm of sufficient standing, that is entirely suitable for auction. A fixed fee with provision for delayed payment might never be paid. For example, the rent value of an uncertain petroleum prospect might be greater than the value of the assets of most of the world's mining companies individually; yet if delayed payment were allowed other companies might be tempted to bid large sums for the lease, knowing that they must either fail to pay or go bankrupt if the value of the lease after exploration were much less than its expected value before exploration. Bidding high to prevent other investors coming in is also a possibility. Hence allowing conditional or postponed payment of an auctioned FF might unduly postpone exploitation of seams other than the richest. If the parameters of forms of rent tax other than the FF were auctioned, there would inevitably be an element of conditionality, and the winning bids might be such as to deter investment wherever the outcome of exploration was not very favourable. The latter possibility is less on the more profit-based taxes than on the SAVD-type royalty, but nevertheless low thresholds and high tax rates even for the RRT could deter some investments that in the absence of tax would be viable. In the case of the BT, there would seem to be no advantage in the auctioned over the general form. For the auctioned FF, however, there is evidence from the USA (see chapter 18) to the effect that it may be a very efficient revenue-raiser from oil or gas leases, at least in certain circumstances. This is not because there is much knowledge about individual blocks at the time of the auction, but it may be that there are fairly firm beliefs about probabilities based on geological information, and in addition any individual extractor in the US outer continental shelf may spread his risks by holding a large number of blocks. These conditions are not fulfilled before exploration for most important onshore hard-mineral mines. In such cases an auction before exploration would probably yield little in relation to the ultimate value of any successful project, while auctioning after private exploration would either risk robbing the exploring company of the fruits of its enterprise or else be fairly meaningless because of that company's monopoly of information. Even for oil and gas, the USA may not be typical, in that confidence in the stability of fiscal arrangements may have been greater there than in most other countries, and this of course favours the revenue-raising power of the auctioned FF.

It is generally no solution to allow private investors to explore freely before the auction, since this either fails to break any monopoly of information or dissipates rent in wasteful duplication of expenditure. The ideal role of the auction would seem to be the allocation of mineral leases when much is already known about their value, either individually or stochastically, because the government or other private investors have provided general geological information or undertaken exploration or because the lease is adjacent to known areas of mineralization. Even in this situation, the risk aversion of private investors, together with any sovereign risk, and the fact that actual costs and prices may be consistently different from those expected, would seem to argue for not relying on auctioning for a fixed fee as the sole source of rent collection but rather for applying more complex rent taxes as well. Indeed, this combination could be considered an ideal form of lease allocation and taxation when there was interest by a sufficient number of investors to create competition and so to justify the administrative costs and delays associated with the auction.

The negotiated form of rent taxation absorbs considerable administrative resources (which can be very costly to a developing country in which they are scarce), leaves governments vulnerable in particular negotiations to the superior information of mining companies, is open to the reality or the allegation of corruption, and usually leads to delays in investment while negotiations proceed. These are large disadvantages of negotiated *vis-à-vis* general forms of taxation, although the negotiated form is, in principle, capable of generating more revenue without deterring investment when the government is sufficiently well informed. It is notable that this hypothetical importance of ideal negotiation diminishes as we move to the right across the taxes in tables 7.1–7.3.

The following discussion of problems of imposition is concerned with the negotiated and general forms of taxation.

Imposition of Parameters of Negotiated or General Taxes

The amount of information that the authorities require to set the parameters of taxes at optimal levels (with respect to neutrality and revenue-maximization) is very great with fixed fees and falls progressively to much more modest requirements for the profit-based taxes. At one extreme, the BT requires the authorities to make judgements only about the maximum rate of tax that is consistent with economizing behaviour. The RRT requires the authorities to make judgements about the maximum feasible rate of tax and ideally also about the supply price of investment, which itself supposes knowledge of the degree of risk and investor's attitude to it. With the PPT, further judgements are required about the likely time profile of cash flows. The HRIT requires judgements to be made on all of these matters and also on the expected level of profits as assessed for income tax. *Ad valorem* royalties require judgements as well about likely unit costs of production and product prices. The imposition of the fixed fee makes possibly the most severe demands on the authorities; its imposition in an optimal manner may be regarded as requiring judgements on all of these matters and also on the likely level of production.

The following section of this chapter discusses considerations that should enter into decisions on the rate of tax. Subsequent sections deal with the delineation of the 'resource-mining phase' of mineral-related activities, that is to say, the part that should be subject to special rent taxation arrangements; with the management of problems introduced by inflation and exchange rate variation; and with the setting of thresholds for the PPT and RRT. The concluding section addresses problems of administration after the imposition of the various taxes.

Setting the Rate of Tax

Sufficiently high tax rates may remove the incentive to economizing behaviour by private investors. It seems reasonable to suppose that there is some maximum marginal rate of tax on profit or cash flow such that, if tax rates are raised above it, there will be significant dissipation of rent. The term 'significant' is meant to suggest a trade-off, as tax rates are raised, between government revenue gain on the one hand and government and social loss owing to wasteful expenditure on the other. However, there is probably no single percentage rate of tax beyond which there will be substantial waste in expenditure across all countries, industries, projects, and types of expenditure within projects. One likely relationship is that the maximum rate of tax consistent with economizing behaviour is higher for projects that generate profits that are large in relation to the size of the investing firm or in relation to the necessary costs. This is because there may be some presumption that the propensity to generate unnecessary costs varies directly with the size of the firm or of the necessary costs, and hence, where profits are very high in relation to these variables, the gain to revenue from setting the rate high may to a higher limit be held likely to outweigh the loss (to revenue and otherwise) through wasteful outlays. Thus, marginal tax rates around 90 per cent on profit may be sufficiently consistent with economizing behaviour in petroleum investments with their large cash flows, while rates of 80 per cent are commonly thought to be high, and possibly conducive to wasteful expenditures, in other minerals. But there are no fixed rules and each government must form its own judgement in its own circumstances. Clear definitions of what are mining outlays for tax purposes, as discussed below, may help to eliminate the incentives that rent taxes might offer to what are really consumer expenditures for mining executives ('gold-plating').

Delineating the Resource Extraction Phase

When profit-based or *ad valorem* rent taxes are applied to resource development, it is important that there should be a satisfactory delineation of activities that are to be subject to them. This is necessary to prevent avoidance of taxation and also to prevent unintended and undesirable effects in other sectors of the economy.

Problems of delineation may possibly arise in relation to downstream processing and to the allocation of overheads between resource development and other activities. Financial transactions of resource investors are excluded by definition from assessment for RRT and BT,[1] but they can raise difficult problems of delineation with the HRIT and PPT if typical income tax rules are used for defining profit under those taxes.

The distinction between 'mining' and 'manufacturing' activities can be fine and arbitrary. A minimum level of processing is sometimes required to convert mineral ore into a tradable form (for example, the concentration of copper ores). At other times, some processing operation may not be absolutely essential to trade but may add greatly to the economic rent available from a resource. It would be desirable in principle to define both of these types of processing operations as resource-related activities for purposes of rent tax.

The standard corporate income tax usually treats mining separately from manufacturing for depreciation and some other purposes. Definitions relating to the termination of the 'resource production' phase have been developed to exclude higher levels of processing. In jurisdictions in which the income tax is well established, no important harm would be done if the existing definitions distinguishing resource extraction were used in assessment of profit-based rent taxes.

Clear definition of the resource extraction phase is required to remove opportunities for tax avoidance: *appropriate* definition, excluding non-mining activities, is required to prevent economic distortions. In some circumstances, the inclusion of non-mining transactions in assessments for rent tax could cause the higher tax rate to deter otherwise intra-marginal investments in those activities. This effect is likely to be very important for HRIT, much less important for PPT and RRT, and unimportant for BT. In other circumstances, the inclusion, in assessments for PPT and RRT, of transactions concerned with much less risky manufacturing activities could cause wasteful over-investment in those activities, as discussed further below. It is possible that the investment-inducing distorting effect of the PPT and RRT would completely offset the investing-reducing distorting effect, but this would be an accidental result and is not to be expected.

[1] In our original article (Garnaut and Clunies Ross, 1975:279) we described procedures that could be used to separate out financing from activities that should be assessed for RRT. We now think that these procedures are unnecessary, so long as separate accounts are kept within each company for all expenditures and receipts that relate to resource development and extraction.

The most important requirement in the allocation of shared overhead costs between resource-based and other activities is clarity, which is needed in order to remove opportunities for avoidance. Accounting conventions in this area have developed in recent years, mainly to handle problems that arise in joint venture arrangements between private investors. These conventions are suitable for application to the assessment of rent taxes, so long as they are described unambiguously in the law or in the practice of the law.

The standard corporate income tax, and so the HRIT and PPT which derive from it (but not the RRT and BT), allow interest payments as a deduction against revenue. If income from resource activities is being taxed at a higher rate than income from the other activities of a diversified firm, there will be a tendency for a disproportionate part of the firm's total borrowing to be ascribed to its resource operations. This could be avoided by requiring the formation of a separate corporation for all activities subject to HRIT or PPT, and by limiting financial transactions between the new entity and its affiliates. Alternatively, arbitrary rules could be used to limit the amount of interest deductions allowed in assessment for PPT and HRIT. In either case, the treatment of financial transactions is a difficulty of PPT and HRIT, and may provide some scope for avoidance. However, either a tax of HRIT or one of PPT form could be defined quite separately from income tax and with different rules, and deductions for interest could be excluded.

Assessment on a Project or on a Company

Any of the profit-based rent taxes can be applied either to the total resource activities of a company or to separate activities by project. The distinction has no practical relevance to the BT, but is potentially very important for the HRIT, PPT, and RRT. The company basis of assessment is relevant only to the general form of rent taxes, and not to the negotiated and auctioned forms (as the terms are used in chapters 6 and 7).

The company basis of assessment for HRIT, PPT, and RRT allows the risk of failure to be shared across various projects. This possibility is likely to lower the supply price of investment and also to reduce the danger that rent tax may deter marginal investments. These advantages are not important with the BT, where the loss offset plays the same role in risk reduction as risk-sharing across projects with the other profit-based rent taxes. They are likely to be less important for RRT and PPT, which tend to have a stronger tendency to reduce risk, than for HRIT.

On the other hand, where the risk of investment varies across projects or industries, and where, for this or other reasons, the supply price of investment varies across projects or industries, the company-wide tax parameters are bound to be inappropriate (from the perspectives of neutrality and revenue maximization) for some of the investments that might possibly be undertaken by a company.

The project basis of assessment can be applied with general parameters, across all projects, or with parameters that are negotiated or determined by auction separately for each project. When tax parameters are set separately for each project, the project basis of assessment for HRIT, PPT, and RRT allows adjustments to avoid the deterrence of otherwise intra-marginal investments. This possibility is not present, of course, when general rules are announced in advance.

The project basis of assessment limits the incentive towards dissipation of economic rents through wasteful over-investment, an incentive that is present with PPT and RRT whenever threshold rates exceed investors' discount rates by more than certain amounts determined by the risk and irregularity of outcomes (see chapter 8). This is because the project basis limits the range of additional investments that can be undertaken to depress the net cash receipts of a successful corporation that appears likely to be assessed for RRT. This does not by itself eliminate the possibility of wasteful over-investment, but it significantly reduces its scope. It allows the authorities greater freedom in setting high thresholds (to avoid deterring otherwise intra-marginal investments) without adding greatly to the likelihood of losses in efficiency through over-investment.

Other considerations are also relevant to the choice between the project and the company basis of assessment. The company basis contains strong biases among investors, improving the relative competitive position of those who are already paying PPT or RRT and those who, because of poor management or bad luck, have accumulated such large negative cash flows in their mining operations that they are unlikely ever to be liable for PPT or RRT. These biases could be reduced by allowing inter-company sale of 'losses'.

There are strengths and weaknesses in both the project and the company basis of assessment for profit-based rent taxes. It is our opinion that the balance of advantages will often lie with the project basis of assessment. However, the choice between the project and company basis must always depend on judgements about the characteristics of the environment in which the tax is to be applied, and especially about the scope that is present for wasteful over-investment.

When the project basis of assessment is applied, it is necessary to determine which of the exploration expenditures that preceded a resource development are to be attributed to the project. The logic of the profit-based rent taxes is probably preserved best by allowing all expenditure undertaken by a company on exploration within the country before the signing of a project agreement to be treated as a deductible payment against income generated in the first project operated by the company. This procedure (which is roughly that applied in the case of the UK's petroleum revenue tax) reduces the risk associated with the early general exploration that is necessary to locate sites for initial mineral exploitation. The limits of the first project would be defined at the time that a development agreement was signed, and further exploration expenditure unrelated to the first project would be deductible only against future projects. This restriction is desirable because companies operating existing projects would receive a competitive advantage in exploration if they were permitted deductions for subsequent unrelated exploration. There should be no restrictions, however, on deductions for expenditure on further exploration within the geographical limits of an existing project.[2]

The Setting of Thresholds for PPT and RRT

The imposition of PPT and RRT, but not of other rent taxes, requires authorities to decide on appropriate thresholds for triggering higher rates of tax payment. The neutrality and revenue yield of PPT and RRT depend critically on the choice of thresholds. As discussed in chapter 8, the PPT and RRT approach neutrality closely over the choice of projects only if the threshold rates are set correctly in relation to the discount rates applied by investors on the various possible investments, to the probability distributions of the outcomes of these investments, to the tax rates, and in the case of the PPT to the regularity of cash flows over time. This list sounds formidable but possibly gives an exaggerated impression of the difficulties. There is, however, an additional problem in approaching neutrality over subsequent investments and techniques within a project. This may in principle require different threshold rates for accumulating different types of outlay or those incurred at different stages in a project's life. In particular,

[2] An alternative approach would be to allow deductions of exploration expenditure against income from *any* project and to allow 'exploration losses' to be traded among mining projects, so that the incentive to new exploration was no greater for corporations with established mines than for others. This would be the effect, but only if the market for such losses was competitive and if profit that was subject to HRIT, PPT, or RRT exceeded exploration losses in the market as a whole. This approach would effectively introduce loss offsets (subsidies) into the exploration phase of mineral development. There would be no further deductions of exploration expenditure in assessment for rent tax. The device should prevent rent taxes from inhibiting high-risk exploration, and so should remove the most important source of non-neutrality in the HRIT, PPT, and RRT. The reduced risk of mineral investment in these circumstances would argue for lower PPT and RRT thresholds. A further alternative, discussed at the end of chapter 9 above, is for the government to undertake or to subsidize exploration.

exploration, being very risky, might require a higher threshold than other outlays.

In an earlier article (Garnaut and Clunies Ross, 1979) we tried to show how in principle tax authorities might maximize revenue in the application of RRT without introducing incentives to wasteful overspending. The method involved a tailor-made combination of tax rate and threshold for each project. It did not avoid the hazard of under-investment in the spending decisions made once the investor had determined to undertake the project concerned. It also depended on unrealistic assumptions about the knowledge available to the authorities. Hence we do not reproduce that argument here.

If the incentives of the tax system to over- or under-investment on a project once established could be ignored, revenue from a RRT or PPT could probably be maximized in principle by fixing the tax rate as high as is consistent with maintaining incentives to economy (at T_{max}) and setting the threshold, by negotiation or auction, as low as each project could bear. However, incentives to over- or under-investment arising from thresholds that are too high or too low cannot be ignored and will in themselves reduce tax revenue, since the incentives operate by providing means of tax avoidance. Hence a system devised in this way will not necessarily maximize revenue even if the problems of ideal auctioning or negotiation could be solved. Our judgement is that, if a tailor-made system for each project were possible, it would be better to set the thresholds at levels that seemed to be most consistent with neutrality and to negotiate or auction the *tax rates* so as to fix them as high as each project could bear, subject to the ceiling of T_{max}. This in practice is of course only as effective in revenue maximization as the negotiating or auctioning makes it. The fact remains that in principle a range of tax rates would be required with any given threshold, or with neutral or near-neutral thresholds, if revenue were to be maximized.

It was observed above that there is a case for separating the treatment of exploration, which being more risky requires a higher threshold to avoid deterring investment, from the treatment of expenditure on mineral development and mineral recovery, perhaps by allowing loss offets (that is, subsidies) for exploration. This procedure would allow lower thresholds to be used without deterring the more risky exploration investment, and so would reduce the risk that high thresholds might introduce incentives to wasteful over-investment for the less risky potential elements within a mineral project.

The RRT or PPT system that gives rise to the highest expected revenue yield that is consistent with neutrality would thus apply quite different tax rates, as well as somewhat different thresholds, to different investments. Indeed, it seems likely that a RRT system that approaches neutrality as closely as possible both among and within projects must be applied with some variety of thresholds; and, given these thresholds, it must be applied with a range of tax rates if it is to yield large amounts of revenue.

Governments do not have access to such detailed information on the possible outcomes or on investors' objectives as would be necessary to extract the highest possible expected revenue yield within a neutral RRT system by the methods described above. Moreover, administrative ease as well as the reduction of business uncertainty (which itself would adversely influence investors' discount rates) argue for thresholds and rates to be set in advance of investment (including exploration expenditure) and to be common across projects to investments within broad categories.

Where there are clear differences in the degree of uncertainty associated with different classes of investment, however, there is a strong case for accumulating losses at different rates for these different classes of investment. For example, a higher threshold could be used for petroleum investments, of which a higher proportion is likely to comprise high-risk exploration, than for metal mining, in which a higher proportion of investment outlays comprises development expenditures incurred after exploration risk has been greatly reduced.

If a single threshold is set appropriately for one investment within a broad class, it will be too high, or too low, or both for some others. If the threshold rate is set so high as to be certain that it will deter no investment that would otherwise have been undertaken, it is likely to provide incentives to wasteful over-investment by firms operating near the threshold of liability for RRT. It is also likely to yield relatively little in revenue. Alternatively, if a threshold is set so low that it is certain to provide no incentive for wasteful investment, it is very likely to deter otherwise profitable investments unless the tax is applied at negligible rates.

The use of progressiveness in the PPT or RRT, with the rent tax rate rising with realized profitability, may facilitate the setting of thresholds to yield high revenues with relatively little distortion of investment decisions. The argument for progressiveness in PPT or RRT itself, in addition to any progressiveness that results from the introduction of one slice of PPT or the combination of RRT and normal corporate income tax, depends on the assumption of risk aversion (as set out in Garnaut and Clunies Ross, 1975), or alternatively on the impossibility of setting a single threshold exactly right for all investments and investors, not only because different projects require different thresholds, but also because of the authorities' ignorance as to what investors' discount rates really are.

To pursue the second of these possibilities, suppose that the authorities are not perfectly knowledgeable about investors' discount rates and assessments of outcome, and that they are concerned with neutrality without being separately concerned with revenue maximization. Consider three separate RRT regimes that are designed so as to give similar NPVs of expected revenue over all relevant projects taken, together, on the assumption that producer decisions are unaffected by the RRTs' imposition. In the first regime, the threshold is set at x per cent, the authorities' lowest estimate of the discount rate appropriate to any likely project, and the RRT rate is set at a per cent. In the second case, the threshold is set at y per cent, the authorities' highest estimate of the discount rate appropriate to any likely project, and the RRT rate is set at b per cent ($b > a$). In the third case, there is a lower threshold x per cent and an upper threshold y per cent, and cash flows in excess of those necessary to yield positive PV at a discount rate of x per cent are taxed at c per cent (less than a per cent), while cash flows in excess of those necessary to yield positive PV at a discount rate of y per cent are taxed in addition at d per cent ($c + d > b$).

The tax in the third case is less likely to deter profitable investment than in the first, because the upward step in the tax rate is smaller at the x per cent threshold. The tax in the third case is less likely to induce wasteful over-investment than the second, because the upward step in the tax rate is smaller at the y per cent threshold. The fiscal authorities' search for the most nearly neutral system of rent taxes must involve some balancing of the hazard of 'investment-deterring' distortions against that of 'investment-inducing' distortions. According to their judgement about factors affecting the supply price of possible investments within their jurisdiction, they may find it easier to reach a proper balance with two RRT thresholds than with one. An alternative, not obviously worse and possibly better, is to use three thresholds, one at the authorities' best guess for the investors' discount rate, one nearer the lower end, and one nearer the upper end, of the likely range.

It should be stressed that this device of using more than one discount rate is presented as a *pis-aller* because of the administrative impossibility of finding an ideal solution to the problem of fixing a threshold. The Industries Assistance Commission (1976a), Swan (1976), and Summer (1978) have argued that an RRT applying two thresholds for the same outlays to the same project cannot be strictly neutral, and we believe that they are right. Dowell (1980) has argued that in a certain special case this need not be so, but the particular rule that he derives seems paradoxical, and he himself attaches no importance to it in practice. We rely here simply on the not very rigorous argument set out above to suggest that this solution might be tailored in such a way as to reduce the practical importance of the inevitably distorting effects that result from government ignorance about investors' discount rates.

A further problem for threshold-setting in the PPT and RRT is posed by the fact that investments made at later stages in a project's life (for example, after production has begun) may incur lower risk than those made earlier, and consequently, if a premium on the discount rate is used as the method of allowing for risk, have a lower discount rate applied. It would be possible to reflect this difference in the choice of thresholds for PPT or RRT by applying one threshold or set of thresholds to investments or negative cash flows incurred before production begins and a lower threshold or set to later ones. Exploration might also be given a higher threshold than developmental outlays, making a third tier. If simple divisions such as these are made, the administrative complications need not be great.

Inflation and Exchange Rate Variations

The real incidence of HRIT, PPT, and RRT can vary with the rate of inflation. Exchange rate movements can also affect the incidence of PPT and RRT in the currency of the investor, raising problems that are akin to those associated with inflation.

Inflation reduces the real value of depreciation deductions in assessment for standard corporate income tax. This well-known effect increases the incidence of the tax in ways that discriminate among industries and projects (Brown, 1948; Swan, 1976). It has its most severe effect on the incidence of taxation on (and so tends to deter investments in) relatively long-term and capital-intensive activities. Because of the higher rate of tax applied, this effect of inflation in increasing the real incidence of taxation is even more important in activities that are subject to HRIT and PPT (which are derived from income tax) than in those that are subject to income tax alone. There is less of a problem with RRT, because capital expenditure is written off against income in the year in which it is incurred, but still inflation would affect the real value for tax purposes of the outlays that are carried forward. The BT, however, is immune from this failing.

Various expedients have been suggested to make the standard corporate income tax more nearly neutral under inflationary conditions (Swan, 1976). This is not the place to evaluate the various proposals, except to note that the most commonly discussed among them would be difficult to administer and would in practice provide imperfect solutions to the problem. Nevertheless, reforms of the standard corporate income tax that reduced the effects of inflation on its real incidence would reduce the distorting effect of the HRIT and PPT if as we have generally supposed, they are based on corporate income tax.

Inflation affects the incidence of PPT and RRT, if their thresholds are set in nominal terms, by reducing the real value of income or cash flow that an investor receives before becoming liable to higher rates of tax. This is a familiar effect of inflation on progressive taxes. For RRT it can be regarded as another variant of the problem already stated. For the PPT it is separate. It may cause the real incidence of the PPT and RRT to vary from the intended incidence, which has presumably been selected for good reasons related to neutrality and revenue maximization.

If the rate of inflation were constant and known in advance, it would present no problem for the setting of PPT and RRT parameters, since it would require merely that the threshold rates of return beyond which various rates of tax are charged be set so much higher than if prices were stable. Since, however, expected rates of inflation do vary over time, it is likely that the 'supply price of investment', and with it the nominal thresholds that are appropriate for reasons of neutrality and revenue maximization, will vary accordingly.

With RRT, this effect could be allowed for by relating the threshold rates to some international long-term lending rate, which could be expected to function as an index of the effects of varying inflationary expectations on expected nominal yields on investment. The threshold rates for new investment could be varied from year to year but, for each project, fixed for the life of the project at the time of the agreement or of the start of major capital expenditures. Alternatively, the threshold rates at which the previously calculated cash flows are discounted could be varied from year to year in line with movements in international lending rates. The former method would be more advantageous to the government in a period in which actual inflation rates exceeded expectations at the beginning of the period and to the investor in the opposite circumstances. However, since such biases would not be anticipated in advance, either approach would be satisfactory.

Another alternative would be for past cash flows to be raised by some actual price index (as well as by the 'real' threshold rate) to determine the current year's tax liability. This would be extremely easy except for the problem of determining the supply price of investment in real terms for the purpose of setting the threshold. Market indicators such as long-term interest rates would be of only limited use, since they would take some account of expected, as well as current, inflation.

The suggestions in the preceding two paragraphs would be simple to administer, involving a change only in the number by which the accumulated value of net assessable receipts is multiplied each year.

For PPT, however, raising the threshold to account for inflation in the year in which tax is assessed will not serve, because an adjustment in a single year would not compensate for the continuing effect of one year's high inflation on liability for PPT. Instead, the value of accumulated capital outlays would need to be raised each year by the rise in the relevant price index.

In general, setting thresholds in real terms reduces perceived risk to the investor without any corresponding cost to the government. The only case against it is the difficulty of deciding appropriate real rates when the markets deal mostly in nominal rates.

For the foreign investor, the relevant supply price of investment is set in his own or an international currency. Thus, the *optimal* thresholds in assessment for PPT and RRT are set nominally in these currencies. If *actual* thresholds are set nominally in other currencies, and especially if they are set in the currency of a country with a doubtful record of monetary management, the investor will discount nominal threshold rates by his expectations about exchange rate changes. As with other reactions to uncertainty, it is likely that the investor will give high weight to possible unfavourable outcomes.

It often happens that exchange rate movements compensate with varying time lags for differential rates of inflation. If the investor is confident that approximate purchasing power parity is likely to be maintained between the currency of his own country and that of the host country, he may be content if satisfactory arrangements are made to adjust nominal thresholds for variations in expectations of inflation in the host country. In this case, if variations in long-term lending rates are used as an index of changes in inflationary expectations, the relevant interest rates are those on securities denominated in the currency of the host country.

Alternatively, if the investor lacked confidence that purchasing power parity would be approximately maintained, or if the government were unwilling to allow automatic adjustments for inflation, all magnitudes used in assessment for PPT and RRT could be denominated in the currency of the investor or in a major international currency. This could be achieved through the application of a simple formula to values denominated in a currency of the host country.

The logic of adjustment for exchange rate movements requires that the same currency be used in assessment for PPT or RRT throughout the life of the project; and that, if adjustments are made for exchange rate movements in one year, they should be made in all years. If adjustments were made for variations in the relative foreign exchange value of two currencies, it would be inappropriate (incorporating double counting) to adjust as well for variations in inflation in the country in which the investment is made. Rather, if any separate adjustment for inflation is to be made, it should be related to variations in prices denominated in the currency in which the PPT or RRT is effectively assessed.

Let us say that the investor is taken to be interested in the real values of returns translated into the currency of his country of origin. Then, if exchange rate changes *and* inflation in the currency of the country of origin were to be taken into account, the capital base for the PPT could be raised each year by $(1+i)$ $(1+e)$, and the discounted cash flows of the previous year for the RRT could be raised each year by $(1+r)$ $(1+i)$ $(1+e)$, where r is the real threshold, i is the rate of inflation in the country of origin, and e is the fraction by which the exchange value of the currency of the country of origin has risen over that of the host country.

Creditability

The profits earned by a foreign investor may be taxed in his home country as well as in the country in which the investment takes place. It is the outcome net of all tax that is relevant to determination of the supply price of investment. The taxing of income in the investor's home country reduces the rent available to the host government.

Most of the major sources of foreign investment for resource development make some provision for crediting taxes paid in the country in which the income is earned against home-country tax liabilities. A small number of countries (including Australia) do not tax income that has a foreign source. Some countries (most notably the USA) credit unilaterally payments of income tax in the country in which the income is earned against income tax liabilities that would otherwise be incurred at home in relation to that same income. Most countries allow tax credit arrangements similar to those of the USA, at least for income earned and taxed in other countries with which they have bilateral double-taxation agreements.

The USA is by far the most important single source of foreign investment in resource development. At the same time, its tax credit arrangements make possibly the most severe demands on the taxpayer if he wants to establish that taxes paid overseas are accepted by the home government as taxes on income. It is worth examining arrangements for tax credits in some detail, both because that country is very important in itself, and because some of the elements in the US arrangements are present in other jurisdictions as well.

The US Internal Revenue Service draws a clear distinction between income taxes and royalties. Income taxes paid to foreign governments in relation to income earned overseas are creditable against US income tax obligations. Royalties are *deductible* in assessment for income tax in the USA, but are not *creditable*.

The difference between creditability and deductibility can be described with an arithmetic example. Let us say that a US company earns $100 income in Indonesia. The Indonesian government imposes a charge of $50 on that income, and the US government taxes all incomes earned by its citizens at the rate of 50 per cent. If the Indonesian levy were classed as a *royalty* by the Internal Revenue Service, the IRS would assess the company for taxation on income of $50 (gross income net of *deductions* of royalty payments to the Indonesian government). The company would receive $25 after payment of $50 to the Indonesian government and $25 to the US government.

Alternatively, if the Indonesian levy were classed as an *income tax* by the IRS, the company would be assessed on income of $100. US income tax obligations in respect of the $100 income would be $50. However, in this case the IRS would credit the $50 income tax payments to the Indonesian government against obligations to the US government. The company need make no additional cash payment to the US government.

In the above example, company income after payment of all royalties and taxes would be $25 greater, or twice as large, if the IRS treated payments to the Indonesia as an income tax than if it treated them as royalties.

Until 1977, US companies were able to combine all of their overseas income and income tax before bringing them to account for income tax assessment in USA. 'Excess credits' in income subject to taxation at rates higher than those ruling in the USA were thus available to balance 'credit deficiencies' (perhaps associated with large royalty payments) in others. The US oil companies in particular had an abundance of excess credits from their Middle East operations, and so were not affected by 'credit deficiencies' in some other countries. However, after changes to US law, it is no longer possible to apply excess credits arising in certain foreign countries to tax obligations arising in other foreign countries. Hence the creditability of fiscal levies (up to the proportion of income assessed for income tax in the USA) in each country individually is not relevant to US income tax liabilities and so to the supply price of investment from the USA.

The distinction between income tax and royalties has evolved largely through case law over many years. There is some uncertainty about the creditability of a new tax applied by a foreign government until the IRS has provided a ruling on the matter, and it is not always prepared to do this until the matter requires a decision for purposes of US income tax assessment.

However, a few general principles are fairly clearly defined. To be an income tax, a levy must be based on some concept of *profits* (gross revenue minus costs). This eliminates FF and SAVD rent charges, and some of the forms of production-sharing. It is helpful for the levy to be called a 'profit tax' or an 'income tax'. The chances of a levy's being creditable are greater if it is based on the whole of a company's income in a country, rather than separately on various projects undertaken by the company. Dividends on equity held by the government of the foreign country, which may have been obtained on concessional terms and so contain a tax element, are clearly not creditable. Additional and excess-profits taxes are fully creditable if they are of general application across an industry, but if they are applied at different rates project by project they are more likely to be treated as royalties. There is a general presumption that a resource-owning government will extract some levy in addition to normal income tax, so that an additional profit tax is more likely to be accepted as an income tax if there is also a substantial element of subsidized government equity or royalty.

Although the probability that a levy will be accepted as an income tax by the IRS is raised if all of a company's expenditure in some foreign country is summed and deducted from all of its revenue in order to assess income tax in that country, there is no absolute requirement that income from all projects be assessed at the same time. It may, for example, be sufficient for losses (for example, exploration losses) on one unsuccessful project to be allowed as deductions against income from successful projects in the year in which the unsuccessful project is finally abandoned.

Production-sharing arrangements which split production and do not automatically allow full deduction of costs are not creditable at the present time. HRIT, PPT, RRT, and BT probably are creditable (even if assessed project by project, but with some provision for ultimately deducting losses on unsuccessful projects from profits on successful projects), so long as they are carefully designed with a view to the creditability requirements. There is recent evidence that the likelihood of US creditability of the whole of a direct-tax combination is greater if any separate resource tax is imposed before regular income tax rather than after, but this is obviously a matter on which up-to-date legal opinion should be sought. (The British petroleum revenue tax is creditable only because of a bilateral agreement between the US and the UK governments.) It may be an extremely important advantage of the profit-related taxes over the corresponding methods of equity transfer that the former can be made creditable while the latter cannot.

The supply price of foreign investment is likely to be very much lower, and the resource rents correspondingly higher, if a government ensures that a sufficiently large part of its fiscal charges are creditable against overseas income tax so as to avoid to the maximum possible extent the payment of taxation to foreign governments in respect of income generated by resource developments within its boundaries.

Problems of Administration

In most forms of taxation there are problems of ascertaining the tax base correctly, especially given the taxpayer's incentive to obscure the true situation.

The administration of the most simple rent tax, FF, requires no information beyond the law or agreement itself. That of the specific duty requires information on the volume of sales. That of the *ad valorem* royalty requires information on both the volume of sales and the prices at which they are made. Taxes based on profit, however, require information on capital expenditure and operating costs, which greatly complicates their administration. The additional information requirements are similar for HRIT, PPT, RRT, and BT, although they are more complicated in the former two than in the latter two.

In most countries (although not in all developing countries) there is now an established tradition of income taxation, and there are meticulous, if not always clear, definitions of the receipts and expenditure that are to be brought to account in assessment for income tax.

The definitions of receipts and expenditure that are used in the assessment of company income tax can be used in the assessment of HRIT and PPT and, with a few exceptions, in that of RRT and BT as well. The fact that RRT and BT are based on cash flows rather than on accounting definitions of income means that depreciation need not be provided for separately and that the valuation of inventories is irrelevant, so that the two most difficult areas of administration of the standard income tax are obviated. Similarly, interest is not allowed as a deduction, and this has the effect of removing an important source of distortion and avoidance in the operation of the standard income tax.

There is, however, at least one new problem of definition for the RRT and BT. Increases in stocks of unsold raw materials would not be assessable receipts for these taxes, and it would be necessary to specify the time at which sales became assessable receipts. Either the time of physical dispatch or the time of receipt of cash would be appropriate.

The opportunities of using transfer pricing to avoid ordinary income tax are also present in the various profit-based rent taxes, and usually in a more severe form on account of the higher marginal tax rates and the possibility of shifting income from higher to lower marginal tax rates within the same jurisdiction. Postponement of tax liability through early payment under longer-term intra-firm contracts for services, and other transfer pricing problems, are important with the income tax, but will need to be policed more carefully with the profit-based rent taxes, especially PPT and RRT, where the rewards in tax avoidance of income postponement are likely to be higher. It might be useful to limit the definitions of deductible payments that are made in respect of services to those services rendered in the relevant accounting period.

On balance, problems associated with the calculation of depreciation, with inventory valuation, and with deductions for interest, probably make the standard income tax (and the HRIT and PPT, which require the same information) more complex administratively than the RRT and BT, at least for a government that is designing a taxation system *de novo*. However, familiarity with the concepts applied in assessment for standard income tax, in jurisdictions in which it is well established, may mean that the application of the HRIT and PPT is easier in the early stages.

Combining Rent Taxes and Income Taxes

The requirements of creditability may introduce overwhelming considerations in favour of combining rent taxes with standard corporate income taxes. With FF and SAVD, the rent taxes are applied first and deducted from income in assessment for income tax. HRIT and PPT, since they apply to the same base, may be treated simply as elaborations imposed on income tax.

It is of little importance whether RRT and BT are applied to pre-tax or to post-tax cash flows provided the tax taken first is allowed as a deduction against the other. The HRIT and PPT might also be treated as separate taxes deductible against, or giving deductions for, income tax, rather than as extensions of income tax. Table 7.5 gives an example which, in its choice of tax rate, supposes that the PPT will be treated in this way. In the case of RRT and PPT, lower threshold rates must be allowed (if they are to have a similar fiscal effect) when they are applied after, and allow deductions for, income tax than when they are applied before, and are deductible against, income tax.

Where an income tax and an additional profit-related rent tax are applied together, it is of course their combined top marginal rate of tax on cash flows that must be limited by the need to maintain economizing behaviour.

It was observed earlier in this chapter, as well as in chapter 8, that the BT is more nearly neutral than the RRT with respect to exploration expenditures (unless the RRT has a very high threshold, which would distort subsequent decisions and collect too little revenue), while the RRT has possible advantages outside the exploration phase because it may have a greater effect in reducing sovereign risk. This argues for a hybrid system of rent taxes, in which loss offsets, that is subsidies (or their equivalent, deductibility against income in established mines, combined with freedom to sell exploration 'losses'), are allowed for exploration, together with an RRT system for other outlays and receipts. When loss offsets have been allowed, the exploration expenses will be cancelled. They will not be carried forward with interest as negative cash flows. Threshold rates may then be lower since they will no longer have to reflect the supply price of investment to risky exploration.

This hybrid is equivalent to the giving of a subsidy on exploration and then treating the rest of a project separately and taxing it by RRT. It goes halfway to full government undertaking of exploration, which, if practicable, has definite advantages, especially in making it feasible to auction mining rights.

Applying Rent Taxes to Existing Projects

Much of the discussion has been conducted as if tax systems were to be devised *de novo* to cover new mining projects. Often, however, countries with longstanding mining industries have recognized the deficiencies of existing systems as circumstances have changed and have looked for improvements. This has been the case with the US and Australian federal governments in the 1970s, and even by 1981 with the UK, in spite of the fact that its current system dated only from 1975.

It is not absolutely necessary that pre-existing projects should be treated in exactly the same way as new ones. Doing so, however, helps to establish an impression that the system is likely to be stable. This sets something of a premium on those methods of taxing that do not require much retrospective information that is not likely to have been registered with the tax authorities. If income tax already exists, the HRIT raises less problems here than RRT, with PPT occupying an intermediate position. All have potential difficulties if projects are to be the entities taxed while income taxation has been levied on persons and companies. The BT is inappropriate for retrospective treatment since its special feature is subsidization of outlays as they occur. If an auctioned FF (cash bonus bid) is to be combined with a profit-based tax for new projects, there is probably no way of making a comparable arrangement for old projects. These have either to be subject to the profit-based tax only at the same tax rate and threshold as new projects, and hence to be more leniently treated than new projects, or else to be given a higher tax rate (or a lower threshold, if that appears to be consistent with neutrality), which might be regarded as some kind of compensation for escaping the cash bonus bid.

If the system introduced includes a RRT or PPT Mark II, pre-existing projects could be made subject to the chosen tax rate for net positive cash flows or income continuing from the introduction of the new system except in so far as they could demonstrate to the satisfaction of the authority that they had (in the case of the RRT) past negative cash flows over the whole life of the project, or of the company's mines within the country, that had not been recovered at the threshold rate, or (in the case of the PPT Mark II) that they had undepreciated capital expenses associated with the taxable entity. Alternatively, the past could be disregarded and only new outlays considered for concessions against a straight tax on positive cash flows or income. This would not be objectionable on grounds of neutrality on the projects directly affected; its only undesirable feature is the apparent unequal treatment that it involves and the doubts that this might raise over the government's consistency in future. If new mines had to face an additional charge such as a cash bonus bid, the force of this objection would be weakened. It would also be weakened if the government could plausibly assert that the companies or projects concerned had covered their mining expenses in the country with a handsome return, but if that were clearly the case the policy of choice might be to leave the opportunity of deductions for past expenses open (probably with an RRT form) and so challenge the companies to prove the contrary.

A special difficulty might arise, however, if it were possible that the current holder of mining rights had recently bought them from another private firm. To ignore the buyer's expenses of purchase in assessing him for PPT Mark II or RRT might be regarded as punitively unfair, while on the other hand treating it like a productive investment might encourage sales of mining rights (perhaps even bogus sales) once the character of the legislation imposing the new tax was known or suspected. At an early stage in discussion over such a new system, the government would need to declare that no future purchase prices paid for mining rights by one company to another would count as the former company's investment for PPT purposes or as a deductible expense under RRT.

Summary

The various problems discussed in this chapter of administration and imposition of rent taxes have led to the following judgements.

Auctioning tax rates or thresholds is appropriate only in the case of the fixed fee paid in advance of detailed exploration; and the auctioned fixed fee itself, though apparently neutral, and also highly successful as a revenue-raiser in certain conditions, is not likely to be effective alone in other circumstances and is most appropriate in combination with a profit-based rent tax. Problems of choosing appropriate rates, if they are either to be negotiated or to be imposed generally across an industry, tell very much in favour of profit-based taxes and most in favour of those furthest to the right in tables 7.1–7.3. (The choice *between* negotiated and general rates is explored in chapter 6.)

The rates of rent taxes should be set no higher than a rate at which unduly large incentives to significant unnecessary outlays set in. This may be higher with projects where rent is very high in relation to necessary costs than with others.

Clarity in definition is required between those activities that will be treated as extractive for tax purposes and other activities.

If rates and thresholds are negotiated by project, the project should be the tax unit. If, on the other hand, there are to be general rules, arguments can be made for either the company or the project as a basis for tax. Various compromises are possible to mitigate the disadvantages of each, such as adopting the project method but either offsetting all of a company's exploration expenditure in the country against receipts in the company's first successful project, or 'trading' exploration losses among projects The difficulties relate particularly to unsuccessful exploration and may be mitigated by the government's undertaking or subsidizing exploration.

There are ideal methods by which, with perfect knowledge, appropriate combinations of tax rates and thresholds could be devised for PPT and RRT so as to maximize the revenue consistent with a close approach to neutrality. In practice, however, sets of rates and thresholds will probably best be fixed for each industry in advance of discussions over particular projects; and, because of uncertainty over, and variation among investors and projects in, investors' discount rates, there may be a case for using more than one threshold and rate of tax as a means of maximizing revenue and approaching neutrality. There is also a case for using different thresholds or sets of thresholds for different stages in a project's life (a lower set perhaps to apply to outlays or investments made after production has begun).

Simple adjustments can be made to all the profit-based taxes to take account of exchange rate variation and of inflation in the currency in which investors are interested. The purpose of these adjustments is to reduce uncertainty for the investor without increasing it for the government. The methods possible include indexation, both for the value of the relevant currency and for the rate of exchange, and connection of thresholds to standard interest rates.

Creditability for favourable international tax treatment generally favours the use of profit-based taxes rather than specific or *ad valorem* royalties or fixed fees, and it sets certain technical requirements on any profit tax.

Creditability may dictate that those rent taxes that are not simply variants on a conventional income tax on companies should be combined with such a conventional income tax, whichever tax is collected first being treated as deductible for purposes of the other. Creditability may favour applying rent taxes before, rather than after, income tax. Where creditability does not require this (because the mineral rent tax is itself creditable against the investor's home government tax), there may be good grounds for not applying conventional income tax to mining projects in countries in which investment in mining comes principally from overseas.

Purely administrative problems of assessment and collection of tax favour the cruder types, such as the fixed fee and the specific or *ad valorem* duty. Among the profit-based taxes they possibly favour the HRIT and PPT where an income tax system is already firmly established, but certainly favour the RRT or BT (which are intrinsically simpler) where it is not.

When a new system of rent taxes is to be introduced where there are mines already in operation, the forms of tax that are most desirable for new projects, such as BT, RRT, and PPT, may give rise either to administrative problems with old ones or to problems of apparent equity between new and old. Various forms of compromise, however, are possible. To preserve equity while avoiding abuse, the authorities would need to adopt special devices for RRT or PPT wherever one private company was able to buy mining rights from another.

11

THE DISSIPATION AND DIVERSION OF MINERAL RENT

Governments do not always receive in their revenues the full potential rent value of mineral resources that they own. Sometimes this is because the resource is exploited in an economically inefficient manner. We describe this as the 'dissipation of rent'. At other times it is because part of the rent accrues to persons or organizations other than the state. This is referred to as the 'diversion of rent'.

The dissipation and diversion of rent can result both from characteristics of the economic environment that are beyond the control of the state (such as monopoly in international factor markets) and from the policies of the state itself. Non-neutral taxation systems are an important source of dissipation of rent, and taxation systems that collect only a part of the mineral rent are an important source of diversion of rent. These circumstances were the subject of chapters 7, 8 and 9 and they are not discussed further here.

The chapter begins with discussion of the circumstances in which rent may be diverted away from the state to other parties. A second section discusses how some of these causes of diversion of rent may also result in its dissipation. The various other ways in which the potential rent value of a natural resource may be dissipated are then considered. A final section covers diversion and dissipation of rent resulting from monopolistic organization in mineral product markets. This is of particular interest in these days when the OPEC oil cartel attracts so much attention.

Diversion of Rent

The diversion of resource rent is likely to result from one or other of three arrangements. In the first, some party other than the state (perhaps another tier of government or the owner of the land upon which the resource is found) must give its approval to a resource development. In the second, some decentralized regulatory agency of the state has the power to enforce some conditions upon the operation of the resource project. In the third, there is monopolistic organization in the supply of inputs to the resource project. All three situations are potentially important and it will be useful to deal with them separately.

First, it is not uncommon for the state's sovereignty over natural resources to be qualified by the implicit or explicit requirement that some other parties, most commonly other tiers of government or landowners, must give their approval before a resource development can proceed. This qualification may be absolute, or may be limited to the raising of reasonable objections (as defined by the state or the courts) and reasonable claims for compensation or reward.

One very important special case of this general phenomenon occurs within federal and quasi-federal systems of government in which there is a constitutional separation of the relevant legal powers. For example, under the Australian and Canadian constitutions, state or provincial governments own and manage mineral resources, but the national government has sovereign powers over income taxation, foreign trade, monetary policy, and many other matters of central importance to the returns that investors receive from resource projects.

Such qualification of sovereign powers over natural resources introduces elements of bilateral monopoly into the relationship among all parties who possess powers of approval over a resource investment. Each party can insist on being 'rewarded' or 'compensated' for its approval, within limits set by the bilateral monopoly relationship. Such insistence leads to the diversion of part of the resource rent to the party providing the approval. (It is possible for disputes among the parties with power of approval, or for poor co-ordination among them, to cause dissipation of rent as well, as discussed in the following section of this chapter.)

It is common, second, for some aspects of the operation of a resource project to fall within the regulatory jurisdiction of a government agency whose operations are more or less independent of central government control.

Though such regulatory agencies do not possess general powers of approval over resource developments and so are not in a position to appropriate for themselves part of the resource rent, nevertheless regulation of the pricing of inputs or products of resource investments, or of the technology applied to them, may reduce the profitability of such investments. This may or may not involve the diversion of rent from the state to other parties or its dissipation. The regulation may simply offset external costs of resource development and so remove the possibility that the investment may generate negative rents for society, as might be the case with some regulation of provisions for employees' safety or for environmental impact. Yet, diversion or dissipation of rent occurs whenever the regulation goes beyond the compensation or removal of external costs.

One important and common example of regulation that causes diversion of rents occurs when an agency with regulatory powers over prices causes mineral products to be sold at lower prices than would be generated in a competitive market. Resource rents in this situation are diverted from the state to purchasers, and perhaps ultimately to users, of mineral products.

There are other examples of rent diversion through the regulation of activities of mining companies. Environmental regulations may cause mining companies to go beyond the compensation of environmental damage, diverting potential rents from the state to those members of the public who make much of the improved environment; or the investor may be required to provide more, or higher-quality, infrastructure than is necessary for the project itself. Other important occasions for diversion of rent are government regulation of labour conditions (which may divert potential rent from the state to wage-earners employed in mineral production); and restrictions on direct foreign investment (which may divert resource rents from the state to preferred, usually national, private investors).

Third, there may be diversion of rent away from the state when there is monopolisitic organization in the supply of inputs to mineral investment and production. Monopolistic organization is potentially important in the supply of private investment, technology, and labour. There may be monopoly in world-wide markets for some necessary input. Alternatively, a few particular suppliers of inputs may have become essential to the continued profitable operation of a project, although they do not exercise monopoly power in world-wide markets. In both cases, the supplier's relationship with the resource owner can be described as one of bilateral monopoly, although the constraints on the supplier's use of monopoly power may be greater in the latter than in the former case.

Monopoly in the supply of world-wide investment in a particular mineral industry may occur because only one firm, or a small number of firms, has had experience in the whole range of activities essential to the profitable production of the mineral. Historically, however, it has been more common for a monopoly or oligopoly to be maintained through restrictive practices imposed on the market for the product by a few large producers; this occurred with copper mining in Latin America and Africa in the first half of this century, with petroleum extraction over most of the world in the 1950s, and with bauxite until recent years. Recently monopoly or oligopoly has sometimes been maintained by a company's exclusive access to some valuable production or processing technology: for example, by the present exclusive ownership on the part of a small number of US firms of advanced technology for offshore natural gas production and deep sea oil production.

Even when there is competition world-wide in the supply of private investment and technology, a local monopoly may be established by the mode of administration of mineral leasing, as discussed in chapter 6, or by a particular firm's special interest in and understanding of the geological, economic, and political environment of some potentially profitable resource investment. The process of diversion of rent to monopolistic suppliers of investment was discussed in detail in chapter 6.

There is rarely monopolistic supply of labour to mining within a country when it is originally established. However, the environment of the mine is conducive to strong union organization. Where there is highly profitable and highly capital-intensive production, a strongly unionized mine work-force is capable of inflicting large losses on a mineral investor through strike action. As a result, it is common for wages to be raised substantially above those ruling generally in the economy in which the resource project is located. The higher wages represent a diversion of potential rent from the state to wage-earners.

Diversion and Dissipation of Rent Together

Some of the sources of diversion of rent described above result also in the dissipation of rent, that is, in the reduction of the total amount of resource rents available to society as a whole. Among the three characteristic causes of diversion of rent that were described above, dissipation is most likely to be present when diversion results from monopolistic supply of inputs.

The first source of diversion of rent described above—the splitting among various tiers of government, and perhaps among landowners as well, of powers to approve natural resource developments—need not cause dissipation of rent. It is in the interests of all parties with powers of approval that it should not. Dissipation does result, however, when the process of bargaining for rent shares among the parties causes delays in investment or otherwise sub-optimal investment decisions. Furthermore, poor co-ordination among various parties that have approval powers, with each levying rent charges in one form or another, may lead to a total fiscal system that is markedly non-neutral and generates less than the maximum possible economic rent. Examples are provided by the combination of federal and provincial or state taxes and royalties in Canada and Australia.

Where the actions of regulatory agencies cause input or product prices to diverge from levels that would emerge from competitive markets, it is also likely that dissipation of rent will accompany the inevitable diversion. The enforced raising of input prices (including wages) is likely to give rise to substitution against the use of the regulated input. In an extreme case, it is possible for the increase in production costs associated with the regulation of input prices to deter investment that would otherwise take place. It is also likely that continued controls on the output prices of some minerals will lead to under-investment in their production, and it is possible for regulation of output prices to deter the establishment of some otherwise intra-marginal projects.

Similarly, where an input is supplied monopolistically and its price is raised above the competitive level, there will be a strong tendency for techniques of production to be modified so as to use the over-priced input less intensively. Various types of labour may be priced in this way. This causes dissipation of some potential rent. Though it will rarely be in the monopolistic supplier's interest to force the abandonment of an otherwise feasible project through insistence upon higher input prices, this result may follow miscalculations in the bilateral bargaining.

One interesting special cause of diversion and dissipation of rent that has been given considerable attention in Australia and Canada is protection of domestic manufactures against import competition. Such protection generally raises the overall cost structure and, in an economy that exports mainly natural-resource-based commodities, reduces the mineral rents available to government. Part of the loss of rents represents dissipation, since ancillary resources are less productive in the protected industries than in those in which they would otherwise be employed. But part represents diversion of mineral rent to other factors of production, especially to labour. In some circumstances this can cause real wages to rise above what they would otherwise be (Garnaut 1979b; Brigden *et al.*, 1929; Copland, 1931; Reddaway, 1937). This possible effect on real wages has been used as a defence of manufacturing protection in Australia and Canada. However, it is not a first-best means of distributing income from mineral rent to wages: the collection of taxes on mineral rent to support some form of redistribution towards labour would lead to higher total product and could raise workers' disposable incomes by as much as, or more than, they would be raised by protection.

Dissipation of Rent without Diversion

Conditions imposed by governments upon the manner in which a mineral deposit is to be exploited can cause dissipation of rent, independently of the various factors described above as contributing to diversion of rent. Dissipation of rent will occur whenever the state fails to use its sovereign powers over resources to prevent duplication of effort in mineral exploration or production. There will usually be dissipation of rent, too, whenever, in pursuance of some objective other than revenue maximization, the state insists that private investors should invest at a time, on a scale, or with techniques, that differ from those that would be chosen by the investor in the normal course of events.

It was demonstrated in chapter 2 that mineral rent is quite likely to be dissipated completely in wasteful duplication of investment in mineral exploration or production unless the state grants exclusive rights to individuals or companies. This is the valid rationale for the granting of exclusive exploration or mining rights, which is now applied more or less universally, except in relation to resources outside national jurisdictions such as parts of the ocean.

It is common, however, for the state to demand, as a condition of mineral rights, higher levels of private expenditure on mineral exploration or processing than would be undertaken willingly by investors. In the extreme case, the state tends to cause potential mineral rent to be dissipated completely if it awards exploration or mining rights to the company that is prepared to undertake the largest investment in exploration or mineral processing (Erickson, 1977).

The intangible and uncertain nature of the mineral rents whose loss is the opportunity cost of excessive expenditure on exploration or processing makes dissipation the more likely. Governments are seldom clearly aware of the costs of rent diversion and dissipation, and so often attribute less value to rents that are lost through them than to more tangible benefits of much lesser value.

A closely analogous situation arises when the state insists on the use of a sub-optimal technique in pursuit of some objective other than revenue maximization. The most common example is the insistence of some governments on high labour intensity in mineral production in pursuit of employment objectives. Such insistence leads to some dissipation of rents, and it is quite likely that the resulting reduction in public expenditure will cause less employment to be provided in the economy as a whole than would have been the case without intervention.

The Special Case of Monopoly in Product Markets

Since many minerals are produced from a relatively small number of deposits in a smaller number of countries, it is apparent that there is at least some possibility of cartel action to raise the price of mineral products. Moran (1975) has argued that monopoly control of copper marketing by a few multinational companies raised copper prices and increased the rents available to governments of developing countries that exported copper before the nationalizations of the 1960s. It is generally believed that the OPEC countries have increased their revenues from oil since 1973 by colluding to raise the world price.

In this concluding section, we examine one special case of dissipation of resource rents: dissipation that results from the use of monopoly power to reduce mineral production and to raise prices. In this case the reduction of resource rents is associated with an increase in monopoly rents. As we shall see, the monopoly rents may accrue to investors or to mineral states, or they may be shared between them. Total monopoly rents may be less than, equal to, or greater than the resource rents dissipated through collusion, depending upon the sensitivity of supply and demand to changes in price. Thus the revenues of mineral states may be reduced, maintained, or increased by the successful use of monopoly power.

It will be helpful to begin the discussion of these issues by reference to a simple diagram. In figure 11.1, SS' is the world-wide supply function for some mineral, reflecting marginal costs of production. D is the world-wide demand curve for the mineral, and MR is the corresponding marginal revenue curve.

With competitive world markets, producers of the mineral maintain output at OQ_1, and it is sold at price OP_1, each determined by the intersection of the supply and demand curves at X. With competition in the supply of investment, the governments owning the mineral deposits receive mineral rents SXP_1, reflecting the difference between the cost of production at the marginal mine and the cost at each producing mine.

Now let us introduce collusion among mineral producers, leaving aside for the moment the question of whether the cartel is organized and enforced by investors or by governments. Production is restricted to OQ_2, corresponding to the point Y at which the marginal revenue curve MR intersects what in competitive conditions would be the supply (marginal cost) curve. This amount of mineral production can be sold at the higher price OP_2. No other levels of output and price will yield greater producer rents.

Let us assume for ease of exposition that the highest-cost mines are retired from production. Resource rents SYP_3 are lower at the reduced level of output, because of the reduced differential between costs of production at the new marginal mine and those at each intra-marginal mine.

The higher price and the lower production give rise to monopoly profits P_3YZP_2. These monopoly rents can be compared with the loss of resource rents P_3YXP_1. The area P_3YWP_1 is common to the areas representing the gain in monopoly rents and the loss in resource rents. Total producers' rents will rise through the use of monopolistic power if the area P_1WZP_2 exceeds the area YXW. This is more likely to be so the greater the rate of increase in production costs for expansions of output beyond OQ_2 and the more inelastic the demand function.

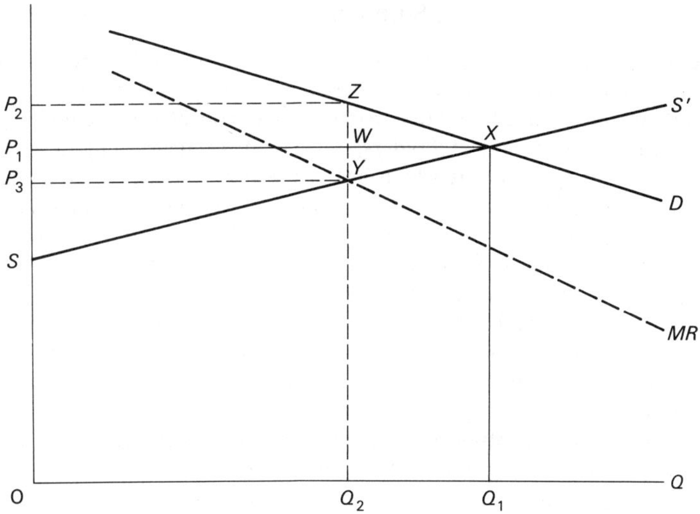

Fig. 11.1 Resource rents and monopoly rents

Who receives the monopoly rents? If the cartel is organized and enforced by governments that enjoy access to competitive markets for mineral investment, the monopoly rents will accrue to the state in the mineral-producing countries. This would seem to approximate the case of the OPEC petroleum cartel in recent years.

Alternatively, if the cartel is organized and enforced by investors who face governments that lack, or fail to use, monopoly power, the monopoly rents will accrue to the investors. This would seem to approximate the case of the copper producers in Latin America and Africa in the first half of this century. In this case, the existence of monopoly in the product market is clearly unfavourable to governments' revenues, the contrary arguments of Moran (1975) in the Chile case notwithstanding. Most commonly, perhaps, there will be elements of monopolistic organization among both governments and investors, and there will be some sharing of monopoly rents.

Thus, the loss of state revenue from the dissipation of resource rents associated with the use of monopolistic power in the product market may be partly or completely compensated, or more than compensated, by the state's receipt of monopoly rents. The state's gain of monopoly rents is more likely to exceed its loss of resource rents (i) the more inelastic is demand for the mineral product in the relevant price range, (ii) the stronger is the collusion among governments of mineral-producing countries, and (iii) the greater is the competition among suppliers of mineral investment.

Summary

Governments do not always receive the full potential mineral rent into their revenues, because it is dissipated in sub-optimal patterns of production, or diverted to other parties. Diversion of rent commonly occurs when private rights holders, regulatory agencies, and tiers of government other than that which has prime responsibility for mineral resource management can negotiate some of the terms upon which mining can proceed, or when investors or workers hold veto powers over smooth production at the mine. Poor co-ordination of the demands made by monopolistic suppliers of labour and investment or by various governments and regulatory agencies (such as occurs in tax conflicts between state or provincial and federal governments) can also dissipate rent, although it is not in the interests of any of the parties that it should do so.

Dissipation of rent often results from government regulation directed at objectives other than revenue maximization, for example, requirements for exploration or for processing, which impose extra costs on the investors that are not matched by their own additional receipts or by external benefits or by both together.

An interesting and important special case of diversion and dissipation of rent occurs when producers use cartel action to raise mineral prices. There is some loss of mineral rent through the fall in production. This loss may be offset by an increase in monopoly rent. The extent to which the monopoly rent is available to governments is likely to be greater the more competition there is in the supply of investment. Such competition is likely to be strong if governments themselves are organizing and enforcing the price cartel.

12

ECONOMIC STABILITY AND THE TIMING OF MINERAL INVESTMENT, PRODUCTION, AND REVENUE

Problems of Timing

The value to a government of revenues from the exploitation of mineral deposits within its control depends on the timing of mine construction and the rate of production from existing mines. This reality causes governments to concern themselves with private investors' decisions on timing, and sometimes prompts government intervention in the investors' decisions.

Here we analyse problems for governments that might arise out of private decisions on the timing of mine construction and mineral depletion, and comment upon various expedients that might contribute to their alleviation. Some of the possible problems are affected by the taxes that are applied to mineral income and so, in principle, are relevant to the choice of rent tax regime.

It was suggested in chapters 3 and 4 that governments typically discount future income at a lower rate than private investors. This chapter begins by discussing how these divergent time preferences of governments and private investors affect the timing of mine development and production and so the real value to the government of mineral revenue.

The chapter then describes how the distribution of a government's mineral revenue over time tends to be highly uneven, analyses problems associated with the uneven distribution, and suggests expedients for managing these problems. Four distinct types of problems with uneven revenue flow are identified and separately discussed: (i) uneven inter-generational distribution of revenue flows; (ii) temporary revenue booms associated with high levels of general economic activity during the construction phase of major resource investments; (iii) large stepwise increases in revenue when new mines begin production; and (iv) cyclical fluctuations in revenues in response to variations in demand for the products of established mines on international markets.

The concluding sections of the chapter discuss the relative merits and shortcomings of several possible responses to the various problems that derive from the timing of mineral production and revenue flows: varying the form or severity of fiscal charges; keeping expenditure stable while revenue fluctuates; and controlling the timing of mineral production and investment.

Divergent Discount Rates and the Amount of Mineral Rent

It is convenient to begin our analysis by assuming that the government applies rent charges that do not affect investors' decisions at all; that is, that are strictly neutral. In practice, as we have seen in earlier chapters, the timing of production may be influenced by the rent charges, and this possibility is examined later.

Decisions on the timing of mineral investment or production by mining companies to maximize their own after-tax profits also maximize the present value of government mineral revenue so long as the investor and the government apply the same discount rate to future cash flows. But as was demonstrated in chapters 3 and 4, the discount rates applied by private investors reflect differing perceptions of risk from those applied by governments, so that they are commonly higher even when private investors and governments have access to international capital markets on similar terms.

Private investors' use of higher discount rates has several important consequences for the timing of investment and production. Fewer mineral investments qualify at any time than if the government's discount rate were applied. This in itself leads to less exploration and to the construction of fewer new mines than would be required to maximize the present value of mineral rent at the government's discount rate. On the other hand, the higher discount rate tends to accelerate production at projects that have qualified for investment, in particular through the earlier exploitation of higher-grade deposits. This distortion of the rates of depletion of deposits that qualify for mining may lead to less complete utilization of lower-grade ores. These effects on the rate and extent of production from those mines that are established also reduce the net present value of mineral rent at the government's discount rate.

It is not clear whether the net effect of private investors' applying higher discount rates will be to increase or to reduce the rate of total mineral production, and therefore the amount of mineral revenue, at any time. It is possible that the government's revenue receipts could be either brought forward or retarded. But it is clear that the divergence in discount rates reduces the net present value of government mineral revenue. It is possible, in principle, that some time-profile of mineral exploitation other than that which resulted from the unconstrained decisions of private investors would generate larger total government rent. The practical importance of this possibility is discussed later in the chapter.

Inter-Generational Distribution of Mineral Revenue

The government of any one country might judge that there was something unsatisfactory about the distribution of its own revenues between the living and the unborn that would result from private investors' unguided decisions. For example, the government of a country with large petroleum reserves might judge that its revenues in the immediate future were likely to be very much higher than in later years, and might want some of the benefits of use of the early revenues to be shifted back in time. Conversely, a government with few mineral reserves but with promising geology might hold the view that it was desirable to bring forward in time some of the benefits of use of the mineral revenues that were expected to be available to it in the longer-term future.

It would be sensible, however, for a government to concern itself with the equitable inter-generational distribution of total government revenue and income, rather than the distribution of only that part of revenue and income which derives from mining. Thus it is only in countries in which mineral rents contribute a high proportion of public revenue that the inter-generational distribution of the use of mineral revenue is a serious issue.

In a country for which government revenue and incomes per capita were expected to rise over time, that is, in the normal situation, there would seem to be no reason for shifting the use of mineral revenue back in time even if receipts from this source were expected to be concentrated heavily in early years. Even an expectation of falling per capita revenue and income in a country with large mineral reserves would not in itself argue conclusively for shifting the use of some mineral revenue back in time: it is possible that the early use of mineral revenue for productive investment would do more to stem the decline over time in per capita income than would delay in the use of some mineral revenue. Judgements must be made on alternative patterns of change in per capita income over time, allowing for the effects of different rates of investment in early years.

The examination of projections of total per capita revenue and income over time, allowing for the effects of variations in the level of early expenditure on economic growth in the economy as a whole, may yet suggest that there are advantages in shifting across generations some of the use of resource revenues. The relative merits of various instruments for securing such shifts are discussed in a later section of this chapter.

The Construction Boom

The construction phase of a major mining project, and its cessation, can generate considerable instability in the wider economy. Problems of instability may suggest a case for government intervention in the timing of mineral investment in order to moderate fluctuations in economic activity. Here, as a prelude to assessment of policy responses, we analyse the effects of instability resulting from large-scale mine construction in various types of economy.

Construction expenditure at a new mining project is often large in comparison with the total economy of a developing country, or with the economy of the particular region within which the project is located. In addition, there is some tendency for the construction phases of different mineral projects to be bunched together at times of buoyant conditions in world markets. The large expenditures on the goods and services that are required in the construction phase of a large mine generate boom conditions in the immediate region or even throughout a small economy.

It is common for the price of labour and other factors of production to rise during the construction boom. Government revenue rises with the general level of incomes and economic activity. If the general increase in revenues leads to greater public expenditure, the boom conditions are intensified. Supply bottlenecks in the general or regional investment boom often introduce inefficiencies into private and public programmes.

The construction boom is often followed by lower levels of economic activity, at least until the government begins to receive and to spend large amounts of mineral revenue. If wages do not fall with receding aggregate demand, unemployment may appear. The wage rigidity causes real incomes and output to recede even further below the levels of the construction boom.

The strength of the tendency for a mineral investment boom to generate unemployment in this way varies with the nature of labour supply in the economy in which it occurs. It is useful to distinguish three separate types of labour market conditions. Type 1 is the dual economy comprising a village sector and a modern sector, within which the supply of labour to the modern sector is infinitely elastic over the relevant range of demand for labour. This is the case assumed in the Lewis and subsequent models of the surplus labour economy. Type 2 is the dual economy within which the village sector is highly differentiated, with labour having a positive but highly variable opportunity cost. More labour can always be induced to migrate from the villages to the modern sector, but only by raising the expected level of modern sector incomes. Type 3 is the fully monetized economy, in which the supply of wage labour is fixed independently of labour market conditions. These highly stylized types do not fit any economy exactly, but the three types have something in common with conditions in Indonesia, Papua New Guinea, and Australia, respectively.

The effects of an expansion in aggregate demand on the domestic economy depend very much on whether wages are set in the market so as to equate the supply of and demand for labour, or whether minimum wages are established institutionally.

When wages are set in the labour market, increased domestic expenditure raises the real wage level in types 2 and 3, but not type 1. It induces migration from villages to the modern sector of the economy in types 1 and 2, but there is no opportunity for such an expansion of modern sector labour supply in type 3.

In the alternative situation, where institutionally set minimum wages prevent the clearing of the labour market (which we refer to hereafter as the 'constrained labour market' case), increased aggregate demand induces migration to the modern sector in type 1 but does not affect permanently the wage level or the rate of unemployment. In type 2 increased expenditure induces some additional migration to the modern sector (because of the increased probability that a migrant will find employment), but also reduces the rate of modern sector unemployment. In type 3 there is an even greater reduction in unemployment, because there is no increase in total labour supply to the modern sector. In types 2 and 3 it is possible that the increased expenditure will lead to the elimination of unemployment and to an increase in real wages in the market place beyond the institutionally determined minimum levels. Increased aggregate demand reduces modern sector unemployment only to the extent that there is no offsetting increase in minimum wage levels. The feeling of prosperity that accompanies the construction boom creates an environment within which wage-setting institutions are more likely than usual to be sympathetic to large wage increases, for reasons of 'equity', or in accession to pressure from wage-earners.

Of course, labour is not homogeneous in any economy. Even in the dual economy in which the general supply of labour to the modern sector is infinitely elastic (type 1), the scarcities of some skills will increase average labour costs to some extent, at least temporarily. There will be opportunities for firms and governments to eliminate scarcities of particular skills over time, through training and through the accretion of skills 'on the job' (see the discussion of the Indonesian oil boom of the 1970s in Garnaut, 1979a).

The nature of the relationship between scale of production and costs, together with the changes in real wages and other domestic resource costs, determines the effects of increased expenditure on prices and on the sectoral composition of economic activity. Let us trace through these effects.

Total demand for modern, non-traded goods rises with increased aggregate demand, the more so since the income elasticity of demand for services tends to be high. Economies of scale are relatively unimportant in these industries, except perhaps in very small and underdeveloped economies. The price of non-traded goods is thus likely to rise with the increased cost of domestic labour in (unconstrained labour market) types 2 and 3. The general price level rises with the price of non-tradables. In all cases, output of modern non-traded goods and services is likely to expand, and with it employment in that sector.

The increases in wages in (unconstrained labour market) types 2 and 3 may cause continued production to be unprofitable in some established export- and import-competing industries in which economies of scale are relatively unimportant over the relevant range of output. There is no similar tendency towards contraction of established industries in (unconstrained labour market) type 1, or in (constrained labour market) types 1, 2, or 3.

The increase in demand is likely to cause investment to be profitable for the first time in some industries in which economies of scale are important. This effect will be most important where there is no increase in the wage level, as in (unconstrained labour market) type 1 and (constrained labour market) types 1, 2, and 3, and least important when the increase in the domestic wage level is greatest, as in (unconstrained labour market) type 3.

The overall impact of the increase in expenditure on sectoral economic activity and employment is as follows. Output and employment in the modern non-tradables sector rises in all cases. Village population falls in all variants of types 1 and 2. Village production of goods and services falls in both variants of type 2, and probably also in type 1, although not if the labour of migrants had been unemployed in the village. Activity and employment in the modern tradables sector rises within some industries and falls within others. Total output and employment in the modern tradables sector rises in both variants of type 1, in (constrained labour market) type 2, and in (constrained labour market) type 3. Employment (and possibly output) in the modern, tradables, sector falls in (unconstrained labour market) type 3, and may rise or fall in (unconstrained labour market) type 2.

There may be some contraction of employment and output in the modern tradables sector if the 'resource boom' induces increases in the institutionally set minimum wage, even in type 1, which would otherwise be immune from contraction of employment in any part of the modern sector. The pressure towards contraction of established tradable goods industries is commonly seen as an important problem in countries with rapidly expanding natural-resource-based industries (McKinnon, 1976; Gregory, 1976; Snape, 1977).

These processes are put into reverse by the post-construction slump in economic activity. Some new industries, encouraged by the boom conditions, are no longer profitable. Unemployment appears or intensifies in all cases in which real wages have risen (through government decision or in the market place) during the mineral investment boom and are inflexible downwards. Even in (unconstrained labour market) type 1, where there is no increase in the wage level, the labour market is severely disrupted by the fluctuations in labour demand, and the maintenance of full employment in the modern sector requires the reversal of the labour migration that occurred during the boom, a response that has its costs and may be slow to emerge.

Stepwise Increases in Revenue

Government mineral revenue rises sharply at the time at which rent charges first become payable in new projects. Where mineral rents represent a large proportion of total revenues, and where governments spend revenues as soon as they become available, this stepwise increase in revenues can lead to inefficient expenditure in both private and public sectors. Manpower and other bottlenecks may prevent the efficient implementation of programmes. New revenues may not be allocated carefully in line with government priorities. The inefficiency in the use of public funds that may be associated with sharp increases in public expenditure in these circumstances may have corrosive effects on development beyond the immediate financial waste associated with them.

Cyclical Fluctuations

The prices of mineral products typically fluctuate widely on world markets, generating large fluctuations in revenue from profit-based rent taxes. If revenues are spent immediately they are received, the cyclical fluctuations in prices generate economic instability in much the same way as the construction boom and slump. It is unnecessary to repeat the above analysis.

Remedies for Undesirable Timing of Production and Revenue

There are, in principle, three types of remedy for undesirable timing of mineral production and revenue. First, the government can control administratively some private investment decisions. Second, it can structure the fiscal arrangements to vary the timing of investment and production or the timing of revenue receipts. Third, it can accept the time pattern of revenue receipts that emerges from private investors' response to the objective conditions and to a fiscal regime selected on other criteria (such as neutrality and revenue maximization), and shift the use of mineral revenue through time by varying the levels of fiscal reserves and indebtedness.

The characteristic strengths and weaknesses of each type of remedy are discussed in turn. The final section of the chapter then discusses the possible role of each in an overall approach to remedying problems arising out of undesirable timing of mineral production and revenue.

Administrative Controls

When there is evidence that private investors' higher discount rates are causing investment decisions to diverge from those that would have maximized mineral rent at the government's discount rate, it is open to a government to apply administrative controls to investment and production decisions. It would be costly for the government to maintain the administrative capacity to 'second-guess' investors in this way, and there would be some risk that the intervention would be poorly judged. Even if the intervention were well conceived to maximize rent, the investor would not find it worthwhile to co-operate, unless the government were prepared to share the mineral rent with the investor. If the private investor were granted the necessary incentives, the well judged intervention would increase the amount of economic rent, but the government's share of that rent would fall. It is not clear *a priori* whether the national benefit from the resource development would increase as a result of the official intervention.

There may be a stronger case for the government to intervene with administrative controls in order to reduce the intensity of a mineral investment boom. Sequential development of a series of mineral deposits, or phased development of a very large deposit, can reduce the impact of the construction boom at some cost to the revenue (Campbell and Scott, 1979). The cost to the revenue may be large or small, depending on the characteristics of the deposits. Whether or not it is worthwhile to incur this cost depends on the deadweight cost of instability and inefficiency associated with the unabated construction boom. Sequential development of mineral deposits could also be used to reduce inefficiency associated with large stepwise increases in mineral revenues, but the deadweight cost of administrative controls would normally exceed those of the main alternative, the maintenance of more gradual expansion of expenditure through variations in fiscal resources and government indebtedness.

Cyclical variations in revenue might also be reduced by administrative controls on the level of mineral output. For example, the authorities could require that a higher proportion of high-grade ore be mined when mineral prices were low and a higher proportion of low-grade ore when prices were high. Such measures would exacerbate price instability in world markets if they were adopted by many countries. If adopted by a single country they would substantially reduce the present value of that country's mineral rent.

Varying the Fiscal Regime

The fiscal regime affects the timing of investment and production through its effects on private investors' discount rates and through the various sources of 'non-neutrality' discussed in chapter 8. It affects the timing of revenue receipts from a given pattern of depletion of mineral deposits more directly, with some forms of rent taxes generating earlier or more stable revenue flows than others.

The most productive means of reducing the dissipation of rent associated with private investors' application of higher discount rates than the government is to reduce the divergence between private and public discount rates. This argues for the use of fiscal charges that not only have their greatest incidence on the profits that represent a high rate of return on investment, but also are stable over time, as discussed in chapters 4 and 6. Unfortunately, where some divergence between private and public discount rates remains after the potential of these expedients has been fully exploited, there is no simple fiscal measure that, when introduced into the rent tax regime, can be relied upon to reduce the loss of rent associated with the distortion of the timing of investment and production.

The choice of fiscal regime has little effect on the economic instability and inefficiency associated with the construction boom, since it is a project's impact on demand for goods and services in the economy as a whole and not government expenditure of mineral rents that is important in this phase of mine development. But the extent of stepwise increases and cyclical variations in mineral revenue depends importantly on the rent tax system.

Stepwise increases in revenue early in the life of a new mine are likely to be greater with BT than with other rent taxes, since at some time there will be a shift from large negative to large positive rent tax receipts. The RRT in hybrid form with normal corporate income tax and with more than one threshold provides for a more gradual increase in revenues in the early years of a project's life than either HRIT or PPT. But even with RRT, large stepwise increases in revenue could be avoided only at the cost of possible distortion of investment and consequent revenue loss, through raising the rate of tax in many steps with the realization of higher after-tax rates of return on outlays.

Cyclical fluctuations in mineral revenue deriving from variations in world market conditions are most severe with the profit-based rent taxes, which, as we observed in chapter 7, 8, and 9, are the most suitable in terms of neutrality and revenue maximization. Among the profit-based taxes they are more severe with PPT than with HRIT, RRT, and BT. Cyclical fluctuations in mineral revenue commonly lead to severe economic instability in small countries that are heavily specialized in mineral exports, but this would not provide an argument for the use of less efficient rent taxes with more stable revenue yields (such as SAVD), unless there were an exceptionally weak system of public finance. Otherwise more stable economic conditions can be maintained, without the deadweight loss from inefficient rent taxes, by varying levels of fiscal reserves and government indebtedness.

Maintaining Expenditure Steady While Revenue Fluctuates

Many problems associated with uneven revenue flows over time can be solved if the authorities set expenditure at levels (or at rates of growth) that are thought to be sustainable on average over time and finance associated budgetary deficits in low-revenue years either by running down reserves accumulated from the surpluses of high-revenue years or by borrowing for repayment in high-revenue years. Financing remedies are often preferable to the main alternatives because they do not distort investment and production decisions and so in themselves have little economic or revenue cost.

Very large inter-generational shifts in the time at which the benefits of mineral revenue are enjoyed can be effected through variations in the rates of accumulation of financial assets and liabilities. The use of revenues can be shifted forward in time by increased borrowing and backward in time by the accumulation of financial assets.

In some cases, governments may reach the limit of their creditworthiness before they have shifted as much mineral-revenue-financed expenditure forward in time as they consider to be desirable. In recent years, it has been common for governments to be constrained in their attempts to shift mineral-financed investments backward in time, possibly because the accumulation of foreign financial assets beyond some indefinite upper limit reduces eligibility for foreign aid.

The impact of the construction phase of major mineral projects can be limited if the government avoids spending the increased revenues available temporarily during the investment boom, and this may well be possible with reasonable political will since it helps to avoid disruption of government programmes. In principle, the effects of the boom can be further offset if the government is able actually to reduce its own expenditure, but this will be politically difficult and will tend to reduce the efficiency of government spending. The increase in fiscal (and foreign exchange) reserves associated with fiscal restraint in these circumstances enables the government to maintain a higher level of expenditure than would otherwise be possible during the post-construction slump.

Financing responses can also reduce problems associated with stepwise increases in revenue. The rate of growth in public expenditure can be accelerated before new mineral revenues are actually received, with the increase above levels that would otherwise have prevailed being financed by borrowing. Identical principles for stabilizing public expenditure growth, or overall economic activity, can be applied if the variations in revenues are cyclical and due, for example, to fluctuations in mineral prices.

All four types of problem associated with unsatisfactory *timing* of revenues are amenable to financing solutions. Ideally, stabilization policy should be based on projections of total domestic expenditure and government revenue. To the extent that the principal stabilization objective is steady expansion of public expenditure for the purpose of facilitating efficient planning and implementation of government programmes, then the projections should incorporate a trend rate of public expenditure growth that can be sustained over the medium and long term. Where the principal stabilization objective is steady expansion of total economic activity, counter-cyclical variations can in principle be built into forward projections of sustainable levels of public expenditure. In either case, the public expenditure restraint in times of buoyant revenue would lead to increases in reserves (or reductions in government indebtedness), which would be balanced by falls in reserves (increases in indebtedness) in times of low revenues and low private sector economic activity.

The maintenance of steady expenditure while revenue fluctuates is described diagrammatically in figures 12.1 and 12.2.[1] Figure 12.1 describes a situation in which the level of real spending power, that is, the level of real expenditure that is consistent with balance in external payments, varies cyclically around a steady upward tendency. We define balance in external payments as a situation in which the net overseas indebtedness of the government and monetary authorities is increasing or decreasing at a desired rate. It might, for example, be defined as zero change in the overseas indebtedness of the authorities in the country under consideration over specified periods. Figure 12.1 also identifies the levels of net accumulation of reserves or net borrowing that must be undertaken in various shorter periods to maintain real expenditure on a steady trend.

Figure 12.2 describes a situation in which the level of real expenditure that is consistent with balance in external payments is subject to large once-for-all increases and decreases. These might be intepreted as large stepwise rises and falls of revenue at the start of exploitation and at its end.

How can a mineral-rich country effect the large net changes in reserves required to maintain stable growth under the conditions described in figures 12.1 and 12.2?

[1] A number of the following paragraphs and figures 12.1 and 12.2 are drawn from a paper prepared by Ross Garnaut for the Australian National University Development Studies Centre seminar on small island economies, 1979.

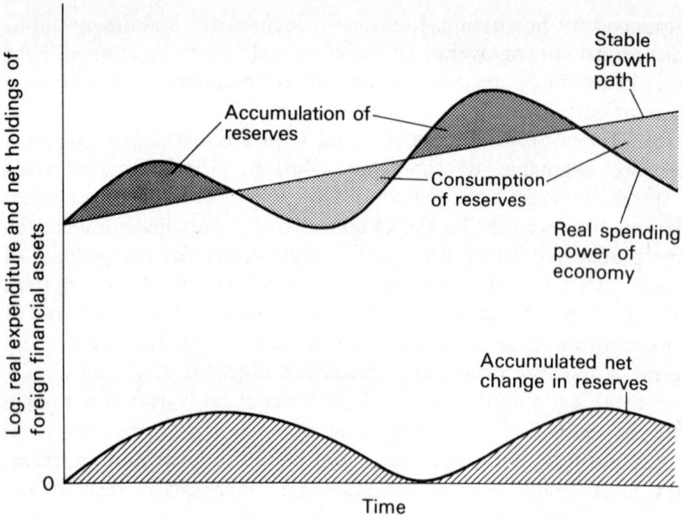

Fig. 12.1 Changes in reserves under counter-cyclical stabilization policy

The first requirement is foresight.

For many minerals, very large price fluctuations on world markets occur around a general tendency for average prices in real terms to be fairly stable over long periods. For some commodities, estimates of future prices by various international agencies provide a reasonable indication of the behaviour of average price in real terms over the medium and long term, although not of actual prices at any particular time. It is, of course, the average of future prices in real terms that determines the sustainable rate of growth in real expenditure. So long as reasonable estimates of this average are available, domestic expenditure can be maintained on a more or less steady path even though there is great uncertainty about prices over short periods.

It sometimes happens that a country is subjected to large and permanent changes in its terms of trade that bear no relationship to historical experience and are not foreseen in the usual estimates of average commodity prices. In these circumstances, the mineral-rich country may have no alternative but to adjust its expectations about growth, and perhaps to accept considerable instability in levels of economic activity in the transition from one growth path to another.

The discovery of a major natural resource may change expectations about the rate of growth in real expenditure that can be sustained simultaneously with balance in external payments. But after the discovery, it takes many years for the resource to be evaluated and

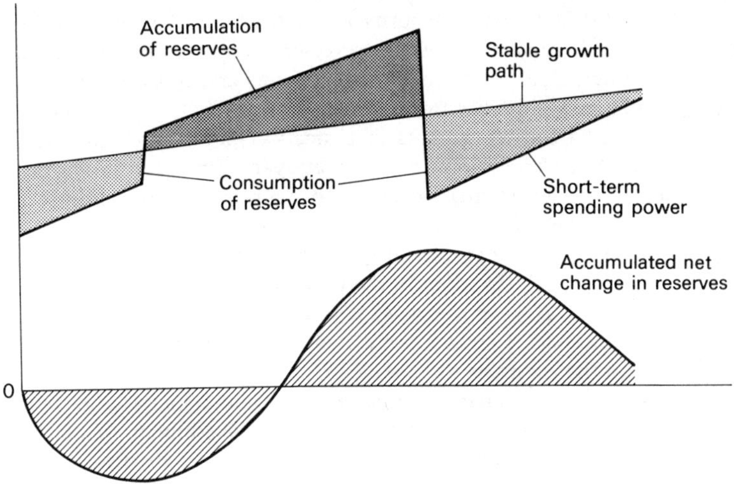

Fig. 12.2 Changes in reserves under policies to offset large variations in rates of growth over longer periods

the feasibility of economic exploitation to be established. These long lead times allow problems resulting from fluctuations in the economy's real spending power through the construction of a mine and the depletion of a mineral deposit to be assessed some time in advance of their occurrence.

Early predictions about the rate of steady growth in real expenditure that can be sustained over long periods are improved as more information becomes available on the behaviour of commodity prices, investment, and production, and this may require the modification from time to time of the steady rate of growth that is judged to be consistent with balance in external payments.

Once the desired steady rate of growth has been calculated, the stabilization policy problem becomes how to maintain expenditure on this path over time in the face of the fluctuations in the economy's spending power. There are a number of 'natural stabilizers' operating in a small open economy that moderate fluctuations in domestic expenditure in the face of large increases in capital inflow or export income, and some of these also assist in accumulating foreign financial assets (reserves) in times of exceptional prosperity. A large proportion of abnormally high profits earned by foreign corporations when export prices are high tend to be remitted overseas, and so do not increase expenditure in the local economy. The local trading banks tend to allow the ratio of lending to deposits to rise in buoyant times, lest they be embarrassed later when deposits are withdrawn. During a cyclical downswing these factors operate in reverse: there is a decline in remittances by foreign corporations; domestic firms and households reduce savings; and banks' liquidity ratios decline.

Government taxation receipts in a small economy are sensitive to external economic conditions, rising in buoyant times and falling with recession. If the government resolves to maintain its expenditure on a steady course, as an alternative to raising expenditure when taxation receipts are high and being forced to make large cuts when receipts are low, counter-cyclical variations in the budget surplus have an important stabilizing effect. A government that does no more than maintain its own expenditure on a steady trend makes an important contribution to stability in the economy at large.

In most small economies, fluctuating mineral incomes accrue almost exclusively to foreign corporations or to the government so that the 'automatic' and 'government expenditure' stabilizers discussed above are effective in an economy in which fluctuations in mineral income are the main source of domestic instability.

Although the prescriptions for maintaining stability in economic activity are easy enough to define, it is seldom a simple matter for a government to maintain expenditure on a steady growth path when there are large fluctuations in its revenues. Even when a government has a clear view of the rate of growth of real expenditure that is sustainable over time, there are strong pressures to raise growth in expenditure above this level when surplus cash appears to be available. The challenge is to build political consensus about the need to stabilize expenditure and about sustainable levels of expenditure. As discussed in chapter 14, Papua New Guinea has found it useful in building political consensus to receive taxation revenue from the minerals industry into a statutory fund, the Mineral Resources Stabilisation Fund, and to transfer it into consolidated revenue only at a rate that is judged to be sustainable over the medium term. To assist in building consensus on the sustainable level of expenditure, the managers of the Fund report annually to Parliament on expectations about mineral prices and incomes. The Mineral Resources Stabilisation Fund institutionalizes the counter-cyclical variation in budget surpluses that is required alongside the steady growth in public expenditure.

Appropriate Responses to Timing Problems

The application of fiscal charges that reduce the dispersion of the probability distribution of investment outcomes, and that are more likely to be stable over time, reduces the loss of mineral rent that results from private investors' discounting future cash flows at higher rates than the government. This reinforces the arguments in favour of the profit-based taxes, and especially PPT, RRT, and BT, that also perform best on neutrality and revenue maximization criteria.

If private discount rates substantially exceed public discount rates after the adoption of the superior mineral rent taxes, it may sometimes perhaps be advantageous for the government to choose an alternative investment pattern that increases the present value of mineral rent at its own discount rate, and to induce the alternative investment pattern by forgoing part of its normal revenues, or through some other subsidy. Such interventions must be based on careful assessments of the profitability of various investment sequences. They require lengthy and costly negotiations over fiscal arrangements, which should leave some of the risk of any expansion of investment with the private investor.

It is always sensible for a government to make as much use as is practicable of financing as a response to problems of unsatisfactory distribution of revenues either between generations or otherwise over time. In most cases, with careful management, the problems of inter-generational distribution, stepwise increases, and cyclical variations of revenues can be more or less eliminated, and problems of the mineral investment boom can be reduced.

Occasionally, a government may face real revenue loss through reduced access to foreign aid and if it tries to go far in the direction of accumulation of budgetary reserves and the reduction of foreign borrowing in times of buoyant mineral revenues. In some of these cases, it will be possible to negotiate arrangements with foreign governments for aid based on medium- and long-term prospects for mineral revenue. In rare cases, however, a government may find it worthwhile to accept an unsatisfactory time path of expenditure rather than to forgo foreign aid.

In some cases, a government may feel that it lacks the political strength to enforce expenditure restraint in times of buoyant revenue. A government in this position may find it worthwhile to modify directly the timing of mineral revenues, through changes in fiscal charges or direct controls on production. But it must be recognized that this is a counsel of despair. It is associated with large revenue losses. And it is doubtful whether a government that so badly lacked the capacity to implement a firm policy on public expenditure would have the political and administrative strength to avoid destabilizing patterns of expenditure through other means.

Even with carefully stabilized public expenditure, the construction boom may bring major macroeconomic problems. It is worth the government's while to assess the costs, in terms of structural adjustment and perhaps unemployment, of an unconstrained cycle of boom and slump in mineral investment. Where it is possible to substitute more smoothly phased investment patterns at relatively low cost to the amount of rent, there is a case for direct government control of the rate of investment. This is the major exception to the general rule, which is that the most efficient government responses to problems of the timing of mineral investment, production, and revenue involve, first, the reduction of divergence between private and public discount rates and, second, the systematic stabilizing of government expenditure over time by financial means.

Summary

The most efficient rent taxes from the point of view of neutrality and revenue maximization are likely to generate highly unstable revenue flows. These create problems for general economic management which are usually best resolved by varying the level of government fiscal reserves and indebtedness. The best solution will seldom be to use administrative controls or compensatory changes in rent charges as the means of maintaining steadier expenditure levels.

13

MINERAL RENT TAXATION IN INDONESIA

The mineral sector of the Indonesian economy has contributed a high proportion of exports and gross domestic product since colonial times and has provided the main impetus to Indonesia's strong economic growth since 1967. The petroleum industry has been most important, and the Indonesian authorities have developed effective instruments for appropriating a high proportion of the petroleum rent. Other mineral industries contribute much less to national development, mainly because the known endowment of resources is poorer, and also because taxation mechanisms are less efficient. The growth of the mining industries over the past decade has placed considerable pressures for adjustment on other sectors of the Indonesian economy, but these have diminished with the accretion of modern labour skills within firms and industries that have been favoured by the mineral boom.

Minerals in the Indonesian Economy

Mineral production is very much more important to Indonesia than to most countries. The mining industry contributed 19.4 per cent of gross domestic product at current market prices in 1977, or more than twice the total contribution of the manufacturing sector. The proportion increased substantially in the two years after 1977 as a result of the rupiah devaluation of November 1978 and the increases in world petroleum prices through 1979.

Most of the output of the Indonesian mineral industries is exported, although the availability of cheap petroleum has spurred the growth of domestic energy-using manufacturing industries in recent years. The mineral industries contributed about two-fifths of total exports in the mid-1960s and about three-quarters in 1979 (Garnaut, 1980). The growth in mineral exports has been prodigious: petroleum and gas, from $230 million in 1966 to about $10,000 million in 1979; other unprocessed minerals, from $37 million in 1966 to an annual rate of $137 million in the first half of 1979; and processed minerals, from $5 million in 1966 to an annual rate of $430 million in the first half of 1979.

Mineral production has grown more rapidly than total output in the period of strong Indonesian economic growth since 1967 and has generated much of the growth in other sectors. The output of mining and quarrying at constant prices increased by 13 per cent per annum from 1966 to 1968, by 14 per cent from 1968 to 1971, and by 12 per cent from 1971 to 1977. This compares with growth of mining and quarrying output of only 1 per cent per annum from 1960 to 1966, and with expansion in gross domestic product at constant prices at an average rate of about 7 per cent over the period 1966–77. The share of mining in gross domestic product at current prices increased from 4 per cent in 1968 to over 20 per cent in the late 1970s. The expenditure of resource rents accruing to the government spurred the expansion of other high-growth sectors, notably construction, public utilities, transport and communication, domestic market-oriented manufacturing, and other services (Arsjad, 1980).

The growth in mineral exports after 1967 had three main causes. The more relaxed attitude to direct foreign investment of the new Soeharto government, symbolized in the Foreign Investment Law of 1967, facilitated rapid growth in the productive capacity of the mining industries; but it was changes in the region's economic environment, especially with rapid industrial growth in Japan, and later elsewhere in north-east Asia, that made these changes in policy so effective. The huge expansion in north-east Asian import demand for industrial raw materials through the 1960s and 1970s transformed export specialization towards natural-resource-based commodities in all western Pacific countries that were richly endowed with these resources relative to their own industrial production: Malaysia, Thailand, the Philippines, Australia, and Papua New Guinea (Garnaut and Anderson, 1980). Then the changes in the approach of the oil-exporting developing countries (OPEC) to setting levels of production and prices after 1973 caused the value of petroleum exports to continue to increase strongly despite slower growth in their volume.

Petroleum production expanded through the 1970s (mainly up to 1973) to about 1.6 million barrels per day in 1977. Lower exploration from the mid-1970s, promoted by renegotiation of agreements and doubts about the creditability of payments under production-sharing arrangements and the tendency towards over-supply of world petroleum markets, caused production to remain on a plateau at that level for some time. Domestic consumption has risen rapidly at an average annual rate of 13 to 14 per cent to about one-quarter of total production.

Indonesia also has substantial reserves of natural gas, but these were little used before the mid-1970s. Two large natural gas fields have recently been developed for the export market, and increasing volumes of gas more than outweigh in energy value the decline in petroleum exports that derives from temporarily static production and steadily rising domestic consumption. The value of natural gas exports rose from negligible levels in 1977 to about $1 billion in 1979, and volumes are expected to rise strongly from 1979 levels. Several small natural gas deposits have been developed through the 1970s to provide low-cost energy or feedstock to domestic manufacturing activities.

Indonesia has been one of the world's major exporters of tin since colonial times (at first mainly ore, but increasingly metal). Production is now concentrated in several large dredging and mining operations. Exports of other non-ferrous metal ores, principally copper and nickel with smaller values of bauxite, have been of some importance through the 1970s. A large mine in Irian Jaya has been exporting ore with copper metal content of almost 100,000 tonnes per annum since 1973. Facilities were established in the mid-1970s to produce about 50,000 tonnes per annum of nickel matte in South Sulawesi. Production has been at less than one-quarter of installed capacity because of technical problems, but strong expansion to near full capacity in the early 1980s is expected. Nickel production at a similar rate from a large deposit in Irian Jaya is being contemplated. Several other deposits of non-ferrous metal ores are being evaluated.

Indonesia also has substantial coal deposits, which are likely to become economically important. Large deposits of coal in South Sumatra are exploited on a minor scale to generate electricity; this activity is in the process of being expanded to replace oil in meeting a major part of the electricity requirements of Java. The Indonesian government is currently seeking foreign investors to develop for export large coal (mainly steaming coal) deposits in Kalimantan. The increasing movement towards generation of electricity from coal in north-east Asia makes it likely that a large export industry based on the Kalimantan coal deposits will eventually be established.

The Institutional Framework for Mineral Development

Ownership of Indonesian mineral resources is vested by the constitution in the state. The national government has unfettered sovereign powers, and so is not constrained constitutionally in policy-making on natural resource developments. However, a substantial proportion of royalties from hard-mineral mining is earmarked for sub-national governments (49 per cent to provincial and 21 per cent to local government) (Gillis *et al.*, 1980: 52).

Indonesian governments have long been ambivalent about direct foreign investment of conventional forms in the mining industries. Formal ownership of most mining operations has been vested in the Indonesian government since the nationalization of the larger foreign enterprises in the late Soekarno period (Hunter, 1965).

Petroleum production is formally vested in a state corporation, PN Pertamina. Most production is financed and managed by foreign enterprises working under one of two types of agreement with Pertamina.

The successors to the old, nationalized foreign enterprises operate within what are known as 'contracts of work', under which the foreign enterprise continues to perform management and financial roles under contract to Pertamina. More recent entrants operate production-sharing agreements with Pertamina, under which they undertake management and financial roles in return for recovery of costs and a share of production beyond what is necessary to recover costs. The contracts of work and production-sharing agreements leave foreign enterprises effectively in the same financial position as direct investors, taking the risk of failure in return for the possibility of high reward, with Pertamina and the Indonesian state jointly playing the role of taxing authority. On occasions upon which Pertamina undertakes an entrepreneurial role, perhaps engaging in exploration itself before reaching agreement with foreign partners, this is reflected in production-sharing arrangements covering only part of the subsequent operations, with part being owned more directly by Pertamina in the manner of a participant in a jointly owned enterprise.

Indonesian coal production in Sumatra is owned and managed by a state corporation. However, a production-sharing agreement with a foreign corporation would have brought this corporate form into Sumatran coal production if the foreign enterprise had not withdrawn for commercial reasons. Production-sharing agreements with foreign corporations are expected to provide the corporate vehicle for coal production in Kalimantan.

The Indonesian state tin mining company owns and operates all tin production except for a small mine that is owned and operated by an Australian company. Other non-ferrous mining is operated by two foreign enterprises established under the Foreign Investment Law of 1967: Freeport Indonesia, which operates the copper mine in Irian Jaya, and Inco Indonesia, which operates the Sulawesi nickel mine. The Indonesian government negotiated a small equity in the former after the start of production, in 1975. Feasibility studies for a possible second nickel mine and mineral exploration are being conducted by other foreign enterprises.

Indonesia has had no restrictions at all on international financial movements since 1967. Immigration laws are liberal with respect to employment of foreign personnel in foreign mining companies in Indonesia. Foreign companies are free to market their own 'share' of petroleum production under contracts of work and production-sharing agreements and are commonly asked to market Pertamina's share as well. Domestic oil output is required to be provided *pro rata* to Pertamina to meet domestic consumption requirements up to a limit of 25 per cent of production, at a price which exceeds the cost of production by only US20 cents a barrel. There are no restrictions on disposal of non ferrous mineral ores produced by foreign corporations.

The Indonesian government makes no special investment in infrastructure required by large-scale foreign-operated mines.

Petroleum Rent Taxes

Indonesia operates general fiscal rules for petroleum production. They are commonly explained as being stable rules as well, although they have been subject to one extended period of uncertainty and change since the 'New Order' government came to power in the mid-1960s.

The general rules for production-sharing provide for 85 per cent of production after deduction for full costs to be made available to the government or its agent. Capital expenditure is generally deducted at one-seventh of the balance of undepreciated expenditure per annum. The rules applying hitherto for contracts of work are very similar, with the same 85–15 split of profits, except that assets are written off over their expected lives. The previous contracts of work were to expire in 1983, and the two companies operating under these arrangements were then placed on the normal production-sharing agreement. The general rules have a similar fiscal effect to an 85 per cent rate of income tax. (In combination with the requirement that 25 per cent of production be made available at cost to the domestic market, they have fiscal effects that are very similar to a uniform 89 per cent income tax from the point of view of the investor.)

The general rules are sometimes qualified by three leading types of negotiated arrangements. The first, which varies the general rules to make them more severe on the investor, involves Pertamina's taking an additional 'share' of the operating enterprise, for which full payment in cash is not made. The second, which also increases the fiscal severity of the arrangements, involves Pertamina's being paid a fixed fee (FF) at the time an exploration licence is issued. The third, which makes them less severe on the investor, involves reduced payments to the state from oil produced with secondary recovery techniques.

The first arrangement is negotiated prior to the issue of an exploration licence when Pertamina judges that an area that it is making available for exploration is exceptionally favourable—that it has positive residual rent value even when the investor expects to make large payments to the state under the general rules. These circumstances might arise because the area covered by the licence is adjacent to an established field, or because Pertamina, or another foreign firm that has since abandoned its interest, has undertaken exploration which has raised the expected value of the licence. In these circumstances, Pertamina may insist on holding directly some proportion of the operating enterprise without making further cash payments. The normal production-sharing arrangements would then relate only to the remainder of the enterprise. The fiscal effect of such a 'free carried interest' is similar to an HRIT with a rate equal to the proportion of the operating enterprise that is held directly by Pertamina.

The second negotiated arrangement is settled at the time exploration licences are issued. There generally is no formal fully competitive tendering process, although competitive pressures are important in establishing the level of the FF established by agreements. The FF component of the fiscal arrangements became more important in 1979 after the oil price increases of 1978–9.

The third qualification is negotiated when a field is at an advanced state of exploitation, and when further large capital expenditure is required to allow recovery of part of the petroleum in the field. This occurs, for example, when secondary recovery, or the artificial augmentation of pressure within the petroleum reservoir, is required for full exploitation of the resource. The Indonesian government completed negotiations in 1978 to allow Caltex to retain an additional US50 cents per barrel for production by secondary recovery techniques from their Sumatran operations. The agreement established through negotiations a base line of production that would have been expected to be achieved without secondary recovery, and any production above this line qualifies for the incentive. The government has indicated a willingness to negotiate secondary recovery agreements with other producers case by case.

The most important exceptions to the stability of petroleum rent tax arrangements emerged in the mid-1970s in response to the increase in after-tax profitability of oil production in the wake of the OPEC price increases, to the government's need for more revenue following the financial collapse of Pertamina in 1975, and to changes in US taxation laws affecting the creditability of overseas tax payments.

The Indonesian authorities thought that changes in the terms of the earlier production-sharing agreements, which provided for a 65–35 split of profits with a 40 per cent limit on the amount of deductions that could be claimed in each year, were justified by the windfall gains accruing to investors from the large oil price increases of 1973–4. Renegotiation of petroleum agreements was common elsewhere in the world at this time, and the justifications suggested by the 'bargaining model' were well known to senior Indonesian officials. Firm proposals for movement to the 85–15 split were put before operating companies in the context of proposals to restore national financial stability after the Pertamina crisis. The renegotiation had not been completed, however, when the US Internal Revenue Service indicated that payments under Indonesian production-sharing agreements were not creditable against US income tax obligations. The issue of whether or not Indonesian production-sharing was a creditable tax was of little importance until US income tax law was changed at about that time, to prevent the use of excess credits from overseas countries in which creditable taxes were applied at high rates against income tax obligations in third countries: most US oil companies operating in Indonesia had substantial excess credits in the Middle East. The change in the law made the creditability of Indonesian production-sharing very important indeed. In the event, the Indonesian government conceded changes in the form of the production-sharing agreement which made it more nearly like the conventional income tax in its fiscal effect and therefore creditable. The main changes were the removal of the limits on deductions, and faster rates of depreciation, which, apart from the creditability issues, made the system somewhat less onerous.

Uncertainty about Indonesian policy prompted by renegotiation of the agreement, the more severe terms of the agreement, the over-supply in world petroleum markets, and anxiety about the stability of the Indonesian financial and political system in the aftermath of the Pertamina crisis together caused the nominal value of expenditure on petroleum exploration to fall from about $400 million in 1974 to about $200 million in 1977. The authorities have said that a stable basis for long-term development has been established with the new arrangements. Aided by the tightening of the international petroleum market, nominal expenditure again exceeded $400 million and was on a rising trend in 1979.

Large-scale natural gas production for export is subject to production-sharing in the same form as for oil but with a 65–35 split of profits. The higher investor share is thought to be justified by the much larger capital expenditure in relation to the value of production: the 85–15 split is considered to leave too little return to private investors to justify new developments. There have been no changes in taxation arrangements since the first agreements for the production of natural gas were negotiated.

A large part of state revenues from petroleum accrue as profits to the state oil corporation, Pertamina. Until the mid-1970s, Pertamina operated with a high degree of fiscal autonomy, with its management virtually deciding how much of its income should be made available to the state. But the financial collapse of Pertamina in 1975 caused the government to insist on close supervision of the corporation's financial affairs (McCawley, 1979).

Rent Taxes for Other Minerals

Indonesian rent taxes for other minerals are exceedingly complex. They are different and generally more onerous for domestic investors (mostly state corporations established by the nationalization of foreign interests before the mid-1960s) than for foreign investors under the 1967 Foreign Investment Law.

State-owned and private domestic firms pay corporate income tax at the standard rate (45 per cent in the late 1970s). However, special lower rates are negotiated for the first ten years of the operation of foreign mining enterprises.

All enterprises are in principle subject to export tax, generally at the rate of 10 per cent for unprocessed materials, 5 per cent for semi-processed, and zero for processed. The rate of export tax has been set at 10 per cent for all mineral products except iron sand and ferro-nickel. Foreign firms operating under 'first generation' or 'second generation' contracts are not required to pay export tax (Gillis *et al.*, 1980, 44–52).

There are general rules establishing a system of royalties that is specific in form, although the government as a matter of policy varies the rates with mineral prices to approximate the effect of an *ad valorem* royalty. The general royalty system is of little practical relevance, since state corporations are required to pay export tax or royalty, whichever is greater (and the export tax is generally greater), and foreign enterprises operate under specially negotiated royalties.

Indonesian land taxes and import duties are significant sources of revenue from mining enterprises. Foreign mining corporations are exempt from the Iuran Pembangunan Daerah (IPEDA, or regional development levy) land tax, assessed at the rate of 5 per cent of presumed profit for domestic enterprises. Similarly, foreign mining investors, and most foreign investors operating under the 1967 law, are exempted from, or qualify for drawbacks with respect to, import duties.

Arrangements with foreign investors have tended to be general for a particular mineral and a particular period of time, but this leaves considerable scope for variation across agreements. Virtually all foreign firms insist on a detailed agreement covering the fiscal terms and conditions of eventual development before undertaking significant amounts of exploration.

The Indonesian government and the Shell Company agreed on a production-sharing arrangement for the large-scale exploitation of Sumatran coal deposits which was identical in its terms to the pre-1977 petroleum arrangements, with the 40 per cent limit on the amount of deductions in any one year, the 65–35 division of profits, and problems with creditability in the USA. There has been no suggestion that the production-sharing arrangements were important to the investor's decision to abandon the project in 1978. Rather, it seems that doubts about the quality of coal and the external market for it, and associated expectations that pre-tax profitability would be low, were the critical factors.

The Indonesian authorities have been unsuccessful in attempts to launch major projects at any of several licence areas in Kalimantan on the basis of the Shell production-sharing agreement. US companies are deterred by the creditability problems, and the limitation of deductions for costs to 40 per cent of the value of sales in any year substantially increases the risk of inadequate returns in an industry that commonly operates on much lower profit margins than petroleum. Negotiations were continuing in late 1979.

The higher investor share in coal, as with natural gas, relative to that in oil, also allowed for the expectation that capital expenditure would be larger in relation to revenue: a higher share of profits could be expected to be required by the investor to justify investment. The main problems with the arrangements seem to derive not from their severity on a typical intra-marginal mine, but from their form. The old form of oil production-sharing agreement has been applied because it had been current in oil when the first coal agreement was negotiated, and the investor had raised no strong objections. There was an element of inertia in the government's attempted imposition of the first agreement, with the same profit split, to subsequent coal licences.

The non-ferrous mining agreements with foreign investors fall into three categories, described as 'generations' by the Indonesian authorities, depending on the period during which they were negotiated.

These agreements have explicitly incorporated some insights of the 'bargaining model', with its assertion that the terms of agreements should shift more strongly in favour of host governments over time, as the government's bargaining power increases. This is no accident, since Harvard economists were advisers on mining agreements to the Indonesian government in the period from 1969. They have made the content of some of their advice available to the general public (Smith and Wells, 1969). Contrary to what is normally taken to be the implication of the 'bargaining' or North American model, however, the Indonesian government generally has not revised the terms of old agreements in line with the changed circumstances that are reflected in the more severe terms of later agreements.

The copper mine in Irian Jaya operates under a 'first-generation' agreement. This agreement established fiscal obligations that were less severe than those on foreign firms operating non-mining enterprises. There was a low *ad valorem* royalty, a three-year income tax holiday, and then a special rate of income tax that was lower than the general rate. In response to extraordinarily high profitability in the year after production began in 1973, the terms were renegotiated to provide somewhat improved terms for the government. The tax holiday was shortened and the government purchased 8.5 per cent of the equity on possibly concessional terms.

Both the established nickel project in Sulawesi and the proposed nickel project in Irian Jaya are based on 'second-generation' agreements. The fiscal terms are slightly more severe than for typical non-mining investments: there is a low *ad valorem* royalty, and income tax is applied at lower rates for ten years and then at the general rate.

There are no mines actually established under 'third-generation' agreements, although major exploration agreements have been negotiated. The first, with the Rio Tinto Zinc Corporation, provides for a low *ad valorem* royalty, income tax at a rate of 35 per cent for an initial period of ten years and then at the general rate, and a form of PPT. The PPT is payable in any year in which *average* profit for the three preceding years, including the year in which PPT liablity is being assessed, exceeds 15 per cent of capital expenditure up to and including the year in which the assessment is being made, less depreciation that has been allowed for items of equipment that are actually being replaced.

The Indonesian authorities believe that the failure of more investors to come forward with exploration projects under the third-generation agreement suggests that it is excessively severe. The period since 1974, however, has been characterized by excess supply on world markets for metallic minerals. The reactions of investors in more buoyant market conditions would provide a better test.

Declared after-tax profits of state hard-rock mining corporations have been low, and have not been an important source of state revenue.

Evaluation of Indonesian Rent Taxes

Indonesian petroleum levies are effective in collecting a high proportion of the mineral rent with relatively little distortion of investment. However, the various taxes applied to the other mineral industries seem to rate poorly on both neutrality and revenue maximization criteria.

The government's share of profits from oil under the general arrangements could not be increased much without introducing incentives to wasteful expenditure. The element of additional income taxation that is introduced in highly promising licence areas through the government's requirement of a free carried interest and a negotiated fixed fee further raises the government's share of the rent from these areas.

However, the Indonesian oil tax system is not well designed to accommodate the requirements of secondary recovery and other occasions for large capital expenditure late in the life of an oil field that might be required for full recovery of the resource. They are built around an arbitrary distinction between oil flows resulting from established investment and those that follow from secondary recovery. The government rarely has the information necessary to decide what level of production would have resulted from the original investment, or how much pecuniary incentive is necessary to induce capital expenditure designed to allow full economic recovery of the resource.

These problems would be avoided with the RRT or BT, which adjust automatically to provide incentives for economically efficient investment. If the Indonesian government were not prepared to place the whole rent tax system in the RRT or BT form, it would still be possible, through minor adjustments to the established arrangements, to introduce their benefits for investments after production had begun. The simplest change would be to allow capital expenditure after the start of production to be written off against income in the year in which it was incurred. If this process caused income to be negative in any year (and this would not often happen, since cash flows from the established investment would normally be very large), the 'losses' could be carried forward at an interest rate thought to correspond to the supply price of investment to the relatively low-risk incremental investments in established projects. These alternative arrangements would be more neutral with respect to resource allocation. The more rapid depreciation of later capital expenditure would result in the postponement of some of the government's revenue flow, the cost of which would need to be set against the revenue gains from avoiding special incentives to higher levels of recovery.

The HRIT, applied at a high rate as in the Indonesian petroleum arrangements, can be expected to deter some highly risky investments (for example, high-risk exploration) and some highly capital-intensive expenditures. At the high petroleum prices ruling in the late 1970s, this was not a barrier to active exploration anywhere in Indonesia except for less promising areas in inland Kalimantan and Irian Jaya. Some small oil fields, which would have been developed under more neutral rent taxation, were left unutilized, but the potential rent value of these was probably small compared with that from fields that were being brought into production. The exploitation of these small fields may become commercially viable with further increases in oil prices. The general introduction of RRT could cause them to be brought into production sooner, but the revenue that they could generate might not in itself justify the costs of disruption to established arrangements.

The arrangements for natural gas are less effective in ensuring that a high proportion of the rent accrues to the government, and they are also likely to be distorting in some circumstances. Since capital expenditure is much larger in relation to income in typical natural gas projects than in typical oil projects, the HRIT is likely to introduce distortions at lower rates of tax. However, some large gas projects could be highly profitable under present arrangements, especially with continuing increases in world energy prices. More flexible fiscal arrangements in the form of RRT or BT might possibly yield greater revenue from these fields while deterring less new investment in less economically attractive areas.

The production-sharing arrangements that have been envisaged for coal share the weaknesses of those for natural gas, but in addition have drawbacks arising from limits on deductions for outlays and doubts about creditability. These arrangements seem to be inconsistent with major private investment in coal development. The advantages in moving directly to RRT or BT are even stronger than for natural gas or oil.

The 'first-generation' agreement on which the Freeport mine is operating has generated very little revenue, partly because of the form of taxation that was applied and partly because of the weakness in world markets for metallic minerals from 1974 to 1979. Indonesia was a most uncertain environment for private mining investment in the late 1960s when the agreement was negotiated, and it is understandable that the government felt that it had to offer favourable terms. Given the investor's concern to avoid the risk of failure, as discussed in chapter 4, the amount of rent that could be collected for the revenue from mineral properties was undoubtedly low, especially if rent was to be taxed by a combination of conventional *ad valorem* royalties and standard income tax. However, there is a reasonable presumption that the investor would have been more willing to give up some of the income that represented a high rate of return on investment in the event of the project's turning out to be very successful than to accept higher royalties or normal income tax obligations, which would have added to what must have seemed the high risk of failure. A higher expected revenue to the government might have been achieved with RRT or BT than with the taxes incorporated into the first agreement.

The form of the second- and third-generation agreements has been progressively better suited to efficient rent collection, but further tilting of the schedule of taxation payments towards relatively heavier obligations in later years and on more profitable outcomes would seem likely to raise the expected value of government revenue in the Indonesian environment, where private investors' perception of sovereign risk continues to be high.

The Indonesian strategy of applying progressively more severe terms to mineral developments with the passage of time has had some justification in the tendency for perceptions of sovereign risk to wane as the hazards of private foreign investment in the Sukarno period recede further into the past. It may not have been wise to establish general rules intended to be stable over long periods in the early days after 1967, when perceptions of sovereign risk were higher, and so the amount of rent available for collection by the state was lower, than in later years. But Indonesia was unable to exploit the main potential advantage of *ad hoc* rules, that the fiscal rules can be adjusted to the special circumstances of each mine, because investors insisted on early agreements, conducted before there was substantial exploration expenditure. There would seem to be a strong case now for moving towards general rules that are expected to be stable, since there does not seem to be any reason to expect any further decline in investors' perceptions of sovereign risk.

Dissipation and Diversion of Mineral Rent

Indonesian policy on petroleum pricing in the domestic market has diverted about one-quarter of the potential rent towards Indonesian consumers of petroleum, and has probably dissipated part of that potential rent through the establishment of uneconomic petroleum-based manufacturing industries. The petroleum that investors have been required to supply to Pertamina at cost has allowed Pertamina, which is responsible for marketing petroleum within Indonesia, to maintain domestic prices well below international levels—with some explicit fiscal subsidy compounding the implicit transfers in most years. The relatively low domestic petroleum price has been a principal reason why consumption has grown at about twice the rate of growth in real domestic product through the 1970s.

Wages and salaries are generally higher in the mining sector than elsewhere in the economy. Part of this difference no doubt represents the diversion of rent to employees. Part represents deliberate attempts to reduce labour turnover, as is required for high productivity in a highly capital-intensive and technologically complex industry (Manning, 1980). The scale of possible rent diversion through this mechanism is minor in comparison with total petroleum revenues.

Much of the petroleum rent has been received in the first instance within Pertamina, and this has provided opportunities for both diversion and dissipation of rent. The diversion has often favoured leading individuals in the corporation and in Indonesian society more generally. The most important dissipation in earlier years comprised the promotion of grandiose and uneconomic investments, sometimes but not always dependent on the use of cheap petroleum. This source of diversion and dissipation of petroleum rent was sufficiently large to threaten the financial and political stability of the Indonesian state in the mid-1970s. After the bankruptcy of Pertamina in 1975 as a result of financial mismanagement (McCawley, 1979), much tighter Ministry of Finance controls were introduced in 1976, and opportunities for the diversion and dissipation of rent have been curtailed since then.

Impact on Economic Development

The expenditure of the large amounts of resource rents accruing to the Indonesian government, especially over the past decade, has provided the main impetus to the growth in output in the economy as a whole (Booth and McCawley, 1981; Garnaut, 1980). For the most part, the resulting growth has been concentrated in what might be called the 'modern sector' of the economy, characterized by large-scale production units applying sophisticated technology and supporting relatively high income for persons employed in it. This has had a range of positive and negative spin-offs for the general population, but so far the 'resources boom' and one-and-a-half decades of high growth have not been enough substantially to raise general living standards (Sundrum and Booth, 1980).

The large stepwise increase in revenues after the OPEC price increases late in 1973 generated considerable pressure for adjustment throughout the modern sector of the Indonesian economy. There was temporarily considerable inflation of costs of skilled labour, urban housing, and other inputs into modern economic activity, the supply of which was fixed in the short term. This caused the competitiveness of other industries using these inputs and producing traded goods to decline, and probably delayed the emergence of an internationally competitive manufacturing sector for several years (Garnaut, 1979a; McCawley, 1980). But the expansion of the modern sector of the economy itself was able to generate an increased supply of scarce skills after a few years, given the availability of an abundant supply of qualified recruits, and the loss in competitiveness was gradually removed after 1976. In the labour-surplus Indonesian economy, the resource boom caused a period of rapid growth, inflation, and temporary pressure on other sectors, but it was insufficiently large by itself to make unskilled labour scarce and so to cause real wages to rise permanently. There was room for strong growth of other modern industries producing tradable goods alongside the much larger resources sector and government sector.

The sudden jump in revenues in 1973–4 was absorbed almost immediately into public expenditure, when a more gradual absorption would have had a less damaging impact on other tradable goods industries. The concentration of the increase in expenditure in a short period also reduced the efficiency of public expenditure by a bureaucracy that was not organized to make good use of it. The loss of financial control at this time was economically wasteful and corrosive of moral standards within the Indonesian government It is widely recognized to have been damaging to the development effort. In recognition of the earlier error, attempts have been made to feed more gradually into expenditure the large increase in revenues that followed the oil price increases in the first seven months of 1979, with the government's net indebtedness falling substantially during the adjustment period.

Summary

Mineral rents, especially from petroleum, have come to represent a major component of the Indonesian economy. The large increase in government revenues through the 1970s has provided considerable impetus to general economic growth. The sudden large increase in revenues in 1973–4 created adjustment problems within other sectors of the economy, including the government sector. These problems would have been less severe if public expenditure had been increased more gradually—a lesson that appears to have been learnt, as evidenced by the more careful response to a further sharp increase in oil revenues during 1979.

The Indonesian petroleum rent tax system seems to perform well on the revenue maximization criterion, but forms of tax that are more closely related to the realized profitability of investment might possibly yield greater revenue from natural gas. The oil and gas systems perform less well on neutrality, requiring negotiated arrangements in particular cases if full economic recovery of petroleum is to be secured.

The rent tax system for other minerals is less effective from the point of view both of neutrality and of revenue maximization. It appears to create difficulties for the establishment of an export-oriented coal industry, and to yield less revenue than might have been expected from non-ferrous minerals. A general system of more nearly neutral profit-based taxes would seem likely to be helpful in correcting current problems in the management of these industries.

14

MINERAL RENT TAXATION IN PAPUA NEW GUINEA

Papua New Guinea has a rich endowment of mineral and petroleum resources relative to its small population of three million. The country's isolation from the world's main centres of industrial production, its difficult terrain, and its poor infrastructure combined to delay its emergence as a major supplier of minerals to world markets. A large-scale gold mining operation in the 1930s had pioneered important new dredging and air transport techniques that had an important impact on the rest of the world (Healy, 1967), but the next major development did not occur until Bougainville Copper Limited, a subsidiary of Conzinc Riotinto of Australia (itself a subsidiary of the Rio Tinto Zinc Corporation), began construction in 1969 and production in 1972 at the huge Bougainville copper and gold mine. The Bougainville project was launched under an agreement negotiated by the then Australian colonial administration, which provided for the retention by the operating company of a high proportion of any mineral rent. After the granting of internal self-government in December 1973, the first Papua New Guinea national government successfully renegotiated the Bougainville Copper agreement in 1974 and subsequently developed general fiscal rules for mineral and petroleum production which formed the basis of several important agreements with foreign companies. The Bougainville mine made the mineral sector important to economic development in Papua New Guinea, and subsequent developments suggest that this sector will remain a major source of economic growth. This result has become more likely as a result of the development in Papua New Guinea of general but flexible fiscal rules for mineral development which seem likely to be stable over time and which are well designed to secure a high proportion of the mineral rent for the public revenue.

Minerals in the Papua New Guinea Economy

The construction of the Bougainville mine at a cost of about A$390 million in the prices of the construction years gave a huge impetus to the expansion of modern sector economic activity in Papua New Guinea, being the main factor behind increases in gross monetary sector domestic product at constant prices of about 15 per cent in each of the fiscal years 1969/70 and 1970/1. The completion of construction in 1971/2 caused output growth to fall almost to zero in that year.

Over the first eight fiscal years from the beginning of commercial production in 1972, the Bougainville mine contributed between 37 per cent (in 1976/7) and 64 per cent (in 1973/4) of the total value of Papua New Guinea's exports. The annual value of mineral sales varied with copper and gold prices between K219 million in 1975 and K415 million in 1979. The mine's contribution to government revenue fluctuated even more widely, from K90 million in 1975 (mostly out of 1974 income) to K32 million in 1978 and K124 million payable out of 1979 income.[1] Fluctuation in government fiscal receipts has had the potential for introducing a great deal of cyclical instability into domestic economic activity, but this was to a considerable extent avoided through the use of explicitly counter-cyclical policies, as discussed later in the chapter.

Macroeconomic stability and growth are likely to depend heavily on export income and government revenue from the mineral sector for the foreseeable future. Construction of a second major copper and gold mine at a cost of about K546 million at constant 1979 prices began at Ok Tedi in the Star Mountains during 1981. The scale of impact of the Ok Tedi mine on economic activity, exports, and revenues will be similar to that of the Bougainville mine, although general economic growth over the past decade, partly resulting from mining on Bougainville, means that its proportionate impact on the economy will be more modest.

The Institutional Framework

The national government has ultimate responsibility for virtually all policy matters that are important to large-scale mineral development. Nineteen provincial governments each have the right to receive payments from the national government equal to 1¼ per cent of the value of all exports, including mineral exports, from the province, or to the value of royalties on the export products concerned, whichever is the larger. Provincial governments do not have the power to collect or to levy any major fiscal charges on mining, although minor charges can be applied in a non-discriminatory way, and subject to limits imposed by the national government, on retail sales of some goods purchased by mining companies (for example, fuel) and on the unimproved value of land. Traditional owners of land have the right under national government legislation to receive 5 per cent of the royalties from

[1] This total includes company profit tax and additional profits tax (the main items), dividends, dividend withholding tax, royalties, personal income tax, and customs duties.

mining on their land. The national government has the legal right to acquire land for mining. Landowners may receive rent payments for the use of their land for mining, at rates agreed with the national government.

The Papua New Guinea government has preferred mining to be undertaken by foreign corporations, itself owning minority equity and leaving management to the majority owners.

The national government has developed general taxation rules for mineral and petroleum development which are designed to be stable over long periods of time. It is prepared to reach early agreement on the amount of, and on arrangements to pay for, equity in mining operations. The government is also ready to finance the installation of some infrastructure in return for equity of equivalent value at par, or in return for a user charge or a tariff negotiated before the beginning of construction. So far this user charge has been designed to recoup the investment with interest at a rate equal to the discount rate that the government applies to its own investment decisions (in 1979, 15 per cent).

Exploration is conducted on a prospecting authority until a commercially viable ore body has been discovered, whereupon a special mining lease is issued, subject to the terms and conditions laid out in an accompanying agreement.

For petroleum, all important fiscal arrangements except those relating to infrastructure are established firmly before the start of exploration. The general pattern in the case of petroleum has been for the government to establish the right to a minority equity in the project from the time of development. This is achieved by the government's borrowing from the private investor to buy its shares and repaying the loan with interest out of its share of the profits. The interest is at normal rates plus a substantial premium for risk. Any part of the loan that is outstanding at the end of the life of the oil field is cancelled, so that this 'carried-interest' equity functions as a tier of RRT.

For other minerals, the taxation and royalty arrangements are of general application but with certain matters such as the extent of government equity participation left for specific agreement, which may be made while much exploration is still to be done. Some investors have been prepared to undertake substantial exploration with only a prospecting authority, on the strength of assurance given by the general legislation, without seeking agreement on the extent of government equity. The Papua New Guinea government has been willing to finalize arrangements for its purchase of equity when the investor is about to undertake exploration expenditures in excess of about K1 million per annum, or less in some circumstances. The government so far has secured the right to purchase equity at par on a fully contributing basis at the start of construction.

Arrangements for government provision of infrastructure to petroleum and other mineral developments are to be negotiated immediately before the issue of a special mining lease.

Papua New Guinea is a very large recipient of foreign aid and has given high priority to the development of mechanisms that prevent this dependence on aid from determining the pattern of resource allocation within the domestic economy. Offers by foreign governments to provide infrastructure for mineral developments are accepted only when the infrastructure is included in a list of public expenditure priorities prepared through the normal budgeting and planning process.

Papua New Guinea maintains exchange controls that require the conversion of export earnings into domestic currency, but restrictions on currency movements are not severe. The government has entered agreements to permit mining companies to hold offshore up to three months' earnings of foreign exchange against commitments for payments in foreign exchange, including the servicing of loans and payment of dividends. Papua New Guinea policy on the terms upon which it makes infrastructure available to mining companies is determined prior to, and independently of, the negotiation of concessional funding.

Petroleum Rent Tax

The fiscal regime for petroleum has four elements: a low *ad valorem* royalty; an income tax of 50 per cent with provision for accelerated depreciation in years of low cash flow; an RRT (described as an Additional Profits Tax) on highly profitable investments; and a carried-interest equity, which has a similar fiscal effect to that of another tier of additional profits tax. The royalty and taxation arrangements are set in general legislation and equity is negotiated case by case. Oil and natural gas are treated in a similar way, except that less demanding equity arrangements may possibly be negotiated where natural gas is the main target of exploration and the rates of return on investment in the case of a successful outcome are correspondingly lower. The system is designed to place relatively light burdens on projects that turn out to be marginal but to raise large amounts of revenue from highly successful investments.

The *ad valorem* royalty is set at the rate of 1¼ per cent, which would generate the amount of revenue that the national government was obliged to make available to provinces if petroleum were exported.

The standard income tax rate of 50 per cent was chosen to absorb all or almost all of the rights to tax credits in home countries. Capital expenditure is written off for income tax purposes on a straight-line basis over ten years or the expected life of the asset, whichever is shorter. To reduce the risk of failure in development investments, more rapid depreciation is allowed if in any of the first four years of production cash flow (gross receipts less operating expenses less capital expenditure, with no deductions for depreciation or interest) is less than 25 per cent of the total nominal capital expenditure undertaken before the start of production.

The RRT becomes payable when the accumulated value of net cash flows after income tax, compounded with interest at 27 per cent, is positive. The RRT is assessed at 50 per cent of cash flows net of standard corporate income tax.

The RRT is assessed separately for each licence area, except that exploration expenditure from an abandoned licence area can be brought to account in other licence areas operated by the same investor.

The Papua New Guinea RRT on petroleum was, on the advice of US legal authorities, probably creditable against US income tax obligations at the time of its legislation in 1977. New rulings by the US Internal Revenue Service since then have thrown this conclusion open to doubt. The importance of this issue is reduced by the fact that tax credit opportunities in the USA would usually be exhausted by the 50 per cent rate of standard income tax.

In the first of the petroleum agreements established under the national government's new policy, with a consortium led by Esso, the government negotiated the right to take up 22.5 per cent equity in a project at the time of a commercial discovery of petroleum. If it did so, then a government account with the investor would be debited with 22.5 per cent of exploration expenditure up to that point, at its nominal value. Thereafter, 22.5 per cent of all expenditure would be debited to the account and the same proportion of sales revenue credited. Interest would accrue on the balance of the account at rates equal to prime corporate borrowing rates in the USA plus 5 percentage points. The government would receive no income as an owner of the oil-producing enterprise until the debit in its account was cleared, but it would be under no obligation to meet any residual debit that remained at the end of the life of a relatively unsuccessful oil field. Carried-interest equity in this form would have a fiscal effect that was almost identical to that of an additional tier of RRT applied when the investment had been recouped with a discounted cash flow return equal to the interest rate applied to the debit in the account, and assessed at a rate of 22.5 per cent.

There has been a considerable revival of interest in Papua New Guinea petroleum exploration, and also an increase in expenditure, since the government announced the general and supposedly stable rules for petroleum in place of the colonial government's commitment to negotiate *ad hoc* arrangements case by case. It is impossible, however, to allocate responsibility for increased exploration interest between the more explicit fiscal arrangements and the more favourable world petroleum market conditions. In the early days of establishing the new system, one US investor, which had been undertaking some exploration expenditure for many years, found the exchange control rules and the concept of RRT unacceptable in Papua New Guinea, and its commitments were taken over by an international consortium led by a subsidiary of Exxon in 1976. Since then, the system has been acceptable to all investors who have entered negotiations with the Papua New Guinea government.

Taxes for Other Minerals

The general fiscal system for other minerals has elements similar to those for petroleum, but there are some important differences that reflect the differences in character between the petroleum industries and others. An *ad valorem* royalty is assessed at the same low rate of 1¼ per cent. Income tax obligations are established in general legislation. Mining income is assessed at the rate ruling generally from time to time (currently 36.5 per cent), except that the rate is limited to 35 per cent until the initial investment has been recouped out of cash flow (defined as sales revenue less the sum of capital expenditure and operating costs). The provision for depreciation is similar to that for petroleum, with identical arrangements for acceleration of depreciation in early years of low cash flow. Dividends paid out of mineral income (other than income from petroleum) attract withholding tax at the rate ruling from time to time (currently 15 per cent). Mining income is subject to RRT with a threshold of (i) either 20 per cent or 10 percentage points above prime corporate borrowing rates in the USA (in the Ok Tedi agreement), or (ii) either 20 per cent or 12 percentage points above the US Prime Rate (in the general legislation). The RRT rate is $(70-n)$ per cent of net cash flow after income tax, where n is the rate at which income tax is assessed. Assessment for RRT is in US dollars, with cash flows recorded in Papua New Guinea *kina* converted into dollars by a simple formula. Minority government equity is to be negotiated case by case at some time when the investor has entered commitments to spend large amounts on exploration and before the start of construction. The two equity agreements negotiated so far, in relation to the Ok Tedi mine and a major gold prospect, require the government to pay for its equity pro rata with other investors. Mining operations are subject to normal import duties and mining personnel to normal income tax.

The interest rate at which negative cash flows are accumulated in assessment for RRT under the general arrangements is lower for minerals than for petroleum. This is because high-risk exploration is thought to represent a higher proportion of investment in the latter, so that higher returns are necessary in successful projects both to compensate for a higher proportion of failed investments and to compensate for risk. This effect can be vitiated in particular cases by the arrangements negotiated for carried-interest equity in petroleum ventures, where this is judged to be appropriate.

The general fiscal arrangements for petroleum and minerals were fixed in 1976 and 1977 after the government had obtained a great deal of experience with fiscal arrangements for the Bougainville mine and the Ok Tedi prospect. The first national government was formed in April 1972, the month in which the Bougainville mine began commercial production. The fiscal arrangements at the Bougainville mine presented the first national government with one of its most pressing political problems until the successful renegotiation in 1974 of the 1967 agreement between the investor and the colonial government.

Mining at Bougainville attracted renewed attention soon after the first national government was formed. The 1967 agreement had been exceptionally favourable to the investor. It provided for a 1¼ per cent *ad valorem* royalty and for 20 per cent government equity purchased at par during the construction phase. These two elements of the 1967 agreement survived the renegotiation. The first Bougainville agreement provided for a three-year tax holiday and for the permanent exemption of 20 per cent of income from taxation. The whole of capital expenditure could be written off immediately against income at the end of the tax holiday period. In the first years for which income tax was payable, a special low rate of 25 per cent was to be applied, with the rate rising gradually to 50 per cent ten years after the project had become liable for income tax. (The range of income tax rates was effectively 20–40 per cent, given the permanent exemption of 20 per cent of income from liability for income tax). The rates of import duty on items purchased by the company were frozen at their low 1967 levels, as were a wide range of other taxes and charges. There was a substantial government subsidy to infrastructure for the project. If these fiscal arrangements had been preserved, the project would have returned to its owners the whole of the investment with interest by about the end of 1974, and would not have paid income tax until the early 1980s.

In its first budget, in 1972, the new government introduced a dividend-withholding tax of general application at the rate of 15 per cent. The owners of the Bougainville mine did not challenge the application to them of this tax, although there was some private comment that the tax should not apply to them.

The new mine turned out to be prodigiously profitable: moderately profitable even through the low metal prices of 1972, because of low production costs; and exceptionally so through 1973 and 1974. The 1967 agreement was seen as being unfair, and calls from political leaders for its renegotiation began in the second half of 1972 and intensified through 1973. It was under this pressure that the government began to articulate the principles that were to form the basis of a new mineral policy: mineral exploitation was desirable not for any benefits that it could provide directly, but for the financial resources it could provide for development in other sectors and for the sake of greater independence from foreign aid; the mineral resources of Papua New Guinea belonged to the country's people, and the benefits of exploitation of those resources must be made available to the people; foreign investors were required to provide the management skills and risk capital to exploit mineral resources, and it was understood that they would not co-operate in Papua New Guinea's mineral development unless they expected to receive a reasonable return on their investment; but any profits in excess of that reasonable return, which could be defined as the minimum necessary to induce desired levels of investment, should accrue to the Papua New Guinea state as taxation.

The implementation of these principles in the Bougainville Copper case was constrained by the circumstances of the renegotiation of established arrangements, but the government did succeed in effecting by agreement a very substantial shift in the balance of advantages.[2] The tax holiday, accelerated depreciation, and exemption of 20 per cent of income from taxation were removed from the first day of 1974. The company was to pay import duties at normal rates from the date of the new agreement, in October 1974. Company income was to be subject to a PPT, so that the part of income in any year that was in excess of 15 per cent of an agreed 'capital base' would be taxed at a total rate of 70 per cent. In the process of negotiation, the government agreed to the abatement of half of the PPT liability with respect to the first half of 1974.

The 'capital base' for 1974 was an agreed figure roughly corresponding to capital expenditure incurred up to the beginning of 1974, net of interest capitalized during construction and of depreciation on items being replaced. The sum was to be varied each year according to a formula, which had the effect of adjusting the 'capital base' fully for changes in the average value of the Papua New Guinea *kina* against the US dollar, augmenting it by the amount of new capital expenditure, and deducting from it depreciation on capital items being replaced. The new agreement also provided that the company would have the right to submit a claim to the Minister of Finance for an upward

[2] See O'Faircheallaigh (1980) and Garnaut (1981) for a description of the renegotiation process.

adjustment to the 'capital base' when there was abnormal inflation, and to have the Minister consider the claim. Any adjustment would be made at the Minister's discretion, subject to the amendment of the agreement by the national Parliament.

The Bougainville mine paid additional profits tax on 1974, 1979, and 1980 income, but not on income from 1975–8 because of the effects of low copper prices on profitability.

Taken over the life of the Bougainville mine to date, the fiscal demands on investors have been less severe than they would have been had the project been subject to the new general rules from its inception. The fiscal demands in future under the Bougainville Copper Agreement are also likely to be less severe than they would have been under the general rules. The Papua New Guinea government has been influenced by these comparisons in rejecting claims to date for adjustment to the 'capital base' on account of abnormal inflation. The effect of continued world-wide inflation would be to make the Bougainville Copper additional profits tax steadily more severe in comparison with arrangements in the general legislation if there were no adjustment of the 'capital base' for inflation, but is is unlikely that this would cause the company to be treated less favourably in any typical future year under its own agreement than it would have been under the general rules.

The renegotiated agreement provided for a review in 1981, after seven years, and it was anticipated that the arrangements might then be varied by agreement between the parties. The review began in 1981 but had not been completed by late 1982.

The Ok Tedi mine is to be developed by a consortium of Australian, German, and American companies under fiscal arrangements that closely resemble, and provided the model for, the general rules. The one important difference is that the Ok Tedi consortium is able to choose at the start of construction between having net cash flows accumulated for additional profits tax at prime corporate borrowing rates plus 10 percentage points, or at a fixed rate of 20 per cent. The recent experience of high interest rates would seem to favour choice of the former, which would bring Ok Tedi into line with the general arrangements when prime rate stood at two percentage points below prime corporate borrowing rates, as it had done at various times in the past.

The agreement with the consortium came only after the breakdown of negotiations with the American corporation Kennecott, which had discovered the ore body and established substantial copper reserves, at a cost of about A$12 million, between 1969 and 1972. This work was undertaken on a prospecting authority, which carried no legal right to mine. No further exploration was undertaken after 1972 as Kennecott watched the outcome of the mounting pressure to renegotiate the

Bougainville Copper agreement. Serious negotiations on the fiscal regime began immediately upon the conclusion of the new arrangements for the Bougainville mine, but Kennecott and the government remained a long way from agreement when the prospecting authority was due to expire in December 1974. The authority was extended for three months to allow negotiations to continue.

By March, the company's best offer on additional profits tax had the following features: an annually progressive profits tax on the Bougainville model, to be applied on that part of income in any year that represented more than 20 per cent of an 'adjusted capital base'; this base to be the sum of all capital expenditure less depreciation on items being replaced, raised by the percentage increase in the general price level between the year in which each part of capital expenditure had been incurred and the present; and no assessment for additional profits tax to be made until after the initial investment had been recovered with interest at a low rate.

There was also an important disagreement on the choice of law, with the government insisting that Papua New Guinea law should govern the agreement and the company insisting on a clause purporting to have the agreement interpreted according to 'the law of developed countries'. It is doubtful whether Kennecott's clause would have had any legal effect even if it had been written into an agreement (Brown, 1977), and it was simply unacceptable to Papua New Guinea leaders.

Faced with this stalemate, the government had to assess whether it would be able to negotiate successfully with other companies if it failed to reach agreement with Kennecott. If it could not, the government, needing the financial benefits of the mine, would be forced to reach agreement with Kennecott even on unfavourable terms. However, if there were reasonable prospects of reaching agreement with other companies, the proper course was to insist on terms that were judged to leave expected after-tax returns to the company no higher than would be necessary to attract the necessary investment to the project from some source, if need be to the point of breaking relations with Kennecott and seeking new investors.

The government at first sought to postpone the choice by offering, while negotiations on an agreement continued, to finance in return for equity the further exploration that the company thought was necessary before feasibility studies could begin. This course had the added attraction of re-establishing momentum on the project: Kennecott had indicated that it would not spend any more of its own money before the feasibility of mining was established, and that, even if agreement were reached immediately, there would be delays while private partners were found.

Kennecott rejected the government's offer. In retrospect, with the feasibility of a profitable mine now established, the company's stance seems surprising. They had insisted on taxation parameters under which additional profits tax might never have been payable even if the mine turned out to be one of the world's most profitable. They had insisted on a clause on choice of law that was politically provocative and probably had no legal effect. And they had finally rejected the opportunity to continue negotiations while having the government finance the next phase of exploration.

There would seem to be two main aspects to the explanation: the company did not place very high value on the prospect; and it thought that its bargaining position was strong, so that it could reach agreement on its own terms if it maintained pressure on the government. It may also have believed that any agreement was likely to be renegotiated at some time in the future, and that an initial agreement needed to be sufficiently favourable to the company to provide for expectations of further fiscal concessions without a strong possibility that they would reduce expected returns below the minimum that the company considered acceptable.

Kennecott negotiators said that there was only one chance in five that a mine would ever be established at Ok Tedi. This may have been only part of the dissembling that occurs in negotiations, but it is possible that they thought Ok Tedi was still a highly speculative prospect. Kennecott had recently been expropriated in Chile, and no doubt gave considerable weight to the risk of changes in government policy in a new developing country. The company was also in severe anti-trust difficulties with the Justice Department at home in the USA, and this may have further reduced the attractions of taking a risk with a large investment in Papua New Guinea.

The second of the two aspects mentioned above was probably more important: it may have appeared to Kennecott that the government needed to reach agreement with them, because it was economically and politically important that the project be advanced, and because they thought it would not be easy for the government to bring in other investors. There were certainly some weaknesses in the government's bargaining position. The government would have been politically vulnerable if it had allowed relations with Kennecott to rupture without having established alternative means to continue work at Ok Tedi. To deal with other companies, the government would need to risk adverse reaction to its putting aside the implied rights of the discoverer of a mineral deposit, and in these circumstances it could not be certain how other investors would react.

In the event, the government did not renew Kennecott's prospecting authority when it expired in March 1975. In announcing this decision,

Prime Minister Somare was able to inform the Parliament that arrangements had been made to maintain community services previously made available by Kennecott in the remote Ok Tedi region, and for a respected firm of geological consultants to take charge of a further programme of exploration designed to take the project up to the feasibility study stage.

Over the next year, a government-owned company, the Ok Tedi Development Company, employed the New York geological consultants Behre Dolbear to supervise a drilling programme based on the further exploration that had been planned by Kennecott. Meanwhile the government discussed the project with most of the major metal mining companies in the world, until the Broken Hill Proprietary Company Ltd (BHP) agreed to take over the project on terms acceptable to the government. The attempts to return the initiative in the project to private hands had been complicated by Kennecott's representations to other companies that it had some continuing implied rights in the project. The government found it expedient in these circumstances to make Kennecott an offer of an *ex gratia* payment of its own bonds of face value equal to Kennecott's audited expenditure at Ok Tedi, with the bonds to be issued at the time of commencement of construction of a mine. The agreement with BHP stated that the Papua New Guinea government would be rewarded with equity for any such payment and also for exploration expenditure by the Ok Tedi Development Company. BHP secured as partners in a consortium in the course of 1976 a German company, Kupferexplorations Gesellschaft mbH, and an American company, Mt Fubilan Development, a subsidiary of Amoco Minerals, itself a subsidiary of Standard Oil (Indiana).

Optimality of Rent Taxes

The Papua New Guinea general systems of taxes for both petroleum and mineral development are well designed to achieve a high degree of neutrality and also of revenue yield. They seem likely to be perceived as being fair over a wide range of circumstances and therefore to be stable over lengthy periods. They provide a high degree of certainty as a basis for exploration and mine development. They require a minimum of discretionary decisions and are generally simple to administer.

The renegotiated Bougainville Copper agreement is less nearly ideal than the general arrangements in some of these respects, reflecting the pressures of the renegotiation process. Under the new Bougainville agreement, the real burden of the progressive profits tax becomes more severe at an arbitrary rate determined by the rate of inflation, unless the Minister for Finance uses his discretion to adjust the 'capital base'. This is undesirable in principle. However, in the course of the renegotiation, the company had traded off its insistence on an automatic adjustment for inflation against other concessions, and the overall effect of the agreement has not been excessively severe. At the review that began in 1981 but had not been completed late in 1982, the company and the government discussed the possibility of applying the general taxation arrangements to Bougainville Copper. This would provide more certain protection against the possibility that the real incidence of additional profits tax might rise excessively with inflation, while possibly giving the government some additional revenue.

Impact on Economic Development

The extraordinary cycle in economic activity generated by the construction and completion of the Bougainville mine contributed to political tensions in both the boom and the subsequent recession. There was considerable resentment of the large increase in numbers of foreign skilled workers in the boom, and even more resentment in Bougainville itself of the inflow of unskilled workers from the New Guinea mainland. Later there were difficulties and hardships because of the loss of employment and the need for many Papua New Guinean migrants to the towns either to return to their home villages or to lower their sights on town incomes. But the purely economic aspects of the adjustment took place with surprisingly little waste of resources: real wages did not lock irreversibly at a higher level as a result of the period of high demand, and generally there is little evidence that other sectors of the economy have been damaged in a lasting way by the short and intense construction boom.

The contribution of the Bougainville mine to the revenue, through royalties, dividends, and dividend withholding tax, rose quickly with the high world copper and gold prices to K28 million for 1973–4 even before the renegotiation of the tax agreement, and the start of the renegotiation process promised increased revenue. This was a period of high prosperity in Papua New Guinea during the world-wide commodity boom, and the increased mining revenue in the absence of an effective stabilization strategy contributed to the raising of real government expenditure well above levels that could be expected to be sustainable. This in itself helped to create an environment in which large wage increases were contemplated without alarm. The permissive attitude to wages policy at this time was exaggerated significantly by political reaction to the high untaxed profits at the Bougainville mine before the renegotiation of the agreement had been completed. The combination of these influences led almost to a doubling of the real value of statutory minimum wages in the large towns as a result of a single decision of a minimum wages board. The increase in levels of real government expenditure and wages in the boom was not easily reversed in the world-wide recession that followed.

The Papua New Guinea government learnt the virtues of a strong stabilization strategy from the experience of 1973–4. Soon after the completion of the renegotiation of the Bougainville Copper agreement, which promised much larger fluctuations in revenue as well as higher average levels, legislation was introduced to establish a Mineral Resources Stabilization Fund, which was designed to secure a much more even flow of funds from the mining industry into consolidated revenue. The managers of the fund, senior officials acting *ex officio*, were required to report their expectations of future mineral revenue annually to the Parliament, together with their recommendations on the level of transfer from the Fund to consolidated revenues that, in their judgement, was sustainable in real terms for the following five years. The government was constrained to accept this advice within narrow limits. The establishment of the Fund appears to have been successful in building widespread understanding of the need for restraint in expenditure in times of high metal prices, and in achieving support for such restraint among leaders in politics and the public service.

The Papua New Guinea stabilization strategy was later developed to incorporate steady growth in public expenditure in real terms, independently of annual revenues, at a rate judged to be safely consistent with feasible and desirable rates of public revenue growth over the medium term. This approach, if implemented effectively, would be useful in limiting the destabilizing impact of the Ok Tedi construction boom in the early 1980s. The continued operation of the Mineral Resources Stabilization Fund could be seen as an aid to the general stabilization strategy, mainly through its effect in helping to maintain political support for large counter-cyclical variations in levels of fiscal reserves (or net public indebtedness).

The expansion of the mining industry since the late 1960s has substantially increased the scale of modern economic activity in Papua New Guinea, and this, together with its important direct contribution to revenue, has facilitated the establishment of an independent state in Papua New Guinea with normal sovereign powers. The increase in revenue has allowed real public sector expenditures—vital to welfare and growth in Papua New Guinea, and yet financed to an extraordinarily high degree by foreign aid before the formation of the national government in 1972—to be maintained for the most part, and in some high priority areas increased, at the same time as the real value of foreign aid transfers has declined. The freedom to accept a steady reduction in real foreign aid without disastrous consequences for development placed Papua New Guinea in a much stronger position to insist on forms of aid that were consistent to the greatest extent possible with the exercise of normal sovereign powers over public sector resource allocation and economic policy in general.

Summary

Papua New Guinea has preferred general fiscal rules for mineral investment rather than negotiations of all fiscal terms on each project individually. It has preferred early rather than late agreement where matters still have to be negotiated, and stable arrangements rather than rules subject to periodic change. In these circumstances, the tax regimes for petroleum and other minerals are well designed to raise large amounts of revenue with a minimum deterrent effect on new investment. Some flexibility across projects is permitted by the (usually early) negotiation of the terms upon which the government takes up equity and by the (usually late, but without additional burdens beyond those associated with the absence of government infrastructure investment) negotiation of the terms upon which the government might provide infrastructure.

Some features of the Bougainville Copper agreement are not ideal, reflecting the constraints imposed by starting from unsatisfactory arrangements. But the renegotiated Bougainville agreement incorporates principles that have been crucial to the building of domestic political consensus, a prerequisite to the maintenance of stable rules, about the management of the mining industry. The renegotiation appears to have been well judged in the amount of revenue drawn from the mine, in terms of what later proved to be acceptable to other large investors *de novo*.

Papua New Guinea is vulnerable to huge fluctuations in economic activity with the ebb and flow of mining investment and with the variations in metal prices on world markets. Effective mechanisms have been developed to counter the effects of the latter and at least to reduce the amplitude of those of the former. But there would seem to be no way of removing completely a tendency towards boom conditions during construction of new mining capacity and recession in its immediate aftermath.

Papua New Guinea, as a small, newly independent country with an inexperienced government, might have been expected to have relatively little bargaining power in negotiations with large mining companies and to be in a weak position to withstand domestic economic instability deriving from the mining sector. That it has managed to achieve favourable terms against the opposition of powerful mining companies seems to establish that there is competitive supply of investment, at least to large-scale mining of certain metals. That it managed for a number of years to stabilize the nation's expenditure in the face of drastic fluctuations in copper and gold prices shows that the characteristic economic problems of countries with large mining sectors are amenable to considerable amelioration, with foresight, and with consistent and wise political leadership.

15

MINERAL RENT TAXATION IN AUSTRALIA

Australia has an unusually rich and diverse per capita endownment of mineral resources. Although importing between one-quarter and one-third of its liquid petroleum requirements, it is a major exporter of coal and virtually all economically important metallic minerals. Soon it is expected to become a large exporter of natural gas. The exploitation of its mineral endowment has made a substantial contribution to domestic economic production and to government revenue, although the industry has operated within a fiscal framework that is unstable and must be judged unfavourably on neutrality and revenue maximization criteria. Part of the inefficiency in the fiscal system as it relates to minerals results from a complex federal constitution and political system, which is a significant obstacle to the reform of mineral rent taxation.

Minerals in the Australian Economy

Australia's location, remote from the main centres of world industrial production before the recent industrialization of north-east and south-east Asia, delayed exploitation of many mineral deposits. Early mineral developments were concentrated on high value-to-weight commodities like gold, copper, lead, zinc, and silver. There were substantial coal exports to south-east Asian markets at about the turn of the century, but the availability of cheap petroleum until the 1970s stopped the expansion of this trade. Exports of the bulkier minerals, coal, iron ore, and bauxite, awaited the establishment of large steel and aluminium industries in Japan in the 1960s. Since then, continued rapid industrialization in several western Pacific economies has provided opportunities for the rapid expansion of the bulk mineral trade, which has been further augmented by the increased world and regional demand for energy minerals that followed the large increases in oil prices of 1973–4 and 1979–80.

The mineral industries have played the role of leading sectors, providing the main impetus to growth in the Australian economy as a whole at various times in the past, and also over the current period of strong growth in mineral production and exports that began in the mid-1960s. Mining's share of gross domestic product at current market prices increased from 1.9 per cent in 1960 to 3.6 per cent in 1970 and

4.5 per cent in 1978–9.[1] Large-scale oil production began in the late 1960s, and since then about two-thirds to three-quarters of Australia's oil requirements have come from domestic sources. Other mineral industries are strongly oriented towards export markets. Minerals and metals account for over 40 per cent of Australian exports in the early 1980s, compared with only 10 per cent in the first half of the 1960s. These ratios are expected to continue to rise over the next decade.

A characteristic of recent growth in exports has been the rapidly increasing proportion of processed metals. In particular, one important development in the Australian minerals industry since the second world oil crisis in 1978–9 has been an upsurge of interest in aluminium smelting that would use electricity from coal and Australian alumina. This seems likely to become one of the major growth areas of the Australian economy through the 1980s, during which Australia will become the largest supplier of aluminium metal to international markets.

Royalties, taxes, and other charges on the mineral industries contribute about 15 per cent of total federal and state government revenues. The greater part of this accrues to the federal ('Commonwealth') government as a levy on crude oil production.

The Institutional Framework

The constitutional, legal, and fiscal framework within which the Australian mining industries operate is extraordinarily complex. Ownership of minerals is vested in the Crown. Part of the coal resources of New South Wales,–which were granted to private landowners in the nineteenth century, represented an exception to this general principle until private rights were extinguished by state legislation in 1981. Ownership of minerals on-shore lies with the six state governments and the Northern Territory government. Offshore mineral resources are vested in the Commonwealth, although recently the Commonwealth government has chosen to exercise some of these powers jointly with the states. The Northern Territory, which was previously entirely under the jurisdiction of the Commonwealth, has recently been granted some of the powers of a state government, including authority over mineral development except in relation to uranium.

The states' powers cover mineral leasing, royalties (in any form favoured by the state), and such matters as provision of infrastructure

[1] These data, from the national accounts, value petroleum at market price less the crude oil levy, following the national accounts convention on treatment of 'excise taxes'. The levy is treated as an excise for these purposes. If petroleum were valued at international prices, the mining share of gross domestic product would be similar to that shown for 1960 and 1970, but several percentage points higher in 1978–9.

and the imposition of requirements for local processing. But at the same time, mining investments are subject to a wide range of Commonwealth government levies and approvals, relating to income tax, export tax, production tax (excise), pricing in export and domestic markets, foreign investment, financing, and exchange control. Both state and Commonwealth governments intervene on the environmental impact of mining.

The legal framework has been further complicated in recent years by legislation granting land rights to aboriginal groups in the Northern Territory (legislated by the Commonwealth government) and several states. The land rights legislation purports to give traditional aboriginal landowners a right of veto over mining developments, and therefore the right to negotiate fiscal terms and conditions. The land rights cover a majority of known economically valuable mineral areas in the Northern Territory and also highly prospective areas in South Australia. The extension of aboriginal land rights to Queensland and Western Australia, which is on the contemporary political agenda, would also have big implications for mining.

The fiscal system within which the Australian mining industry operates has been developed with little co-operation or even communication between the various states or between the two tiers of government. The result has been a complex, unstable, and distorting system, that appears to fare poorly on both revenue maximization and neutrality criteria. There has been considerable public discussion over the past five years of the need to rationalize the fiscal arrangements, but so far to little effect.

Slightly more than one-half of the value-added in the Australian mineral industries is owned overseas. This ratio rose from about one-third in the seven or eight years of rapidly expanding mineral output that began in the mid-1960s. Since then, the share of production owned overseas has stabilized, under the influence of government policy requiring majority Australian ownership in new mining projects. Most of the Australian equity in mineral production has been privately owned, but there has been substantial government ownership in two uranium mining projects and, through the Electricity Commissions of state governments, in coal mining. One feature of the institutional environment of mining in Australia is that mines are often the major source of employment in isolated locations, and there are strong political pressures to keep mines in production even when they are operating at a loss. Thus taxes and royalties are readily abated, and in some situations cash subsidies paid (as in the gold industry over long periods and the copper industry during the period of low prices in the 1970s), to prevent the closure of mines.

Petroleum Rent Taxes

The main production of liquid petroleum in Australia is in Bass Strait, offshore from the State of Victoria, and mostly in areas administered by the Commonwealth government. The producing company is a 50–50 joint venture of the Australian company, BHP, and Esso Australia, a subsidiary of the Exxon Corporation of the USA. There is much smaller production by largely Australian-owned producers on-shore in Queensland and Western Australia. Natural gas is produced in the Bass Strait and Queensland oilfields and also in South Australia and Western Australia.

There were major changes in the petroleum rent tax system annually from 1975 until 1979. Many of the changes have been introduced at the same time as variations in the controlled price at which domestically produced crude petroleum is sold within Australia.

All petroleum production is subject to an *ad valorem* royalty at a rate between 10 and 12.5 per cent of well-head value. The proceeds of the royalty from Bass Strait production are divided in the ratio 3–2 between the Commonwealth and the Victorian government, under an agreement between the two governments.

All income from petroleum (and other mineral production except gold) is liable to income taxation at the normal rate, which is currently 46 per cent. Income tax from mining was treated substantially more favourably for taxation purposes than other income before major revisions of the income tax in 1974 which removed all preferential arrangements for mining. A minor part of the preferential treatment of mining income was restored through further revisions of the Income Tax Act by a new government in 1976. Dividends remitted overseas by subsidiaries of foreign corporations are subject to a 15 per cent withholding tax. A special rate of income tax, at 53 per cent, is applied to the earnings of branches of foreign corporations, which are not required to pay withholding tax on overseas remittances. Most depreciation and amortization provisions for capital expenditure on petroleum production are similar to those applied in other industries. Expenditure on petroleum exploration can be written off immediately against income from any source.

In addition to income taxation and royalties, all oil production is subject to a production excise which is levied and collected by the Commonwealth. The production excise, or 'crude oil levy', was first applied in 1975 at the rate of A$2 per barrel, at the same time as the price of domestically produced crude was raised by the same amount in partial and delayed response to the increases in world petroleum prices in 1973–4.

In 1965 the Australian government set the price of crude oil, which was then produced only at the small Queensland field, at A$3.14 for the next five years, which was substantially above the price of imported petroleum. The higher price was meant to encourage exploration. Elaborate arrangements were made to ensure that all domestic oil refiners took a share of the high-priced Australian crude pro rata with their total requirements. When the Bass Strait fields came on-stream in 1969, a lower price was set for their production, but still above the (then declining) import parity level. A new agreement in September 1970 set a price of A$2.06 per barrel, again above current import parity prices, for crude from the Bass Strait fields, and slightly higher prices for fields producing less than 10,000 barrels per day (in practice, the Queensland and West Australian fields). The new prices were to be maintained for five years. This new agreement, executed at a time when world crude petroleum prices were declining, caused Australia to fall well below world prices when the latter rose strongly through 1973. Since the expiry of the agreement in 1975, the price of Australian crude has been raised, through a series of unconnected and sometimes surprising decisions, to a version of 'import parity pricing' from late 1978. Each increase in the price of domestically produced petroleum has been accompanied by an increase in the crude oil levy. The amount of each increase in the crude oil levy has been determined *ad hoc* without any attempt to articulate the principles by which it was determined.

In 1981 crude oil levy was applied at different rates depending on whether petroleum was produced in liquid or gaseous form. On liquids, the levy was applied at three different rates depending on whether production was classified as 'new-oil', 'import parity old oil', or 'old oil'. 'New oil is from fields discovered after September 1975 and attracts a zero levy rate. No 'new oil' was being produced in 1981. 'Import parity old oil' includes a specified amount of production from each of a number of defined 'fields', all discovered before September 1975, an amount that has been rising over time at a pre-announced rate to total about 50 per cent in 1980. The remaining production from those fields is 'old oil'. In mid-1979, the levy rates were set at import parity prices less A$9.59 per barrel for 'import parity old oil', and import parity prices less A$2.38 per barrel for 'old oil', with provision for raising the Australian dollar amounts per barrel left in the hands of the companies in line with increases in the consumer price index. There is provision for all or part of the levy to be abated at the discretion of officials of the Department of National Development, and this has been used in an attempt to remove the deterrence to high-cost marginal production that would otherwise be introduced by the levy.

The crude oil levy does not apply to natural gas production, nor to petroleum condensate marketed separately from a liquid petroleum stream. In 1980 the levy was applied at a rate of A$2 per barrel to liquified petroleum gas (LPG) from fields in production before August 1977, but was not applied at all to LPG from fields brought into production after that date.

The form and rates of the crude oil levy were changed in major ways in each of the years from 1975 to 1979. The financial fortunes of producers have been influenced more by the annual changes in the crude oil levy than by any conceivable difference between good and bad management.

Rent Taxes for Other Minerals

The production and sale of other minerals is usually subject to a royalty levied and collected by the relevant state government. Coal attracts a Commonwealth export tax as well. The whole system has been established *ad hoc*, so that there is little obvious rationale for the wide variation in types and rates of royalty within and between states, and across different minerals.

The Commonwealth export tax applied to coal is a specific duty, at a higher rate for coking than for steaming coal (in 1981, A\$3.50 per tonne for coking coal and A\$1 per tonne for steaming coal). Australian coal exporters' profits increased sharply after the 1973–4 oil crisis and the subsequent increase in coal prices on world markets. The introduction of the levy in 1975, originally at the rates of A\$6 per tonne for coking coal and A\$3 per tonne for steaming coal, was motivated by the Commonwealth government's desire to claim as taxation part of what it saw as windfall profits when the Queensland government had not been willing to increase state royalties. The specific export duty was undoubtedly a deterrent to the opening up of new marginal mines or to the expansion of high-cost output from established mines, and was much criticized on these grounds. A new conservative government came to office in later 1975 having promised to eliminate the export levy. The levy was reduced in the 1976 budget, whereupon the Commonwealth warned the states that they would lose financially in other ways if they increased royalties in the wake of the Commonwealth tax reduction. The tax was reduced again in 1977. The promise to phase out the levy completely was not repeated after 1977 and was revoked in 1981.

State royalties are specific, *ad valorem*, and profit-based, the three forms often being applied in the same state. In addition, several states have obtained large amounts of revenue by providing infrastructure (such as rail transport) at charges substantially in excess of cost. This has been especially important to the State of Queensland.

The royalty system is fairly uniform across most of the mining industries in Tasmania, the Northern Territory, and South Australia, which are the least important states as mineral producers. Tasmania levied no royalties on a general basis until 1975; now, royalty is levied on the most important mining products (including concentrates) at the rate of 2 per cent of net proceeds of sales or 5 per cent of profits from sales, whichever is the less, with no royalty being paid in any quarter in which proceeds of sales do not exceed A$50,000. The general royalty rate for established mines in the Northern Territory is 1¼ per cent of proceeds of sales, but there are low specific rates for bauxite and higher profit-based rates for gold. In the case of the Northern Territory's uranium, which is still administered by the Commonwealth government, the *ad valorem* royalty is supplemented by charges negotiated by, and payable to, aboriginal landowners. These charges are an annual lump sum of A$1 million plus 2.5 per cent *ad valorem* in the most important case of the Ranger mine, with higher lump sums and an additional 3 per cent *ad valorem* in an agreement for a new mine negotiated by Pancontinental in 1982. A new Northern Territory Royalties Bill enacted in 1982 applies an HRIT at a rate of 18 per cent, without any deductions for payments to aboriginal groups, to all new non-uranium mines. In South Australia, most mineral production is subject to a 2.5 per cent or 5 per cent *ad valorem* royalty. A 1981 agreement on mining a major newly discovered copper and uranium deposit at Roxby Downs combines a 3 per cent *ad valorem* royalty with an extraordinarily complex PPT that will yield little or no revenue unless the mine turns out to be prodigiously profitable.

Queensland applies a range of specific and *ad valorem* royalties, often at lower rates for minerals processed within the state. Different royalty arrangements apply virtually to every mineral. Of the most important minerals, royalty is levied at the nominal rate of A5 cents per tonne for coal consumed within the state, 4 per cent of free-on-rail (f.o.r.) value for coal mined underground and exported, and 5 per cent of f.o.r. value for coal mined open-cut and exported. Bauxite production is levied on a specific basis that relates to the volume of ore mined, at a rate that varies proportionately with the world price for aluminium ingot: A40 cents per tonne times the ratio of the current Alcan world price to the 1973–4 price; provided the royalty is not less than A50 cents per tonne for ore consumed within the state, and not less than twice that rate for ore consumed elsewhere. Copper, silver, lead, and zinc at the large Mt Isa mine are subject to a royalty that is identical in form to that for bauxite, but levied at the higher rate of A$2 per tonne when prices are at 1973–4 levels, with a minimum of A$1 per tonne. Nickel ore is levied at the low rate of A10 cents per tonne. Sand-mineral concentrates and copper other than from Mt Isa are taxed at the rate of 2 per cent of gross sales proceeds in excess of A$30,000 per annum or 5 per cent of profits in excess of A$30,000 per annum, whichever is the less. In addition, the Queensland Treasury negotiates rail charges mine by mine, which, since mining investors are required to use the state rail system, amounts to an additional tax (calculated at the equivalent of 15 per cent *ad valorem* for some coal mines) when the freight rate exceeds the cost of providing the transport service.

In New South Wales, coal is subject to a low specific royalty, although there is provision for companies to offer to pay a 'super-royalty' when competitive bids are called for new leases. Some use is made of negotiated rail freight charges as a taxation instrument along the lines followed by Queensland, although this is used less vigorously than in Queensland. The important Broken Hill lead, zinc, and silver mines are subject to a profit-based royalty at a rate that is progressive with the absolute amount of nominal profit: 4 per cent of the first A$400,000 'net profit' (calculated on an agreed basis) increasing at the rate of 2 per cent for each additional A$400,000 of profit to a maximum of 50 per cent. Other minerals, and other projects, are subject to different royalty arrangements.

In Western Australia, the important iron ore production generally attracts an *ad valorem* royalty of 7.5 per cent for ore if it is exported and a low, specific rate if it is used within the state. The royalty for nickel is similar in form to the copper, lead, silver, and zinc royalties at Mt Isa in Queensland, although it is applied at a low rate, and relates to the mass of the metal content rather than the mass of the ore. Bauxite is generally subject to royalty at A50 cents per ton, and alumina at 25 cents per ton and rising with the world price of alumina. Each other mineral is subject to its own rate of specific or *ad valorem* royalty. The large gold mining industry is subject to negligible royalties and is exempted from company income tax. A 1981 agreement governing the mining of a major diamond field provides for the investor to pay royalty *ad valorem* at the rate of 7.5 per cent or on profits at the rate of 22.5 per cent, whichever is the higher, with some provision for averaging annual profits.

Victoria generally applies specific duties on non-metallic minerals (most importantly brown coal at A4 cents per ton), and a 1.75 per cent *ad valorem* royalty on metallic minerals.

The large amounts of coal used by state electricity commissions for the generation of electricity to supply the aluminium smelting industries is generally subject to little or no royalty in any Australian state. Electricity is said to be sold at a rate that covers mining and generation costs and generates a small profit.

This bewildering array of royalty arrangements has grown historically through large numbers of *ad hoc* decisions, and has no obvious economic rationale. Arrangements within each state have been unstable over time, with royalty rates often being raised at times of high profitability in the mining industries. For example, changes in royalty arrangements in Queensland between 1973–4 and 1974–5 increased royalty obligations at the Mt Isa mine approximately three-fold, and at the major bauxite and coal mines approximately ten-fold.

Optimality of Rent Taxes

The Australian system of mineral rent taxes is highly unstable, does not on the whole perform well on the criterion of revenue maximization, and is a significant source of distortion of investment and production decisions.

The instability of the Australian rent tax system seems to result from pressures arising out of the common view that highly profitable mining ventures should make substantial fiscal contributions through some form of rent taxation, in a situation in which governments have not articulated principles for the setting of rent taxes. The forms of rent tax applied in Australia do little to compress the probability distribution of after-tax outcomes of mineral investment, sometimes having the opposite effect, and so are poorly designed to promote stability. The instability has been exacerbated from time to time by tensions between the federal and state levels of government. The complex federal system of government has certainly reduced the incentive at each level of government to expend analytic, administrative, and political effort on the rationalization of the rent tax system, because of doubts about whether the necessary co-operation from other governments would be forthcoming.

Perceptions of instability in the fiscal regime are probably most important in relation to petroleum. The high proportion of investment in the high-risk exploration phase in this industry causes the probability distribution of pre-tax outcomes to be exceptionally widely dispersed, introducing a greater than normal tendency towards instability in the fiscal regime. Realization of this characteristic of the industry causes investors to be sensitive to the possibility of changes in the fiscal regime, and the frequent large changes in Australian policy through the 1970s provide produce good reason for concern. In these circumstances, the Fraser conservative government's promise, repeated frequently between 1976 and 1982, that new discoveries would receive import parity prices without being subject to the crude oil levy or some substitute, can hardly be credible. Investors judge that there is some chance that additional taxes will be applied, at least in the case of highly successful outcomes. Only an optimistic investor would expect that a discovery that gave rise to rates of return, before tax, as high as the Bass Strait fields would be taxed much more lightly than they.

A cautious investor might consider the announced fiscal system for 'new oil' to be unlikely to stand the test of time and expect the worst. The perceptions of high probability of changes from the announced fiscal system for petroleum production from new discoveries could probably be dispelled only by the announcement of an alternative regime that did more to compress the probability distribution of investment outcomes and that was less obviously open to political criticism should highly profitable new projects be established.

Although Australian rent tax systems for other minerals have been highly unstable, this has probably been less damaging to efficient resource allocation for them than for petroleum. High-risk exploration expenditure represents a smaller part of total investment in production of other minerals, so that the expectation of very high after-tax returns from successful ventures is less crucial to the maintenance of incentives for exploration. A mineral deposit that is expected to yield returns that would make mining investment eligible in the absence of tax is unlikely to be deterred in Australia by the possibility that future changes in the fiscal regime will cause the project to operate at a loss. There is a long tradition in Australia of alleviating fiscal burdens on mines that are operating at low rates of profitability, even to the extent of ensuring their continued operation through payment of subsidies. But instability in the fiscal regime for hard minerals still has large costs: such uncertainties undoubtedly cause some delays in investment, and divert corporate effort from finding, mining, and selling minerals to managing relations with governments.

Australian rent taxes taken as a whole yield large amounts of revenue, but the absence of a clear rationale based in the concept of economic rent has meant that substantial rents have been left in private hands in some highly profitable mining operations. This is most obviously true in the coal industry, where valuable deposits have been identified with little private exploration expenditure, and in which there is a wide range of profitability across mines, depending on the nature of the deposits upon which they are based. The use of a combination of uniform specific and *ad valorem* royalties has prevented highly profitable mines from being levied at high rates, for fear of deterring investment or production at higher-cost mines. However, the negotiated 'rail freight taxes' and super-royalties in Queensland and New South Wales have loosened this restraint on high state taxation of coal mining.

The reliance on specific and *ad valorem* taxes and royalties more generally means that there has been substantial distortion of mine development decisions. This has been clearly apparent from time to time with the crude oil levy. Even at the lower rate applying to 'import parity old oil', the crude oil levy has removed incentives for high-cost incremental production from established fields. Where the companies have been able to establish the significant presence of this effect, officials have applied their discretion to restore incentives. But the process of alleviating the distorting effects has involved complex negotiations over matters on which officials have necessarily relied on information from the producing companies. Adjustments have been arbitrary, often late and imperfect, expensive in terms of managerial time, and sometimes unnecessarily corrosive of the revenue.

The combination of Commonwealth and state specific export taxes, royalties, and super-economic rail charges appears to have deterred some production of higher-cost coal from established mines in New South Wales and Queensland, and it has been asserted by the industry that they have also deterred the establishment of some high-cost mines. It was this claim that led the present Commonwealth government to promise to remove the coal export levy on several occasions between 1975 and 1977. It would be surprising if the high rates of specific duties related to the volume of copper, lead, zinc, and silver ore production in Queensland had not led to some high-grading effect, or if the high rates of profit tax had not deterred some marginal investment at the Broken Hill mines.

The Resource Rent Tax Debate

The weaknesses of the Australian rent tax system have been much discussed, and there has been considerable legislative action on mineral rent taxes, since the early 1970s. A report by Fitzgerald (1974), commissioned by a recently elected Labor government, argued that the especially favourable treatment of mining income for taxation purposes greatly reduced the industry's contribution to national welfare and served no good purpose. This report inspired a major revision of the Income Tax Act in 1974, removing the preferential treatment for mining income. There were major upward revisions of royalty rates in 1974 and 1975, in response to a recent history of exceptionally high profitability in some mining projects. The large increases in world energy prices in 1973–4 generated windfall profits for coal exporters, and opened the prospect of much higher profits for oil producers should domestic crude oil prices be raised towards international levels, and this prompted calls for higher taxation on these industries. These calls were reflected in the coal export levy and the crude oil levy.

The changes to the mineral rent taxation systems through the 1970s greatly increased the industry's contribution to the revenue, while on the whole increasing the fiscal system's complexity, arbitrariness, and distorting effects. This highlighted the need for a more rational and internally consistent mineral rent taxation system. A major official report on the taxation of the petroleum and mining industries (Industries Assistance Commission, 1976a) recommended that mining income be subject to income tax in more or less the same manner as other income, and that the complex system of state royalties be replaced by a uniform higher rate of income tax (HRIT). Another official report, on crude oil pricing, recommended a gradual increase in the price of domestic crude oil prices towards international levels and the simultaneous introduction of a form of resource rent tax (RRT) on the petroleum industries (Industries Assistance Commission, 1976b).

Much of the Australian discussion has centred on whether a general progressive profits tax (PPT) or RRT should be introduced in Australia. In 1975, the Commonwealth Minister for Minerals and Energy, Mr Rex Connor, said in a submission to the Industries Assistance Commission inquiry on petroleum and mining taxation that a general, Bougainville Copper-type PPT should be introduced. Although the first Garnaut and Clunies Ross article (1975) on RRT was directed mainly to the circumstances of developing countries, it was the subject of considerable discussion in Australia, and there were many suggestions that the RRT proposal should form the basis of a revision of the system of rent taxation either for petroleum or for all minerals (see, for example, Caves, 1978; Crawford and Okita, 1978; Swan, 1976; Garnaut and Clunies Ross, 1979; and several contributions to Smith, 1979). It was discussed at length in each of the two Industries Assistance Commission reports on the mining and petroleum industries (1976a, 1976b), unfavourably in the first and favourably in the second.

The Australian Labor Party policy conference in July 1977 committed the party to implement an additional profit tax for the mining industry upon its return to office. Subsequent statements from Labor Opposition spokesmen have made it clear that the additional tax would take the form of RRT, and that it would replace the crude oil levy and the coal export tax. A Labor government would seek to reach agreement with the states on their abandoning their own royalties in exchange for a share of the Commonwealth RRT revenue (Keating, 1982).

Soon after the 1977 Labor Party conference, in his August 1977 Budget Speech, the then Australian Treasurer, Mr Philip Lynch, announced that the government was considering the application of RRT to petroleum (to replace the crude oil levy) and uranium. Late in 1977 the Minister assisting the Treasurer, Mr Viner, spoke of the possible application of resource rent tax to the north-west shelf natural gas export project, and described the characteristics of RRT that would ensure that it would not deter investments that appeared in advance to be marginal. These statements by ministers prompted a crude and noisy campaign against RRT by parts of the mining industry. In July 1978 the government announced that it had shelved plans for an RRT because of its possible adverse effects on exploration and foreign investor confidence.

The resource rent tax discussion has continued among academics and officials who are interested in the possibility of introducing a less distorting fiscal system. It has survived in political debate on current policy for two reasons. The first is the Labor Party's commitment to RRT. The second is that executives of several of the major mineral producers, and especially the oil producers, believe that RRT offers a superior alternative to the current fiscal systems.

The RRT provides a means to a fiscal system that, in comparison with existing arrangements, is at once more stable and more efficient from the points of view of revenue maximization and neutrality. However, the realization of this potential requires co-ordination between the federal and state levels of government: the imposition of a Commonwealth RRT on top of (with deductions for) existing taxes and royalties might increase revenue yields, but might also exacerbate the distortions in the existing arrangements. It would, however, be possible for the Commonwealth to legislate sensible RRT arrangements for petroleum, where the main production is offshore in its own jurisdiction, in place of the crude oil levy, and probably for uranium, where the main deposits remain under Commonwealth control in the Northern Territory.

For other minerals, it would be necessary for the government seeking to rationalize the rent taxation system, presumably the Commonwealth government, to make it attractive for other governments to co-operate. The current state of federal–state financial relations is in one important respect conducive to such an exercise, since the Commonwealth is under strong pressure to increase its grants to the states. Even generous concessions to the states on access to mineral-related revenue collected by the Commonwealth could possibly be seen as a form of inevitable fiscal concessions which would otherwise need to be made in other areas.

We favour the introduction of RRT on all mineral income by the Commonwealth, accompanied by arrangements for increased Commonwealth grants to each state that agreed not to collect taxes and royalties on the mining industries. There would be no way of policing a commitment by the states not to raise revenue through super-economic charges for the use of infrastructure, so that in the last resort these charges may be a possible source of anomalies in the fiscal system. There is no reason for the grants to each state to be related precisely to RRT collections in that state, and there would be advantages in making grants equal to some fixed *ad valorem* rate on mineral sales. For example, the rate could be set at such a level that no state obtained less revenue than under the existing fiscal system, and that average total payments to states were similar in magnitude to total RRT receipts. The fixed ratio to sales would provide a more stable flow of revenue than would be obtained from remitting the RRT collected from mining operations in each state.

A state that continued to collect royalties would have no entitlement to special mineral-rent-tax-related grants from the Commonwealth. Mining operations within that state would then be subject both to state royalties and to RRT. The state would need to levy royalties at negligibly low rates, which would yield much less revenue than the Commonwealth grants; otherwise the combined effect of state royalties and Commonwealth taxes would introduce a powerful deterrent to production from marginal mineral deposits. Hence there would be strong pressures for co-operation with the Commonwealth.

Alternatively, and with less risk of intractable confrontation between the Commonwealth and the states, certain royalties imposed by the states might be made creditable against Commonwealth RRT commitments without the introduction of any system of additional grants to the states. This arrangement would introduce powerful incentives for the states to introduce royalty arrangements that were similar in form and rate to the Commonwealth's RRT system.

Under present arrangements, or either of the arrangements suggested in the preceding paragraphs, the revenues of the mineral-rich states, especially the lightly populated outlying states of Queensland and Western Australia, and those of the Northern Territory, could be expected to expand through the 1980s relatively to those of the more densely populated south-eastern states in which Australian manufacturing industry is concentrated. This in itself might be seen as undesirable, especially since the growth of economic activity and state revenues from other sources is likely to be most rapid in the mineral-rich states. However, all states are heavily dependent on Commonwealth grants, which could be adjusted to compensate for any undesirable distributional consequences of the suggested arrangements for mineral rent taxation. Such adjustments would be resisted by the mineral-rich states, but they would be in a weak position, since they receive, per capita, much higher than the average Australian level of grants from the Commonwealth government at present.

If, as would be desirable, the Commonwealth vacated the fields of mineral price control, excise, and export taxation at the time it introduced RRT, it would by so doing enhance the power of the states to manage the mineral industries. This might be a useful bargaining coin in seeking agreement with the states.

An Australian government could rely upon large RRT yields from established projects, and so would be in a strong fiscal position to subsidize mineral exploration at the RRT rate (or perhaps the lower of the rates in a two-tier RRT system). This would remove, or at least greatly reduce, the possible distorting effects of RRT on exploration decisions, while allowing revenue yields to be raised through lowering the RRT thresholds.

Diversion and Dissipation of Rent

There have been three major sources of diversion and dissipation of mineral rent in Australia: Commonwealth government controls on pricing, which have kept the price of some minerals in domestic markets at well below world parity levels; state government promotion of further processing; and the demands of mining unions for higher wages and improved working conditions, ahead of those generally available to Australian workers.

From 1973 until 1978, the price of Australian-produced crude petroleum was well below international levels. There was considerable debate on the merits of this approach to pricing (see, for example, Industries Assistance Commission, 1976a), with the result that the advantages of world parity pricing were widely accepted in academic, official, and business circles, although perhaps not in the community at large. A policy of near-world-parity pricing, supported by additional taxation on income from petroleum production, seems now to be well established in Australia.

When international copper prices rose to unusually high levels in 1973–4, the Prices Justification Tribunal, a statutory authority established by the Commonwealth with powerful persuasive but ultimately no enforcement powers, fixed the domestic price of copper for a period at about two-thirds of the peak international levels.

This Commonwealth government involvement in mineral pricing has led to a large-scale diversion of rents to mineral users, and in some instances (especially with petroleum) to substantial dissipation of rent in uses that would not have survived if minerals had been priced at international levels. Price intervention introduces a major additional source of uncertainty into investment decisions in relation to minerals produced for domestic consumption. General Commonwealth undertakings to avoid holding domestic prices below world prices on average over time would, in the context of agreements on the fiscal regime between the Commonwealth and the states, be helpful to the husbanding of mineral rent.

The states have a strong political and economic interest in the promotion of mineral processing within their boundaries, even at the cost of dissipating mineral rents, a major part of which accrue to the Commonwealth government under existing arrangements. Several promotion mechanisms are employed. First, substantially lower royalties are often applied to ore that is processed within the state in which it is mined. Second, companies are sometimes required to undertake further processing within the state as a condition of a mineral lease. Third, states have made energy available for mineral processing at well below international parity prices, especially in the form of electricity for aluminium smelting.

The first of these mechanisms would not be available to states within the context of Commonwealth–state agreements on fiscal arrangements along the lines suggested in this chapter. The second would continue to be available to the states, unconducive as its use is to efficient resource allocation within the Australian economy, since there would seem to be no simple means of ensuring its elimination.

The third of these mechanisms became politically contentious from about 1980. Complex issues arise in the identification of mineral rents in coal production where the coal is used for generating electricity to process minerals.

Where coal has an international trade value, allocative efficiency requires that it be priced at export parity. Economic rents can then be identified in the same manner as in other mineral production. However, a considerable part of the coal used for electricity generation appears to be unsuitable for export because of its quality or location. An efficient system of resource rent taxation in circumstances in which coal had no alternative use would seem to require the provision of electricity to smelters at cost, and the application of rent taxes to income earned in smelting. To avoid arbitrary distinctions between tradable and non-tradable coal resources, it might be sensible for the states to continue with present electricity pricing practices and for the Commonwealth to apply a general resource rent tax to aluminium smelting along the lines suggested above for mineral extraction.

Australian mining unions are strong, and there is a long tradition of bargaining for improved conditions from mines earning high profits. The rapid expansion of coal production for exports following the rise in world energy prices in the 1970s was associated with unusually rapid improvements in the relative incomes of miners. Unions sometimes argued that the low levels of taxes on mineral rents generated in the most productive mines justified especially high wages in those mines. By the end of the decade, average weekly earnings of coal miners were about twice those of other Australian workers, and Australian mining wages were the highest in the world.

Impact on Economic Development

The exploitation of newly discovered mineral deposits, or those that are newly valuable as a result of changed export market conditions, has provided the main impetus to growth in several periods of rapid expansion in the Australian economy, most notably in the middle of the nineteenth century and from the late 1960s. The expansion of the mineral industries is the main reason why economic growth was significantly stronger through the late 1970s and early 1980s in Australia than in other advanced industrial countries.

But despite its effects on growth in the economic aggregates, the expansion of the Australian mineral industries has often been seen as a mixed blessing. Mineral industry growth has forced the contraction of other sectors of the economy producing traded goods (Gregory, 1976). The established exporting and import-competing producers have commonly resisted decline and applied pressure to governments for protection, or, in the important case of the mineral boom of the late 1960s and early 1970s, for the maintenance of an artificially low exchange rate (Smith, 1978; Garnaut, 1979b). Accession to the first of these pressures has led to dissipation of part of the potential mineral rent, and accession to the second has led to inflation. High capital intensity and commercial profitability has made expanding mineral industries a target for claims by labour unions, and there have been strong tendencies for increases in wages and improvement of conditions to flow through to other sectors of the economy. At times this has complicated macroeconomic management directed at restraining inflation and maintaining full employment.

The pressure that rapid expansion of the mining industries has applied to other sectors of the economy, and to macroeconomic stability, has been associated with a rise in political tensions in recent times. The resolution of these tensions has seemed to require measures to distribute more equitably the benefits of the natural resource industries. Demands for heavier fiscal impositions on mining have flourished in this environment, and have been and are likely to continue to be effective. But in the complex constitutional framework of Australian politics, this seems at least as likely to spawn a further proliferation of *ad hoc*, distorting, and ultimately unstable levies, as a reform of the rent tax system that is able to serve well the desiderata of neutrality, stability, and revenue maximization.

Summary

Australian economic growth has owed much to exploitation of mineral deposits for well over a hundred years. The country's recent relatively strong growth and its expectations of more to come are based on the good prospects for expansion of mining.

The institutional framework within which the Australian mining industry must operate is extremely complicated, with federal and state governments sharing regulatory powers, and other interest groups, recently and most notably aboriginal landowners, having recognized claims to participate in decisions on fiscal and regulatory arrangements. The location of the major petroleum deposits off-shore in Commonwealth territory eases somewhat the constitutional problems of policy in this area, but arrangements still need to be conducted within the context of an agreement with the neighbouring state of Victoria.

The petroleum rent taxation system yields large amounts of revenue but is not based on clear and explicit principles, and partly for this reason has been highly unstable. It discourages full economic recovery of petroleum. Rent taxes for other minerals are bewildering in their variety, on the whole being distorting and unstable, and yielding, on average, relatively low amounts of revenue. The Australian system has been one of the less efficient examined in the case studies in this book from the point of view of maintaining incentives to full utilization of the mineral endowment, and, with the possible exception of the levies on crude petroleum, also from the point of view of revenue maximization.

There is a sense in which these inefficiencies matter less in Australia than in the other countries that have been the subject of case studies: Australia's per capita endowment of minerals, and its other economic advantages, are so large that considerable dissipation and diversion of mineral rents is consistent with the maintenance of high and growing average living standards and relatively equitable distribution of incomes in the economy as a whole. But the Australian polity is uneasy about the waste of potential public wealth associated with the current arrangements, and there are likely to be further major changes through the 1980s. There are considerable barriers, associated with the federal constitution, to the introduction of a rent tax system that is more nearly ideal from the point of view of neutrality, stability, and revenue maximization. But unless these barriers are overcome, *ad hoc* responses to demands for greater public revenues are likely to compound the distortions inherent in the system at present, and to dissipate even more of Australia's potential mineral wealth.

16

MINERAL RENT TAXATION IN SOME AFRICAN COMMONWEALTH COUNTRIES

This chapter examines rent taxation regimes for hard minerals in selected African Commonwealth countries. It deals at some length with the important case of Zambia and more briefly with Botswana and Tanzania. The experience of these countries is challenging to theorizing within both the 'Commonwealth' and the 'North American' traditions. Botswana with diamonds and Tanzania with uranium exploration in the late 1970s found themselves in favourable positions to extract from investors in particular negotiations conditions that were far more onerous than would normally be acceptable under general rules as favoured by the 'Commonwealth' logic. And in Tanzania's case, the first major agreement reached with a foreign investor in hard minerals after independence in 1963 was more favourable for the government than agreements for new developments in any of the established hard-mineral producers, in contradiction of the 'North American' logic. The Zambian experience is challenging, too, because it provides an instance in which fiscal arrangements that were too severe to be consistent with substantial amounts of new investment may have been suitable instruments for the pursuit of narrowly national goals. We return to an examination of the implications of these characteristics of the African experience after describing the rent tax regimes in the three selected Commonwealth countries.

Zambia

Minerals in Zambia's Economy

Zambia has been one of the supreme examples of states whose economy is dominated by a single mineral: only in some of the oil-producing countries has a single mineral been so overwhelmingly important. At the time of Zambia's independence in 1964, and for some years after, copper typically supplied over 90 per cent of the country's exports, 50–70 per cent (directly) of government revenue, and nearly a half of net domestic product (Martin, 1972:146). The highly unstable character of these contributions, however, is exemplified by the fact that the industry's share in gross domestic product fell irregularly from 48 per cent in 1969 to 11 per cent in 1977 and 1978, while its contribution to government revenue fell from 59 per cent in 1969 to 19 per cent in 1972 and from 53 per cent in 1974 to 3 per cent in 1976 (Zambia Mining Yearbook, 1978:24).

At independence, Zambia exported more copper than any other country and was the world's third largest producer (O'Faircheallaigh, 1980:156, 161, 180). Copper output varied between 630,000 and 720,000 tonnes per year from 1964 to 1978. Other metals (principally zinc, cobalt, and lead) have also been of some importance, these three accounting for nearly 5 per cent of total export earnings in 1972 and 1973 (Commonwealth Secretariat, n.d.:7.1). Since the mid-1930s most copper has been exported in refined form (O'Faircheallaigh 1980:156). The very small role played by commercial agriculture and manufacturing at the time of independence meant that, apart from subsistence food production, the economy consisted very largely of mining and government services and activities closely related to them or serving their employees.

Direct employment in mining has been important, but much less so proportionately than has output or revenue; in 1971 about 16 per cent of the working population was engaged in mining and quarrying (Commonwealth Secretariat, n.d.:7.2), but the importance of the industry in employment has clearly gone far beyond the number directly employed as miners.

The figures given above illustrate the enormous vulnerability of Zambia's economy to changes in copper prices. A fall in price may reduce government revenue by a half over a very short space of time even without allowing for secondary effects through general reduction in income and spending: total domestic expenditure must either fluctuate enormously from year to year, or there must be large surpluses and deficits in the current account of the balance of payments, reflected in variations in foreign exchange reserves or foreign debt.

Ownership and control of mining was an issue that readily roused political emotions. Colonialism in Zambia's case had been associated with much of the racial discrimination characteristic of South Africa and of what was then Southern Rhodesia. One of the two main copper mining groups, the Anglo-American Corporation, was South African-based, and South Africans have been dominant among expatriate staff on the Copper Belt. Africans had been prohibited until shortly before independence from exercising any but the most menial roles in the copper mines, a limitation only gradually removed between 1953 and 1963 (O'Faircheallaigh, 1980:174–95). Thus, the copper mining establishment could easily be associated with the most resented features of the colonial era. It was an added source of resentment that the mining companies had had to pay royalties for over thirty years to the British South Africa Company, which had been granted mining rights over the whole area in the 1920s but had made no contribution to mining over this period.

Zambia in the 1960s and 1970s thus inherited a mining industry long established and familiar, and the issues raised were inevitably different from those faced on the same period by Indonesia and Papua New Guinea over non-ferrous metals, Botswana over diamonds, or the UK over oil and gas. The strong feelings inevitably aroused were no help to calm and rational deliberation over the government's relationship with the industry.

The Institutional Framework for Mineral Development

There have been four important changes in the arrangements applied to revenue-raising and the control of mining since independence, giving five phases in all.

At independence mining was subject to a royalty and an income tax, the royalty rights having been transferred from the British South Africa Company to the state three hours before. The royalty had been designed so as to adjust the incidence with profitability by being levied at 13.5 per cent of the London Metal Exchange (LME) price *minus* £8 (K16) per long ton produced. Rises in prices and costs, however, had made the adjustment of little significance, and the incidence of the charge was for practical purposes 13.5 per cent *ad valorem* at the margin. Income tax was collected on various mines at a variety of rates until 1968 and thereafter on all mines at 37.5 per cent on the first K200,000 and at 45 per cent on the remainder.

A change was made in April 1966 to tap the windfall profits of a substantial rise in price. For a period from late 1963 the Zambian companies followed the major copper producers in voluntarily charging what was called a producers' price, which was well below the LME price. This was done in order to discourage substitution of other products for copper. By 1966 the Zambian government wanted its mining companies to abandon this system, and it added an export tax which took 40 per cent of the excess of the LME price over K600 per long ton. This had the effect of both encouraging the companies to raise their prices and tapping the windfall when they did so. It did, however, raise the total *ad valorem* levy at the price level then prevailing to about 24 per cent (figures calculated from information given in Commonwealth Secretariat, n.d.:7.16). This would have constituted a fairly large disincentive to production, and instances are given in which royalty and export tax (with the income tax) had converted a profit into a loss (Commonwealth Secretariat, n.d.:7.16–17).

The system was apparently recognized as unsatisfactory but was not immediately revised in any way relevant to this problem. Instead, a second important change was made by President Kaunda's Mulungushi speech in April 1968. This was the rule that allowed the mining companies to remit not more than half their profits abroad in dividends.

This was an unsatisfactory and short-lived provision. It was contrary to international agreements; it annoyed the companies. Without bringing any obvious benefits to Zambia, it did not appear to increase investment in Zambia of mining profits, but instead led to accumulation by the companies of liquid assets in Zambian banks which to some extent were lent overseas by other channels; and it led to confusion and dispute over what definition of profits should apply and consequently to delays in dividend payments (Martin, 1972:147–8).

The most dramatic changes, however, were announced in broad terms in the President's Matero speech of August 1969. A new tax system had been worked out for income tax and 'mining tax' with abolition of the royalty and export tax. The state would require any mining company seeking a prospecting licence to offer it a chance to purchase 51 per cent equity in any new mine. When the exploration was done and an application was lodged for a mining licence, the state would decide whether to participate; but if it did so decide it would pay a 51 per cent share of the exploration costs, and it would also pay 51 per cent of the capital required for the development of the mine. At the discretion of a company, its prospecting expenditure might be offset for tax purposes either against its own subsequent mining income or against the current income of its shareholders (Gordon, 1972:190). In addition, the government wanted the companies to give the state a 51 per cent share in existing mines, to be paid for at a fair estimate of book value. At the same time, the 50 per cent limit on remittance of profits would be abolished (Martin, 1972:154–6).

The precise terms of the partial nationalization of the copper mines were worked out by negotiation with the two groups of companies involved, the Anglo-American and Amax subsidiaries, by October of the same year. The companies would lend the government the sum required to buy the equity, and the government (through its holding company) would repay the loans at 6 per cent interest over a maximum of twelve years in the case of what became NCCM (the Anglo-American-related company) and of eight years in the case of what became RCM (the Amax-related one). Against the initial opposition of the companies, the assets were valued at book value with only slightly upward adjustment in the case of RCM. The faster repayment was also put forward as some compensation to the Amax group, which was thought to have been the more disadvantaged by the book valuation. The repayments were to be made in equal half-yearly instalments, except that, if dividends received by the state in a half year were sufficiently far above the scheduled repayment, part of the excess should be added to the amount repaid.

The agreements required the international companies to provide certain services including those of management. Fees were specified for these services. The agreement gave the companies' directors, who would be in a minority on the respective boards, a veto over a number of classes of decision. The private partners were required to hold at least 20 per cent equity in the respective operating companies (NCCM and RCM), but this requirement was modified in 1972. Broadly, breach of the agreement by either party would relieve both parties of their obligations except that the state would then have to repay immediately the outstanding part of the loan. (These details are summarized from Commonwealth Secretariat, n.d.:7.21–6; Martin, 1972:176ff.; O'Faircheallaigh, 1980:188–92.)

The new system of tax for mining in the wake of the Matero speech consisted of income tax and mining tax, both assessed on profits. The income tax was to be assessed on the balance of profits after deduction of the mining tax and was to be at the standard rate of 45 per cent. The mining tax was to be at different rates for different minerals, ranging from 10 per cent on gold, cobalt, and some others to 51 per cent on copper. The system, however, had a progressive element: companies whose only mining interests were 'new mines' were to be able, at intervals of three years, to recover from the state so much of the value of mining tax paid (to a maximum of the full amount) as would be necessary to ensure that the profit (after mining tax and income tax) did not fall short of 12 per cent of the value of equity. In other words, there was an assurance that mining tax alone would not cause after-tax profits (average over a three-year period) to fall to less than an annual rate of 12 per cent on equity. Because income tax is assessed on profit after mining tax has been deducted, its rate of 45 per cent implies that for copper mines (which pay mining tax at 51 per cent) income tax collects only a little over 22 per cent of profits, and the later remission of mining tax for low returns does not alter that figure. Thus, effectively, taxes on profits may vary for new mines between 22.05 and 73.05 per cent. The principle is similar to that of the progressive profit tax. In addition, however, under the 1969 arrangements capital expenditures could be deducted immediately for purposes of both mining and income tax, and any resulting losses could be carried forward indefinitely. The government also had power to exempt a mine from the mining tax in a particular case, but, as Gordon has pointed out (1972:194), this could lead to a higher rate of tax than without the exemption provided the firm would otherwise have enjoyed a substantial part of the rebate of mining tax. The arrangements agreed with the two major copper producers guaranteed that their overseas (non-citizen, non-resident) shareholders would not be subject to any tax until the bonds on which the government had borrowed to buy its shares were repaid, and there was for the time being to be no tax on any mining dividends paid overseas.

These arrangements for 'new mines' had about them elements of the Brown tax (in the provision for the government to buy equity and to contribute virtually pound for pound to capital outlays and exploration outlays), of the resource rent tax (in the immediate deductibility of all expenditures), and of the progressive profit tax (in the protected annual rate of return). Provisions for NCCM and RCM were somewhat different, lacking the progressive profit tax element. The scheme for the government to borrow in order to buy equity and to repay the loans out of its dividends at an interest rate could be described as a form of carried interest. If it had been used in association with a mine development investment and the interest rate had reflected the investor's discount rate and dividends paid to the state had been wholly and exclusively used to repay the debt, the effect of the set of provisions would have approximated closely to those of a resource rent tax, and the form used may be a not unreasonable one for an old mining enterprise that is agreed to have covered its investments already at an adequate rate of return. On the face of it, then, the 1969 system might have seemed not unattractive to new investors and not grossly unfair to old ones.

The government's sense of grievance which led to the next set of changes seems to have been connected with the view that the major companies reinvested too little in Zambia, that there was too little new mining activity, that there were insufficient guarantees for Zambianization of staff, and broadly that Zambia still had inadequate control of the industry. These general concerns were expressed by the President in a speech given on 31 August 1973 (Commonwealth Secretariat, n.d.:7.38–40). Such worries were understandable but not necessarily remediable in any satisfactory manner.

A change had already been made, in that from April 1972 a dividend-withholding tax of 20 per cent had been added. This could not be imposed on the two major companies until their bonds had been repaid. Consequently, arrangements were made with them after the President's August 1973 speech for the state to repay the outstanding debt quickly and consequently for changes to be made in a number of the provisions agreed upon in 1969. The two companies, NCCM and RCM, were to be free to make their own arrangements over management and other services without being tied to use of the parent multinationals; the veto of the minority shareholders was ended; the dividend tax would be imposed; and immediate depreciation of capital expenditure was replaced for the time being by the rates of depreciation that had operated for various types of investment before 1970.

Optimality of Rent Taxes

Obviously, an important motive in Zambia's developing policy towards the mines was the attraction of a sense of national control. Possibly also there was ingrained suspicion of the mining companies' good intentions. For the present purpose we leave aside whatever subjective benefits may have arisen from satisfying these impulses and turn to what would usually be treated as the economic effects.

The pre-1969 tax system was highly unsatisfactory by the standard of neutrality and hence, prima facie, of efficiency and longer-term revenue prospects. The reforms of 1969–70 conform apparently to some of the principles espoused in this book in that they were closely related to profitability and took some account of required returns. However, the reliance on *ad hoc* arrangements with frequent renegotiation ran counter to the recommendations on general and stable rules in chapter 6. In addition, the version of progressive profit tax applied by Zambia had unsatisfactory features: the suitability of taxes of the progressive-profit and resource-rent type depends on their having appropriate thresholds. To have very high rates with inappropriate thresholds can be highly distorting. As has been shown, with a progressive profit tax the threshold must be higher than the investor's required rate of return if negative effects on investment are to be avoided, and the shorter the life of the mine (or of the investor's time horizon for recovering his capital with interest), or the more irregular the cash flows, the more the threshold would need to exceed the required rate of return. The 12 per cent 'threshold' (unindexed) is almost certainly lower than most new investors would have required even with the comparatively stable price conditions that obtained in 1970, and their requirements are very likely to have risen with increasingly prevalent inflation.

The effect of the 1969–70 tax system on the eligibility of investment is explored by Gordon (1972:192–8). He shows, for example, that if the investor required a 15 per cent return within ten years from an investment financed entirely by equity, in which capital expenditure was spread evenly over three years and thereafter for the next ten years there was a constant flow of annual profit (gross of any depreciation), the annual rate of gross profit to capital expenditure that would just meet this requirement would be 23 per cent in the absence of tax, but 42 per cent under the 1969–70 tax regime (even on the most generous interpretation of the conditions for mining tax refund). It could be argued that an income tax at 50 per cent would bring about a similar gap between pre-tax and post-tax requirements of profitability, and in a case 2 country (see chapter 7) it would be desirable on grounds of neutrality for such a tax to apply to mining in order that pre-tax profitability requirements should be inflated by tax in mining as they are in other industries. Yet Zambia is not a case 2 country, and, if foreign investors would be prepared to invest in Zambian mines for a return of 15 per cent, Zambian national income and potential revenue would be reduced, *ceteris paribus*, by taxing in such a way as to raise the requirement to 30 per cent. A regime approaching neutrality in the appropriate sense for a case 1 country would perhaps dispense with the regular income tax element unless the requirements of overseas creditability overrode other considerations. Then, if an RRT or PPT were applied it should be with appropriate thresholds, representing somewhat more than the rate that the investor was likely to require, and, in the case of the PPT, the shorter the investor's required pay-back period, the higher the threshold would need to be. Gordon, in the same passage, making what seemed in 1972 not unreasonable assumptions about costs and receipts for underground extraction of Zambian copper, shows that the tax might raise the minimum profitable grade of copper from 2.83 to 4.09 per cent, and he notes that none of the mines on the Copper Belt regularly showed average grades over 4.09 per cent. Thus it seemed very likely that the tax system would be at least sufficient (with other existing conditions) to prevent new investment in copper mining. Whether new mines would have been opened by private investors on any reasonable conditions is another question. To add to the sources of discouragement as they appeared in 1971–2, the copper price in the 1970s was to undergo drastic fluctuations which were not an inducement to confidence.

The state-ownership conditions proposed for new mines were probably harmless to general investment incentives and must even have increased them by reducing risk, so long as a firm was really convinced that that state would contribute 51 per cent of expenditure on exploration and 51 per cent of the capitalization of mine development and operation in exchange for its 51 per cent shareholding. But it would be understandable if there were some doubt on this score. The ownership provision, for new and old mines alike, raised the maximum marginal rate of positive cash flow that might go to the state to about 82 per cent, which was high but probably not high enough to pose a serious threat of wasteful management as a result.

The changes of 1972 and 1973 served, if anything, to aggravate the disincentives of the Matero system. The imposition of a 20 per cent dividend withholding tax raised the maximum marginal rate of cash flow going to the state to about 86 per cent and the overseas shareholders' maximum rate of tax on income from their investments to about 78 per cent even if the government paid its full share of capital expenditures. The abolition of immediate depreciation again made the tax system more severe and tended if anything to increase risk of loss.

However, a disincentive to new private mining investment, which would probably have been conclusive even if the tax system eventually fixed had been much more favourable, was the uncertainty produced by frequent changes of the rules and by doubt from time to time as to what the rules actually were. We have recounted here four important sets of changes in just over seven years. There were also periods in which major questions about the tax system were left unanswered—for example, after the time of the Mulungushi declaration, doubt over the remittance of dividends and the mode by which royalty and export tax were to be replaced; and after 1973, over the rules for depreciation. Martin (1972:131–2) records that RST (the Amax company) in the mid-1960s would have gone ahead with at least two new mine developments but abandoned the idea because the government delayed in removing the royalty system. Added to this, the mining companies could have been excused if they had regarded the attitude of the government as needling. They may have felt themselves the prey of unpredictable political forces.

Altogether, it is hardly surprising that no new mines of any consequence were started on private initiative. If there had been any serious prospect, on grounds jointly of copper price, Zambian factor prices, and the nature of Zambian deposits, for new investment, the incentive to provide it might have been preserved if a stable system for new mines could have been announced at the time of Mulungushi, say, and maintained thereafter—a system involving only a Brown tax or a resource rent tax with a suitable threshold or one of their variants, though possibly disguised as a means for acquiring equity. An arrangement like the one announced at Matero for 'buying in' on equitable terms might have been quite tolerable if combined with one of these taxes, or a combination of them, alone, provided the stake left to the private shareholders had not been too small to interest them.

On the most sober calculations, however, there would have been a dilemma for the Zambian authorities that would not have arisen as acutely, where it existed at all, for the other countries considered here. Should they construct their tax system to encourage new mines, or should they milk the old ones to the limit? Some distinction, as we have shown, was made between the two groups, but in form the same system applied to both. Perhaps greater divergences were appropriate, though an RRT could largely have avoided the dilemma or considerably reduced it. It does seem that Zambia has veered to the side of milking the old companies. As we have suggested, there may well have been no new privately initiated mines, however well constructed and lenient the taxes to which they were subject. But if the country were to have remained open to the possibility, a different system of taxation, designed to last, would have been necessary.

We shall not consider Zambia's stabilization problems except to mention that with the heavy dependence on one product, and one with a highly volatile price, they have been extreme. To preserve some stability in the country's growth path the authorities would have needed to be far more conservative in the late 1960s than in fact they were. There was indeed a reluctance among the authorities to spend to the limit during the boom at the end of the decade, but to come close to stabilizing government expenditure they would need to have estimated with some accuracy the average copper price over a period when it was in fact extremely erratic, and to have behaved with extreme restraint in times of high prices. The authorities certainly did not meet these tests, and the forced reductions in expenditure in the mid-1970s imposed huge costs on the Zambian economy. These costs are testimony to the doubtful benefit of large increases in mineral revenues that are not used within a suitable stabilizing framework.

Botswana

Botswana is a land-locked country bordering South Africa. It is the size of France but has a population of less than 1 million. The very low population density suggests a high per capita mineral endowment, but at independence in 1966 the mineral industries were of little importance. At that time Botswana was heavily dependent on foreign aid to cover recurrent public expenditure, and despite high per capita foreign aid had one of the lowest per capita incomes in the world: US$50. Since then, Botswana has enjoyed rapid growth in domestic product, at about 10 per cent per annum in real terms, so that in 1980 per capita product at US$940 was in the middle rank of developing countries. It has at the same time significantly reduced its dependence on foreign aid. This economic progress since independence has been achieved mostly through the agency of the mining industries. (This paragraph, and this section of the chapter, have made considerable use of Johnson, 1981.)

The value of mineral production has increased from less than US$1 million at independence to over US$300 million in 1979. There are two diamond mines, both operated by a subsidiary of De Beers Consolidated Mines. The Selebi-Phikwe nickel–copper mine is operated by the US company Amax and by the Anglo-American Company, the latter being closely linked with De Beers Consolidated Mines in senior management and ownership. A small mine operated by Anglo-American produces coal for domestic electricity generation. In 1980 mining contributed about two-thirds of the value of exports and about one-third (directly) of both government revenue and gross domestic product.

Mineral exploration is undertaken by private firms on prospecting authorities that grant no right to mine any mineral deposits that are discovered. Prospecting authorities are granted, on application, to firms that are judged to have adequate technical, financial, and marketing capacity. Fiscal terms are negotiated *ad hoc* after the discovery of a commercial mineral deposit. The fiscal terms so negotiated have been unstable: there have already been two major renegotiations. There has been considerable government expenditure, largely financed by foreign aid, on basic geological work in recent years. Uncertainty about fiscal arrangements is probably one reason why there has been relatively little private exploration other than for diamonds. Private exploration for diamonds, however, has been maintained at a high level, mainly because of the highly prospective geological environment together with De Beers' anxiety to maintain its world-wide diamond monopoly in the face of challenges from possible new entrants. The search for diamonds became much more active when a second company, Falconbridge of Canada, was issued with authorities to prospect for diamonds, so that two companies were effectively racing each other to locate any remaining large deposits of diamond-bearing mineral.

In addition to generating uncertainty in relation to private exploration investment, the *ad hoc* negotiation of fiscal terms appears to have led to delays of up to a year in the development of new mines. However, given the high and apparently adequate rate of growth of mineral production, it is hard to argue that slight delays in some mineral developments have imposed a heavy cost on Botswana.

The fiscal regime that has been established for diamonds operates to split profits between the government and the investor in the approximate ratio of 75–25. The regime operates as an HRIT, the rate of which has been adjusted to leave the company with an internal rate of return equal to what is thought in the process of negotiations to be the minimum rate expected *ex ante* that is consistent with a decision to invest. There was one major renegotiation of fiscal terms to yield more revenue to the government in 1974 when *ex post* returns to the company were judged by the government to be higher than were required *ex ante* to induce a decision to invest. If such adjustments continued to be made to established fiscal terms, making the terms symmetrically more and less severe as pre-tax profitability rose above or fell below what had previously been expected, then the effect *ex post* would be similar to RRT with a tax rate of 100 per cent.

Investors in mining other than for diamonds have been required to pay a 3 per cent *ad valorem* royalty on sales revenue, normal income tax (35 per cent in 1980), and normal dividend withholding tax (15 per cent). In addition, the private investor has been required to make 15 per cent of the equity available free to the government. The established nickel–copper mine has been a financial disaster, so that the government has received almost nothing from dividends, income tax, and dividend withholding tax. The original fiscal terms had provided for a 7.5 per cent profit-based royalty, and this too yielded no revenue until renegotiations in 1978 replaced it with the *ad valorem* levy. The renegotiated fiscal system now has the effect of an HRIT at about 53 per cent combined with a deductible 5 per cent *ad valorem* royalty, except that payments to government as dividends on the free equity would not be credited against investors' home taxation commitments.

The mineral fiscal regimes in Botswana are in conflict with what has generally been recommended in this book because they have been established in late agreements, *ad hoc*, and unstable. One clear cost of this has been low levels of exploration other than for diamonds. Stability and early agreement under general rules is not required to maintain high levels of diamond exploration, but investors may doubt the stability of arrangements for other mining if the terms of diamond mining are subject to periodic renegotiation. The government has not won spectacularly favourable terms from late agreement: the government share of benefits has been little different from that which would have been obtained under the types of general rules applied in Papua New Guinea. If HRIT or RRT were applied generally in Botswana, the rates could be higher for diamonds in recognition of the government's highly favourable position in that industry. The general imposition of HRIT or RRT would remove a further undesirable feature of the *de facto* fiscal regime for diamonds: the reduction of *de facto ex post* marginal tax rates below 100 per cent would restore incentives to economization in mineral production. But this having been said, the regime applied to diamonds has not stood in the way of rapid development of and high revenues from diamond mining, and no sensible arrangements could have yielded substantial revenue from unprofitable nickel–copper production. The main costs of the non-optimality of Botswana's mineral fiscal system lie in the undue postponement of exploration and the consequent delay in the exploitation of some of the deposits that very probably remain to be investigated.

Tanzania

Tanzania is a lightly populated country with prospective geology, but the mining industries have been relatively unimportant in total economic activity. Minerals contribute less than 10 per cent of the value of exports and less than 1 per cent of gross domestic product, almost all diamonds from one mine.

This mine was built on the world's largest deposit of diamond-bearing mineral, but the deposit was substantially depleted by the time of independence and production has declined steadily since then. At independence in 1963 it was fully owned and operated by a subsidiary of De Beers, but soon afterwards the government acquired 50 per cent of the equity for a negotiated sum based on book value. Diamond sales are subject to an *ad valorem* royalty of 5 per cent, as well as the country's high standard corporate income tax, currently over 50 per cent, and dividend withholding tax at 15 per cent.

The diamond tax regime yields a high proportion of profits to the government, but the advanced depletion of the deposit has meant that there was little cause for concern that major investment that would otherwise have been undertaken at the established mine would be deterred. The relatively high *ad valorem* royalty possibly deters some high-cost recovery of diamonds now in the later years of the mine's life. Tanzania has in place an elaborate system of royalties for all minerals, which has been of little practical significance because of the absence of other mines. Specified royalty rates tend to be high, and vary from mineral to mineral (see Commonwealth Secretariat, n.d.:6.13 and 6.14 for a detailed description). Some are specific and some *ad valorem* charges, while some are progressive with the ratio of annual profits to annual sales value. (The latter formulation is quite different from the PPT, because there is no consistent relationship between capital investment in a mine and the annual accounting cost of operating a mine.) Mining investors would also be subject to the normal high corporate income tax, and to normal dividend withholding tax.

The severe general fiscal rules may have contributed something to the almost complete absence of foreign investor interest in hard minerals in Tanzania, but there were other, more immediate causes. Following the nationalizations of foreign enterprises in the 1960s, and especially the Arusha Declaration of 1967, it was reasonable for foreign investors to doubt whether they were welcome in Tanzania. Persistent balance of payments problems and tight exchange control raised questions about remittance of profits and recovery of investment in foreign exchange. And any investor who had successfully negotiated these psychological hurdles would have then had to find his way through outmoded colonial mining legislation and rusty mining administration.

But the possibility that income from mining could loosen the tight constraint that chronic balance of payments problems imposed on development caused the Tanzanian government to review mining policy in 1977. In late 1977, the Minister for Mines in a budget speech announced that the government would seek to enter joint ventures with foreign investors for mineral exploration and development. Foreign experts under Commonwealth Secretariat auspices provided technical assistance in the revision of the range of mining, taxation, and other policy relevant to the encouragement of joint ventures with large foreign companies.

The first negotiations with foreign investors under the new policies were conducted through 1978, and resulted in a major exploration agreement with the German company Uranertz. The general shape of the agreement can be ascertained from public sources, and it is clear that the terms were exceptionally favourable to Tanzania. The company seems to have accepted the normal taxation arrangements: a low *ad valorem* royalty; 51 per cent government equity as required under Tanzanian national policy, with the private investor carrying the government through the exploration phase; and two tiers of RRT, with a combined effect that would yield more revenue from similar intra-marginal projects than the Papua New Guinea RRT arrangements.

Far from having to re-enter the market for large-scale mining investment through offering exceptionally favourable terms to the investor, as suggested by the 'North American' model, Tanzania in its first major agreement was able to secure terms that were far more favourable to itself than those available to governments in the main established hard-mineral producers in Latin America and Africa. One reason is that the Tanzanian government was able to offer a high degree of political stability and credibility: it had a good record of honouring commitments that it had entered. A second reason was that the government had itself undertaken considerable basic exploration and had located mineral deposits that seemed to justify a much closer look. A third and probably the most important reason was that the company had access to finance for exploration on highly favourable terms from its home government, which greatly reduced its exposure to risk in the exploration phase of mineral development.

Thus the Tanzanian government was able to take advantage of some special characteristics of one potential investor to achieve more favourable terms than could normally be written into general rules with confidence that they would be consistent with high levels of investiment. This early success creates a dilemma for the Tanzanian government: should more foreign investment be required (and it amost certainly will), it may be necessary to introduce less severe general rules to which the first agreement is an exception, or to negotiate *ad hoc* new agreements that are less severe than the first.

Summary

From the considerable experience of African Commonwealth countries with the hard-mineral industries we have drawn from three countries whose mining industries are at very different stages of development: Zambia's massive mining industry, which is unlikely to grow rapidly simply because of its established size; Botswana's mining industry, now very important in the national economy after more than a decade of rapid growth; and Tanzania's, possibly at the beginning of a period of expansion. All have relied upon *ad hoc* provisions that have in practice been unstable. None has had much broadly based private exploration, and this could possibly be seen as evidence of failure in policy. But each has been able to exploit special strengths in its circumstances to obtain favourable terms from investors in parts of the mining industries. In Botswana's case the special arrangements probably have not been more favourable than could have been written into general and stable rules. In Zambia, similar outcomes from established mines probably could have been obtained, given more foresight, from less unstable arrangements. But there is at least a case for advocates of general rules to answer in the experience of the African Commonwealth countries.

17

TAXATION OF OIL AND NATURAL GAS IN THE UK

Minerals in the UK Economy

Offshore oil and gas discoveries from the mid-1960s presented the UK with a new set of opportunities and dilemmas. Private mining had been of no great interest in the country for the two decades that had followed the postwar nationalization of coal. Mining for metals had long been vestigial.

Coal remains a major industry and source of energy in the UK. It employed (in collieries) 233,000 in 1979; the number has fallen fairly consistently, from 724,000 in 1948, but the size of the work-force and the fact that many villages and small towns have depended on coal for their existence mean that local human problems still bulk very large in policy decisions over the industry. Nationalization of coal, though it appealed in principle to the Labour government that effected it in 1947, was partly motivated by the need to ameliorate the social dislocation that arose when pits declined below the level of commercial viability, and the continuance of nationalization has not been seriously challenged. Because of the unpleasantness of underground mining and the risks and deprivations of mining communities in the past, coal miners touch a soft spot in the national conscience. Their bargaining organization is also good. Coal itself is an item of mass consumption, directly and indirectly, and its price is potentially a sensitive issue. Exports and imports of coal are normally not important. These factors have combined largely to rule out the possibility that economic rent should be collected fiscally from the nationalized industry, though with present high energy prices this may not continue to be so. Coal, despite its importance, will thus be ignored in what follows and attention will be directed to oil and gas.

Legislation on oil drilling had been passed in 1934 vesting ownership of all mineral oils in the Crown, and since then small oil and gas finds had been exploited commercially on-shore in various parts of England and Scotland. By the early 1960s about 1,500 barrels of oil a day were being produced (Hamilton, 1978:12). This, however, was nothing to what was to come, and represented little more than a thousandth of the rate that production was to reach in 1979.

Offshore drilling in the North Sea began in 1964, when the government first issued licences for it. Commerical gas production began in the southern part of the sea in 1967, and reserves discovered were so large that the nationalized gas industry proceeded to convert the whole of its supply from manufactured to natural gas. The first attractive discoveries of offshore oil were made in the Montrose and Forties fields off north-east Scotland in 1969 and 1970, and commercial production of offshore oil began in 1975.

Britain became self-sufficient in oil in 1980 and may easily be a net exporter over much of the 1980s, the only one of the large industrialized countries other than the Soviet Union to be in this position Production of natural gas in the late 1970s had settled at around 40 billion cubic metres per year (36 billion in 1982), which is the energy-equivalent of about 32 million tonnes of crude oil. Production of oil was 54 million tonnes in 1978 and 103 million tonnes in 1982. In 1978, sales of oil realized £2.8 billion and the value of gas extracted in oil-equivalent prices was about £2.0 billion (Department of Energy (UK) 1979). In 1979 consumption for all purposes was 94 million tonnes, and for domestic energy use 84 million tonnes, in crude-oil-equivalent terms. UK use of oil for domestic energy had been about 95 million tonnes a year before the Yom Kippur war but had fallen to something like 65 million tonnes by 1982.

In 1980, after the oil price rises of that year and of 1979, actual gross revenues to the extractors from oil and gas production were expected to be about £10 billion in current prices, and the government was expected to take about £3.5 billion in various forms of tax (of the order of 2 per cent of GNP) (*The Economist*, 29 March 1980:35). In fiscal 1981–2, government revenue from offshore oil and gas amounted to £6.4 billion (roughly 3 per cent of GNP at factor cost).

Total UK oil reserves have been thought to be in the region of 2–5 billion tonnes (14–35 billion barrels), though there must be great uncertainty, especially at the upper end of the range. Gas reserves were estimated in early 1977 as about 90 trillion cubic feet or about 2.5 trillion cubic metres (Hamilton, 1978:55), equivalent to about 2 billion tonnes of oil. That has more recently (1981) been cited as the upper limit of gas originally present; and end-1980 figures of reserves still available are given as 0.75–2.15 trillion cubic metres. Thus at the 1980 rates of consumption, if there were no net exports, UK reserves might maintain the country in a state of net self-sufficiency in oil for twenty-five to sixty years, and in natural gas for twelve to forty years, more. Earlier projections, however, supposing a rise in oil consumption and a gradual decline in output and substantial net exports in the 1980s, had predicted a renewed net import of oil by 1990 or before.

Officially, the oil and natural gas industry is estimated as being responsible for the creation of about 100,000 jobs, slightly more than half of them in Scotland. This would represent about 0.5 per cent of the UK work-force and 2.5 per cent of that of Scotland. Such estimates, however, inevitably ignoring both the downward effect on employment resulting from any oil-related rise in real wages and also the upward effect on employment resulting from the greater relaxation in spending that oil earnings have made possible, give only part of the picture. In contrast with the mining industries of countries such as Indonesia, Zambia, and Papua New Guinea, British oil and gas production tends to stabilize the economy, even in the absence of corrective action. This is because it brings the country close to net balance on the world's most important traded primary commodity and hence neutralizes the impact on the terms of trade of the drastic changes in that commodity's price.

As in Papua New Guinea, mining developments have had a considerable political impact, especially in their encouragement of separatism. The Scottish National Party (SNP), which advocates independence for Scotland, made much use of Scotland's oil in its propaganda of the mid-1970s. The Party raised its parliamentary representation from one to seven and then to eleven in the two general elections of 1974 when oil euphoria was probably near its zenith, and it continued (on the evidence of opinion polls and local elections) to rise in popular support until 1977. Nationalist gains in turn were undoubtedly important in prompting the devolution bills for Scotland and Wales which would have set up elected parliaments in both nations and a responsible executive cabinet in Scotland. These bills occupied a large part of the time of the 1974–9 Parliament, but the resulting acts were not brought into effect since referenda in March 1979 gave a conclusively negative verdict for Wales and an inconclusively positive one for Scotland. In Shetland, too, an autonomist movement is undoubtedly a reaction in part to the immense changes that oil has brought to the islands.

The gas and oil bonanza also raised issues for British, as against Scottish, nationalism. Was it necessary to offer open competition to foreigners for extracting the product and for providing the equipment and services? To what extent should British operators be favoured? Nationalist feeling probably helped to determine the way in which blocks have been made available to investors. Entwined with this nationalist issue was that of public ownership, which offered on the face of it one way of exercising public control. The dependence of the Labour governments of 1974–9 on the support of trade union leaders for maintaining their wage policy, and the unpopularity of that policy with their own party's left wing, probably pushed them farther along the road to public ownership than they would otherwise have gone. Over these matters it can be said that politics has impinged on oil rather than oil on politics.

Treatment of economic rent has differed between oil and gas. The economic rent of gas has until recently been largely diverted to those lucky enough to be domestic gas consumers, and it is likely that some has also been dissipated through the 'high-grading' that results from low gas prices; it is possible that some has gone to the British Gas Corporation. In oil, on the other hand, there has been a serious attempt to divert a large share of the economic rent to the Treasury through the combination of world market prices and the tax system that, since 1974–5, has applied to both gas and oil: a moderate and 'removable' *ad valorem* royalty; corporation tax of the normal kind; and an additional profit tax, the Petroleum Revenue Tax (PRT), which has a loose relationship with profitability. A further, ostensibly temporary, levy, the Supplementary Petroleum Duty, was added in 1981.

The Institutional Framework for Mineral Development

Allocation of Leases

The first allocation of offshore leases in 1964 was hurried through by the Conservative government so that it could be completed between the passage of the Continental Shelf Act and the impending general election (Hamilton, 1978:19). The UK continental shelf was divided into rectangular blocks each about 90 square miles in area. Exploration licences, tied to no particular area, were issued, each to run for three years, and production licences, each for one or more blocks, were issued to run initially for six years. After six years a certain part (generally half) of the area allocated for production to a licensee under one licence had to be given back, but the lease on the remainder could be extended for a further forty years at the licensee's discretion (Sibthorp, 1975:179). A licensee could not assign his licence to another person without the Minister's permission, and the authorities aimed to prevent the emergence of a market in licences (Dam, 1976:47–51). Where more than one licensee had blocks in one field, the Minister was empowered to impose an agreement between them on co-operative extraction.

There was a large element of official discretion in determining who got which block and on what terms. Application was open to any UK resident or UK-registered company 'that could prove a reasonable degree of technical and financial competence' (Hamilton, 1978:18), but it was made clear that preference would be given to companies with refineries in Britain and to those prepared to commit themselves to the most exploration and investigation. There was thus in practice an element of preference for British companies, an element of bidding by 'work commitment', an element of preference for the oil majors (which were more likely to have refineries in the country), and a refusal to lease to purely financial companies unless they formed consortia with companies that had experience in the industry.

Hamilton (1978:17ff.) attributes the choice of a discretionary system to the desire for speed in exploration and extraction and for substantial British participation, and to the fact that both Treasury and the then Ministry of Power relied on the co-operation of Shell and BP, which favoured discretionary allocation over auction. The discretionary system has persisted through seven licensing rounds (to 1980), except that in 1971 15 blocks out of the 282 allocated in the Fourth Round were auctioned as an experiment, realizing between them £34 million in successful bids. There has been public advocacy of an auctioning system, but the cause has lacked sufficient political force to counter the natural preference for discretion on the part of officials and of politicians from both major parties.

A peculiar feature of the system has been that the 12½ per cent *ad valorem* tax or royalty on oil and gas has been imposed as part of the agreement with each licensee, and the policy since 1974 has been that the Secretary of State may at his discretion absolve the licensee from paying it and so avoid the 'high-grading' effect that the royalty might otherwise introduce in certain wells at certain stages.

Pricing Policy

Offshore gas depends mainly on fixed submarine pipelines to convey it to market. These facilities are very costly; so is the alternative of freezing the gas and conveying it by tanker. Hence the destination of natural gas from any area is determined heavily by proximity. Oil, which can much more readily be transported by ship, is much less constrained as to its destination.

Legislation has obliged operators to deliver their oil and gas in the UK, except with the permission of the Secretary of State, but otherwise oil may be sold to anyone or used at the operator's discretion, while gas may not (as of 1981) be delivered to premises in the UK or used by the operator himself in the UK, except with the Secretary of State's permission, and then only for non-fuel purposes and after it has been offered to the nationalized gas industry (gas boards and later the British Gas Corporation) on reasonable terms (Sibthorp, 1975:176). This provision has in fact allowed British Gas to operate as monopsonist for the extracting companies and monopolist for the domestic users of gas as a fuel and has thus enabled it to determine the price. In October 1981, however, the government's intention was announced of ending British Gas's right to be sole buyer, and also of ending its right to be sole seller in the industrial (as distinct from the domestic) market.

The method of setting the price of gas, and the level at which it should be set, were the subject of much argument over the years 1966–8, with the operators wanting the price to be related to its market value, which presumably would bear some relationship to the prices of other fuels, and the Gas Council and government wanting a percentage over cost. The companies eventually accepted an arrangement by which the price was set much lower than they had asked and was varied somewhat according to the supposed costs of the different fields (Hamilton, 1978:22–4). The price selected made North Sea gas very attractive to domestic consumers and made it acceptable for the whole of the British gas distribution network and domestic gas appliances to be converted over about ten years to using it in place of manufactured gas. The price remained low in relation to that of other fuels until 1980, since when moves have been made to raise it closer to what are regarded as competitive levels. Gas bought by the Gas Corporation over and above the amounts whose purchase the Corporation guarantees receives a substantially lower price. This 'valley' gas is mainly sold outside the winter months (Dam, 1976:88).

Oil by contrast has been sold at market prices. Despite the formal requirement that it be delivered on-shore in the UK, operators have been free to sell overseas, and the type and quality of oil extracted has made it economic to sell some British oil in countries for whose refineries it has a comparative advantage even while the UK was a net oil importer.

Taxation

Receipts from gas sold to the British Gas Corporation under any contract made before June 1975 were subject only to corporation tax at the normal rate and to a $12\frac{1}{2}$ per cent *ad valorem* royalty or tax on output, the figure being chosen apparently as an 'average' world rate of royalty and as being the minimum rate applied to federally owned oil and gas in the USA. Gas, extracted under later contracts, and oil extracted after November 1974, have been subject in addition to Petroleum Revenue Tax (PRT), which was defined in the Oil Taxation Act 1975 (c.22). From the March 1981 budget oil and gas were subject in addition to Supplementary Petroleum Duty, which was originally to run until June 1982, and later had its life extended until December of that year.

For purposes of corporation tax, operators have been able to offset against profits on one oil or gas field their losses on other fields within the UK's jurisdiction, but (since 1974–5) not losses on other activities (HMSO, 1974:8). This restriction has become known as the 'ring fence'.

The precise form of the PRT derives from arguments between the government and the oil companies that went on between October 1974, when plans for a similar but more rigorous tax on oil and gas profits were communicated to the oil companies, and February 1975, when the final form of the proposals was published. The form first outlined appears to have been essentially a single-rate additional tax on net positive undiscounted cash flow. In response to pressure from oil companies and other interests connected with the industry, this was seriously modified between November and February by providing a number of offsets that would have most effect on the tax due on small fields, and by limiting the tax taken by reference to the annual 'rate' of profit on accumulated investment. The tax as it emerged is thus extremely complex. Patching, rather than radical reconstruction, seems to have been the response to the inadequacies of the original simple form.

The taxable unit for the PRT is the operations of a particular licensee within one field. PRT, unlike corporation tax, does not in general permit expenditure on one field to be offset against receipts from another. There is, as it were, an inner ring fence for PRT within the outer ring fence that applies to corporation tax. It is allowable, however, for a licensee to offset the abortive expenditure or unexhausted allowances (other than the 'oil allowance') that he has generated in one field on the UK continental shelf against the profits of another field, for PRT purposes.

The PRT is perhaps best seen as two taxes assessed on different bases, the lower of which applies in any particular case. The first is in principle a tax on positive net undiscounted cash flow within each six-month period, losses being carried forward indefinitely and in some cases backward. The rate of the tax was originally set at 45 per cent but was raised to 60 per cent in 1979, and to 70 per cent in 1980, and to 75 per cent from the end of 1982, in response to the OPEC price rises of that time. Apart from all actual expenditures, both capital and recurrent, within the field concerned and abortive exploration expenditure (otherwise undeducted) incurred anywhere in the UK sector, the taxpayer can deduct an additional 75 per cent (35 per cent from 1979, making 175 per cent, later 135 per cent, in all) of most pre-production expenditure (other than that on fixed assets) and of post-production expenditure that in general increases the field's productivity. The taxpayer can also deduct an 'oil allowance', which reduces the 'assessable profit' (on which tax is calculated for each licensee in any six-month period) in the proportion that the value of 500,000 long tons (250,000 tonnes from 1979) of oil bears to the value of the field's output in that half-year (or reduces it to zero, whichever is the lesser reduction). The value of 500,000 tons (or later 250,000 tonnes) or of total output, whichever is the less, is described as the 'cash equivalent' of the oil allowance for the field for the half-year. Oil allowance ceases to be deducted, however, when 10 million long tons (5 million tonnes from 1979) have been 'deducted' from the field's output in this way over its whole life.

The 'first' tax is in principle somewhere between an HRIT and an RRT in the sense of these terms introduced above in chapter 7 and is applied to the 'project' rather than to the 'company'.

The 'second' tax, or 'safeguard', is what we have called a progressive profit tax (PPT), also applied to the project rather than to the company. It is assessed on a larger base than the first tax in that exploration and development expenditures (and the extra 75 or 35 per cent of such expenditures) are not deducted, nor is the oil allowance. The base is called 'adjusted profits'. This tax is assessed at 80 per cent of the amount by which the adjusted profit in a year exceeds 30 per cent of the value of accumulated (but without interest) capital expenditure incurred in the field concerned. The rate of 30 per cent is what we have called above the threshold rate of the tax, but, since PRT is collected before corporation tax rather than after, the threshold would amount to about 15 per cent after tax in so far as corporation tax was actually paid. This tax used an assessment period of a year, not a half-year, but shifted to a half-year period in 1981. It is essentially a safeguard to prevent undue exposure to the 'first' tax.

Interest on borrowing is not a deductible expense for either 'first' or 'second' tax. This follows the logic of their application to projects rather than to companies. The extra 75 (or 35) per cent deduction on pre-production, and other, expenditures was thought of as some kind of substitute for deductions on interest.

This 'second' tax was one of the measures introduced to modify the rigour of the first in its original form. If there were no further pre-production expenditures to be deducted and the oil allowance had been exhausted or represented a small fraction of the field's output, the 'second' tax would be the lower and hence the operative one except at very high rates of profit on accumulated past capital expenditure. Where the oil allowance per period represented a large part of a field's output, or where there were pre-production expenditures (at 175 or 135 per cent of their value) still to be deducted, the 'first' tax would probably be the operative one. When the 'second' tax is operating, the top marginal rate of tax on cash flow could in principle be at least as high as 90 per cent. Under the tax system as a whole, Dafter (1982: 23) estimates that the top marginal rate on income would be 76.9 per cent from 1975–8, 87.4 per cent after the 1979 and 1980 changes, 90.3 per cent after the 1981 changes, and 89.5 per cent when the 1982 changes come into effect at the start of 1983. Average rates for 1981–2 he cites as 85–6 per cent on income.

A well informed observer from the industry has allowed us to insert the following generalized account of how the two arms of the tax would probably operate under conditions obtaining before the changes of 1981. If oil prices were constant and the production profile of a field were symmetrical over about twenty years in the form of rise to plateau followed by fall, the initial years to about year 4 would be free of PRT because of the offset of accumulated pre-production and development expenditures. From, say, years 5–10, the liability would be PRT at the full 'first-tax' rate on cash flow but reduced by oil allowance. Years 11–14 would have reduced PRT on cash flow because the 'second' tax would be the lower and hence take precedence over the 'first'. Years 15 and following would be free of PRT under the 'second' tax. If, on the other hand, oil prices were to rise sharply over a period in which the field had the same production profile, the PRT liability would probably still be nil for years 1–4, full 'first-tax' rate (reduced by oil allowance and using the whole of the oil allowance) for about years 5–15, and progressively reduced by the 'second-tax' safeguard to zero thereafter. (The real costs are assumed to be those typical of North Sea fields in the mid- to late 1970s, and the constant prices in the former examples, and at the starting point in the latter example, are assumed to be the real oil prices of the mid-1970s.)

The Finance Act 1981, c.35, introduced an additional tax called Supplementary Petroleum Duty (SPD), which was to apply to assessment periods over 1981 and the first half of 1982 and was later extended until the end of 1982. The new tax was to be collected 'before' PRT and corporation tax and to be deductible for purposes of both of them. It was basically to be a 20 per cent tax on 'gross profit' as defined in the 1975 Act (which meant something very close to gross revenue), but with an 'exempt allowance' (worked out on the same principle as the oil allowance for PRT) deducted. The exempt allowance on a field would be exhausted for any period or for good when the oil allowance was exhausted. If and when that stage had been reached, the SPD would act like an additional *ad valorem* royalty.

This extra tax was accompanied by various changes in PRT that increased its severity. The 'expenditure supplement', the *extra* 35 per cent of certain pre-production expenditures that could be deducted for PRT purposes, was to apply on a field only until the end of the chargeable period in which net profit first accrued from the field, unless under certain circumstances a period of losses followed the period of net profits. Furthermore, the limit introduced by what we have called the 'second' tax would apply only for the number of periods to which the expenditure supplement applied plus half as many periods again.

The motive for these changes was undoubtedly the government's urgent desire for extra revenue in 1981, combined with a realization that the oil price rises of 1979 and 1980 had provided oil extractors generally with windfall gains and possibly also the belief that a delay in extraction on account of the fiscal charges was not necessarily bad. There was, in response, considerable murmuring from the oil companies and threats from the Offshore Operators' Association that its members might not take up blocks in the next leasing round. The ending of SPD, and its replacement by a higher rate of PRT in the March 1982 Budget, was probably a recognition of the criticisms, encouraged by falling oil prices.

State Participation

Perhaps the most unusual element of the arrangements and the most redolent of political compromise is the British National Oil Corporation. The White Paper of July 1974 (HMSO, 1975) proposed participation by the government as the fairest way of sharing profits, and the British National Oil Corporation (BNOC) was set up under the Petroleum and Submarine Pipelines Act 1975, with allowable functions covering virtually every possible activity commonly carried out by an integrated oil company, though for many of them the Secretary of State's permission was required.

BNOC has become an equity participant in a number of the active consortia. It has been the operator in three fields under development, namely Thistle, Beatrice, and Clyde, and in numerous exploration projects. There has been a condition, in all licenses issued in the Fifth, Sixth, and Seventh Rounds, that there should be what is called majority government participation, either through BNOC or through the British Gas Corporation (Department of Energy, 1977). BNOC bought equity in all projects established in the Fifth Round and took a carried interest in all those established in the Sixth. This practice ceased to be general in the Seventh Round. The rule of majority government participation does not mean, however, that the state corporation is necessarily a *majority shareholder*, but in many cases means simply that it has the right to buy 51 per cent of the product at the ruling market price. This option to buy the oil continued in the Seventh Round. These arrangements over the buying of oil are framed with the intention of having no fiscal effect, that is, of not leaving the private operator any worse (or better) off than if they had not been made. The private operator may be allowed to buy the oil back, and in that case does so on similar terms to those on which the oil has been sold. Until 1979, legislation exempted BNOC from payment of Petroleum Revenue Tax. Commercially, BNOC appears to have been successful, and it could be regarded as having fulfilled one possible purpose of its creation in building up a body of expertise.

The British Gas Corporation, the state's gas distributing monopoly, has acquired licences on blocks and is the operator for the Morecombe gas field. Conservative government plans in late 1981 were to privatize some of these holdings.

The UK government has also had a controlling interest in British Petroleum (BP), one of the major offshore operators, but has not apparently used this holding in any way for a policy purpose.

The peculiar position of BNOC seems to have been the result of a compromise between those that wanted a large government stake in all oil and gas extraction and those who were satisfied with the private operators and did not want to rock the boat. The government planned in late 1982 to denationalize Britoil, which comprised the oil production of BNOC. It had in 1979 announced a policy of increasing private-sector participation in the Corporation's offshore assets, together with various changes designed to reduce the privileged position and quasi-governmental role of BNOC: by ceasing to use it as adviser to government, denying it privileged access to government finance, reducing its exploration activities, and ceasing to make it a mandatory equity participant in new leases.

It can be argued, in favour of the existence of BNOC as it stood in 1979, that it had built up for the state a body of expertise in the oil industry, that it had maintained a capacity for ensuring supply to British users of oil, and that its presence permitted the government, if there were a need, to speed up exploration or extraction or otherwise to influence operations for social purposes. Where it had participated in equity, there was no clear evidence that it had done this in order to reduce the risk for the private partners, but there was always a possibility of its participation for this purpose. On the other hand, the sceptic might ask what the government could achieve by means of BNOC that it could not achieve through its holdings in BP. The logic of the arrangement for compulsory buying of 51 per cent of a company's oil *with no fiscal effect* is also questionable. If this would ensure supply to UK refineries at a time when oil companies would not otherwise guarantee it, the provision must involve some potential sacrifice on the part of the oil companies. In 1979–80, BNOC showed that it was able, through forward sales of its oil, to earn interest as its price for guaranteeing continuity of supply to other oil companies. If BNOC had not had rights over the oil, this interest or its equivalent would have accrued to the companies themselves, because they would have been able either to guarantee their own supplies or to sell forward in the same way as BNOC. This is a case in which it is hard to see how BNOC could make a gain out of its rights over the companies' oil if the companies had not correspondingly lost thereby. However, any fiscal effect can probably be regarded as accidental and exceptional.

Optimality of Rent Taxes

The PRT has clearly been designed with a view to taking a large share of the economic rent remaining after royalty and corporation tax while minimizing deterrence to extraction and exploration. What is wanted is a tax that, in face of variation in cash flows and their timing, will leave a company's minimum requirements fulfilled after tax, provided they would be fulfilled in the absence of tax. It seems reasonable to suppose that such a tax would need to be framed with reference to some measure of what the company's minimum requirements are.

The wrong first step was taken when the original draft of the tax was based on undiscounted cash flow. Reduction of undiscounted nominal cash flow to near zero, which might occur if the rate of such a tax were high, would of course not please an investor who must discount his future receipts for time and for expected inflation and allow for risk. Hence it no doubt seemed out of the question for the tax in that original form to be applied at a very high rate, and even with the 45 per cent rate, concessions were thought necessary. The oil allowance was introduced to favour particularly those fields with small annual output and short productive life. Though these will generally be the least profitable, there is no simple relationship, since capital costs, and recurrent costs per barrel, may vary from one field to another. Hence the oil allowance is a clumsy, hit-and-miss way of favouring the less profitable. Ideally, a test more closely related to profitability would be required. The form of the oil allowance probably does not avoid any possibility of distorting the rate of depletion, since postponing production can increase the cash value of the oil allowance when prices are rising, and bringing it forward can increase its cash value when prices are falling (a point made to us by Mr A.G. Kemp). (In such a case it will often be the means of avoiding a distorting effect in the reverse direction which the 'second' element in the tax might otherwise have, but this is no satisfactory excuse.) Kemp's figures (1975, 1976) suggest that the PRT in its 1975 form did indeed make substantially greater difference to the IRR and NPV of fields that, in its absence, would have a high IRR and NPV (at a 10 per cent discount rate) than to those of fields that would have a lower IRR or NPV. Since the more profitable projects considered by Kemp are also projected as long-lived, and since the burden of PRT tends to be concentrated *after* the first few years of a project's life, use of a 20 per cent discount rate, say, to assess the effect of PRT on NPV would probably indicate less of a difference between the proportional reduction on highly profitable and that on unprofitable projects.

A later version (Kemp and Cohen, 1980:9) shows the projected total tax under the 1975 version as taking, in real present-value terms with a

10 per cent discount rate, proportions between 67 and 79 per cent of 'resources generated' on seven of the eight fields studied but 230 per cent on the eighth field, which is left with a negative real present value. However, it is worth noting that the anomalous result in the eighth case (Cormorant South) is due not to the 1975 PRT, which would not have been payable on this field, but to corporation tax and royalty. The 1979 system (Kemp and Cohen, 1980:8) leaves the first seven fields with an even narrower range of proportions of real present value of resources generated taken in tax (80–7 per cent) but leaves Cormorant South with 234 per cent taken and a slightly more negative real present value than under the previous system (because this time some PRT is due from it). Residual real internal rates of return on the 1979 rules range across the eight fields from 4.5 to 33.4 per cent. Under the 1975 rules Cormorant South is prevented from paying PRT, which would otherwise be due by the 'safeguard' or 'second' tax, but only one other field out of the eight has its PRT reduced by this device, and that is the highly profitable Forties field. Forties, Cormorant South, and Ninian have PRT reduced by this means under the 1979 regime. It is notable that the small but reasonably profitable Buchan field would have paid no PRT under the 1979 system, presumably because of the tax's bias in favour of smallness. These conclusions are of course sensitive to assumptions about future oil prices and (where the assessments are in real terms) about other prices as well. Treatment of Cormorant South (under the 1979, and even the 1980, regime) looks much less harsh (Kemp and Cohen, 1980:39) when 1980 prices are taken into account, allowing it to expect a (just) positive real NPV.

What we have called the 'second' tax, the progressive profit tax, that sets an upper limit on the 'first', is more clearly geared to profitability, and in a time of stable prices its threshold of 30 per cent before corporation tax might be a reasonable estimate of the required rate of return or discount rate in the industry. The fact that it is not indexed for general price changes makes its base a more arbitrary and unpredictable one. Moreover, if net operative receipts were likely to fluctuate from one year to another over the life of any field, this method of taxing excess profits would be somewhat arbitrary in its effect, even if inflation were allowed for through indexation. Because it is not impartial between various time patterns of cash flow, it may encourage the speeding or slowing of extraction for the purpose of spreading profits more evenly over time. Nevertheless, oil companies could reasonably expect this second tax, if it were indexed and their real cash flow were steady over a longish period, to allow them a satisfactory return if they would have been able to make a satisfactory return in its absence. (Applied with corporation tax, it would raise the marginal rate of tax on positive cash flow in certain circumstances to about 90 per cent.) But it is not indexed;

and cash flows are not stable. However, the tax is unlikely to be operative on any project, either during the early period of the project's life when no PRT is paid, or during the subsequent ten years or more, in which the first tax is reduced by the oil allowance and is thus likely to be the operative one. Desirably, the combination will probably often prevent the 'second' tax from biting on high annual profits early in the life of a project when those high profits are a purely temporary phenomenon. This will remove the risk that it will have a punitive effect when the lifetime returns on a project will not be particularly high. But this safeguard could also have been produced, more systematically, by an RRT.

Though undoubted ingenuity has gone into preventing the PRT from being punitive, it is hard to escape the conclusion that the result could have been much more simply achieved, and in a more reliable manner, by an indexed tax of RRT form or by one of the arrangements for equity transfer or carried interest that are roughly equivalent to an RRT. A tax whose burden would be related to a fairly satisfactory indicator of profitability might have avoided the need for the many changes in the rules that took place in 1979–82. It would be better to have a tax the rules of which did not need to be amended when oil prices changed.

Harrigan (1977), using Kemp's data and assumptions on three fields, has compared the package as it existed before 1979 with a combination of corporation tax and RRT (and no *ad valorem* royalty), applying two different assumptions about RRT rates and thresholds. Regime 1's RRT has a single threshold of 10 per cent in real terms and a tax rate of 55 per cent. Regime 2's RRT has the same threshold and rate but also a second, higher, threshold of 20 per cent in real terms, above which an additional 25 per cent tax is added. The fields used are the large, rich Forties and the much smaller Auk and Heather, of which the latter then seemed marginal. Regime 1 leaves NPV (at 10 per cent) and IRR somewhat lower on Auk and Heather than the 1975–8 package, but higher on Forties. (The latter result reflects, as Harrigan suggests, the absence of royalty in his alternative arrangements, and it may also reflect the smaller relative importance of the oil allowance in the case of the large Forties field.) Regime 2 reduces NPV and IRR substantially on both Forties and Auk below their levels under the 1975–8 package but leaves all three, even Heather, a real rate of return still above 10 per cent. It is worth noting that the RRT regimes avoid the 'high grading' effects of the royalty and that the top marginal rate supposed under regime 2 is the same as under the PRT 'second' tax.

There is no consistent evidence that exploration and the rate of new operational drilling fell off in response to the mooting and announcement of PRT in spite of the uncertainties over 'participation', which continued probably until 1977. Harrigan's exercise, however, limited as it is, suggests that an RRT regime with quite reasonable rates and thresholds might eventually have raised substantially more revenue than the 1975–8 package if the latter had been continued. The 1979–80 package would doubtless raise more than the earlier one and, initially, the 1981 package more again. It is relevant, however, that these new packages were not adopted when oil prices were at 1975 levels, presumably because they would have been regarded as potentially punitive then for some otherwise intra-marginal fields; and by the same token the rates might have to be lowered in future if the real oil price were subsequently to fall substantially. Also, the 1981 package appears to have led to the abandonment of projects that would otherwise have been undertaken (Dafter, 1981). It would seem that the same limitations would not arise with Harrigan's RRT regime 2.

British experience moreover suggests that, where there are two profit taxes, and the one that can more acceptably be altered is also the one more closely related to profitability, that tax should be collected after, rather than before, the other. Alteration of the tax can then have a more controlled impact on the profitability of projects. There may, however, be legal and institutional reasons for doing the reverse, as Britain in fact did.

If there must be an *ad valorem* royalty, there are undoubtedly advantages in the provision that the Secretary of State may remit it. Given an assurance of probity and perfect knowledge on the part of the administering official, the arrangement would be quite satisfactory. But perfect knowledge is never given, and probity, not always. For these reasons the remittable royalty is far from ideal as a general rule.

On the face of it, the SPD renders the system even less satisfactory in the incentives it offers to efficiency. As pointed out earlier, over considerable phases of the lives of certain projects the SPD will act like an additional *ad valorem* royalty, with all the faults of that kind of charge. It may easily inhibit the more marginal investments and, in certain cases, put a premature end to output from fields already producing. The fact that SPD was initially imposed for only eighteen months (later twenty-four months) would not be enough to reassure potential investors.

The need for heavy impositions makes it all the more urgent that the form of tax should approach neutrality. Distorting effects of a tax structure that are trivial at low rates become serious when the proportion taken from cash flow is high. Thus, flaws of tax structure which were probably of little consequence in the mid-1970s have since become more damaging. The 1981 and 1982 changes, and the reaction of the industry to them, emphasize the need for a rational reorganization of the highly complex patchwork that has been evolved. As we have indicated, we believe that a much simpler combination of taxes may go considerably closer to neutrality: that, by discounting cash flows for tax purposes at an appropriate rate, the rules can approach the touchstone of defining a tax base in true economic rent.

There is persisting doubt over the wisdom of discretionary allocation of leases, which precluded any realization of a part of the rent through auctioning. In general, this book has argued against exclusive reliance on the auctioning of rights as a means of rent collection except where a profitability-related additional tax is administratively out of the question. There is, however, no reason why auctioning of leases should not be combined with an additional profit tax. The experiment in the Fourth Round made it clear that certain blocks could realize quite substantial sums. The fifteen blocks then auctioned realized £34 million. Generalizations from such a small sample are made doubly precarious by the extreme skewness of the distribution: two blocks earned £29 million and the other thirteen, the remaining £5 million. Little more can be deduced from this except that there is money to be earned from some blocks. It might be argued that the companies bidding were doing so without taking into account the PRT, which was introduced only four years later, but it would be surprising if they had not taken some account of the possibility of additional taxation, just as it would be surprising if they *had* taken account of real rises in the oil price as large as those that were to occur over the next decade. In short, with better knowledge of future taxes and prices, they might well have bid at least as much. If we were to accept Hamilton's inference (1978:108, note) that about £250 million might have been realized in 1971 if all the 158 blocks then issued in the northern part of the Sea (where most of the auctioned blocks were) had been auctioned, this suggests that (*if* the companies had *not* taken the possibility of a PRT into account and *had* foreseen the price rises) auctioning would be no adequate substitute for PRT; but, as suggested above, these conditionals are probably not realistic and on that point no useful conclusions can be drawn from this elusive fragment of evidence. In any case, we would propose the auctioning of rights for fixed sums as not a substitute but a supplement for a profitability-related tax.

In May 1980 the government announced that, for the Seventh Round of leasing, about twenty of the ninety blocks to be offered would be sold on application for £5 million each (about a tenth in real terms of what Esso offered for the highest-price block in 1971). These blocks were all in the region of established fields. The test of the *ex ante* value of blocks that this sale will provide is again not a very useful one; for if all the blocks were sold on those terms what would be indicated is merely a lower limit, and because the purchase fee could apparently be offset against profit for both corporation tax and PRT purposes (*Financial Times*, 7 May 1980) the value of the net additional revenue would be very small.

The fact remains that it is possible that much revenue could be gained, and virtually inconceivable that financially much could be lost, by auctioning blocks in or near established fields, whatever other taxes were applied. The question of interest is whether those financial gains are likely to be outweighed by the alleged advantages of a discretionary system: preference to UK operators and to those that will engage UK contractors, work commitment bidding, the inhibition of a speculative market in oil leases, and the capacity to enforce policies on oil companies other than through legislation and legal process. Work commitment bidding is of dubious worth unless it is assumed that any well must have a positive social value: if necessary, a minimum work commitment of, say, one well on average for so many blocks, as *de facto* in Norway's first round (Dam, 1976:55–6) could be required. A speculative market could be prevented by forbidding resale of the block, though again the desirability of inhibiting a market may be questioned. Extra-legal influence on companies is naturally appealing to many officials, but is rather dangerous as a general rule. Finally, the nationalist argument lost much of its force as long as BNOC was in existence and could step up its requirements for participation in the financial sense if, for whatever reason, the proportion of home to foreign operators or contractors and suppliers was considered too low. Economic nationalism is not a costless indulgence. If some concessions to mercantilism are considered necessary, they should be pursued so as to interfere as little as possible with selection by efficiency. Hence it might have been better to let auctioning take its course and to make any necessary corrections in subsequent rounds by stepping up (or down) the participation of BNOC, rather than to give arbitrary preferences to British-owned firms such as would preclude any element of competitive self-selection.

Finally, though BNOC's rights to 'participation' might in certain circumstances have constrained the oil companies as a means of bringing financial advantages to the state or of achieving some other social benefit, the corporation and its rights may largely be ignored in an assessment of the tax system strictly so-called. Its potential, so far unexploited and apparently to be removed altogether, must be seen as that of a backstop in case other policy instruments failed. In particular, it might have helped to justify a less cautious attitude over lease allocation. Clearly, however, direct state enterprise is one way of realizing mineral rents for the nation. There is no reason in the case of BNOC to think it has been an inefficient method. Despite certain privileges, it has been exposed to market prices and costs very much as other oil companies have faced them. The fact that it may have put any after-tax profits into further exploration and production should not count against it.

Summary[1]

The UK oil and gas tax system conforms with the main tendency of this book in applying general fiscal rules and in relying heavily on an additional profit tax. This tax, however, is unduly complex and is not based on any clear principle. This makes it almost inevitably subject to changing rules if a large share of rent is to be culled under markedly changing external conditions; leads almost certainly to significant distortions when rates are high; and makes modifications of its rules and rates difficult to assess.

The *ad valorem* royalty might be dispensed with, at least under a more satisfactory additional profit tax. Its discretionary remission is bad in principle but in this case makes it possible to avoid the distorting effects of the royalty on the extent of extraction.

It seems a mistake to forgo the extra revenue that might be raised by auctioning licences. The objectives of the discretionary system of allocation either might be pursued by other means or were better not pursued at all.

Altogether, there is a strong case, particularly in the light of the changes of 1979–82, for a simpler additional tax more closely related to profitability and for the auctioning of new licences.

Though the BNOC, while state-owned, has been a potentially useful instrument for pursuing certain subsidiary objectives, much of its 'participation' was virtually irrelevant to the fiscal system. It may, however, if it is allowed to survive, or is revived, turn out to be quite a good example of the building of state expertise in mining business and of the realization of mineral rents through semi-competitive state enterprise.

[1] POSTSCRIPT 1982–3. Under the 1983 Budget, various reliefs produced further complications in the patchwork. All exploration and appraisal expenditures were to be immediately deductible for PRT; no royalties were to be collected on certain new fields and their oil allowance was to be doubled; and a start was made on phasing out Advance PRT (PRT collected through dividends). In the debate prompted by the rigorous measures of 1981, a Committee set up by the Institute for Fiscal Studies (*The taxation of North Sea oil*. London: Institute for Fiscal Studies. December 1981) proposed replacement of all taxes on oil and gas extraction by a simple three-tier RRT with a top marginal rate of 85 per cent and thresholds un-indexed at 15, 25, and 35 per cent. Applied on 26 existing fields, this tax, the Committee calculated, could be expected to raise almost as much revenue as the (since relaxed) 1981 regime. They believed that it would be generally creditable overseas and that its application to existing projects posed no great problems of information. Our criticisms of this device are that the thresholds are un-indexed (and therefore unpredictable in their effects) and that (on expectations of 7–10 per cent inflation) the top threshold is too high (encouraging avoidance through over-investment and otherwise needlessly surrendering revenue) and the bottom threshold possibly too low. These faults are easily remedied, say by substituting fully indexed thresholds of 12½, 15 and 17½ per cent.

18

TAXATION OF OIL AND NATURAL GAS IN THE US FEDERAL JURISDICTION

The USA plays a dominant role in the world markets of virtually all industrial minerals and fuels. Because of its affluence and the size of its population, it is the greatest user of all or almost all. Not surprisingly, given its continental extent, it is also a major producer of many of them. An estimate for 1974 (Morgan, 1974:3) attributes about 4 per cent of US gross national product to the production of mineral raw materials. These were supplemented by net imports equal in value to nearly 2 per cent of gross national product.

Mining in state territory is constitutionally under the jurisdiction of the states, an arrangement followed in Canada and Australia. In addition, certain mineral properties in the USA rest with private persons, who consequently receive royalties from them. This chapter will deal with two products only, petroleum and natural gas, and with those two mainly in so far as they come under federal jurisdiction.

Federal concern for the oil industry dates back to the period before the First World War. Over the years it has been directed to maintenance of supplies, especially in wartime, to conservation of the resource, to control of monopoly, and during the interwar years to sustaining local business activity and employment in the producing areas (Nash, 1968: *passim*). From 1971, there have been several phases of price control and allocation control. Federal government activity was greatly increased in response to the Arab countries' temporary oil embargo and the considerable rise in the world price of crude oil that followed the Yom Kippur war of October 1973. Regulation has been exercised by virtue of federal powers over inter-state and overseas commerce. The federal government has also owned mineral rights to certain lands within the territory of the states; and, since the Outer Continental Shelf Act of August 1953, the continental shelf outside a three-mile limit from the coastline (nine nautical miles in the Gulf) has been treated as under exclusive federal jurisdiction (Nash, 1968:194). The proportions of US oil and natural gas production coming from the outer continental shelf (OCS) rose steadily from 1953 to reach about 11 per cent for each product in 1972 (Epple, 1975:25). Epple (1975) estimates that these areas accounted for about 5–10 per cent of total discoveries over the period 1955–69. The federal share of oil resources extends also to various naval reserves which include an immense area along the eastern part of the north coast of Alaska. Since the oil crisis stimulated presidential declarations of intent to increase the rate at which federal oil lands were made available, it is likely that their importance in output has continued to increase. The oil and gas of the OCS has been subject to a distinctive fiscal regime which was not seriously modified until the 1978 amendments permitting and enjoining experimentation and the imposition on all domestic oil producers of the Windfall Profits Tax under the Act of 2 April 1980.

Oil in the US Economy

At the time of the 1973 oil crisis the USA depended on oil and natural gas for about 80 per cent of its energy needs, 50 per cent on oil and 30 per cent on gas (official figures from Keesing: 26226). The USA is the largest user of oil in the world. After the Soviet Union it has also been, in most years, the world's largest producer, though in 1976 and 1977 its output was exceeded also by that of Saudi Arabia. Through the 1960s the USA became steadily more dependent on imported petroleum as demand rose more rapidly than domestic output. When import controls were introduced in 1959, imports were set at no more than 9 per cent of estimated domestic demand. This limit was raised to 12.2 per cent in 1962 and gradually edged up to reach 23 per cent in 1972, shortly before quantitative controls were abandoned (Epple, 1975:20). Imports of oil by 1973 provided about a third of US oil use and a sixth of total energy use. By the end of the decade this proportion was little altered, despite the measures taken to reduce it.

Oil has been concentrated in certain states in the central South and the Far West, notably Texas, Oklahoma, Louisiana, Florida, Kansas, and California. In Texas above all it has played a dominant role. Alaska's economy is likely to be similarly dominated by oil over the remaining years of the century.

American-based companies have been concerned with much of the extraction, refining, and marketing of petroleum in the world at large. All but two of the 'seven sisters', the top international oil 'majors' of the postwar period, are American. Through the 1960s and 1970s host governments have admittedly played an ever larger role in these activities, but, despite changes of title, the expertise of the great companies is still needed, and generally they remain in one guise or another. While in much of the world the industry has been an oligopoly of oil majors acting for considerable periods rather like a cartel, there have, in the USA itself, been numerous smaller independent companies and apparently a considerable element of competition. A report issued in 1977 (McAvoy, 1977:1) described the petroleum industry as a network of more than 300,000 firms, the fifteen to twenty-five largest performing all functions from exploration to retail marketing with many smaller independents each participating in one activity alone. In 1969 the four largest suppliers of crude oil accounted for 31 per cent, and the twenty largest integrated companies between them for 71 per cent, of the US crude oil market, while the four firms with the largest refining capacity provided 32 per cent, and the twenty largest integrated companies 83 per cent, of US refining capacity (Epple, 1975:26). At the time of the imposition of mandatory price controls on the industry in 1973, the twenty-four largest companies were described as doing more than 95 per cent of the gross industry business (McAvoy, 1977:7). There is clearly great disparity in size of firm but no ground on which to suppose effective monopoly or oligopoly. Like other observers, the McAvoy report supposes 'workable competition'.

Politically, oil and gas in America have been subject to conflicting forces and concerns: the apparent interests of producers as against those of consumers; the demands of current as against future supplies; strategic and environmental objectives as against free enterprise ideology and states rights; the majors against the independents. The fiscal regime for federal oil leases, as established in 1953, was the product of a fairly tranquil period in oil politics (Nash, 1968:190–4). It was a happy compromise which continued for a number of years without serious challenge. In the troubled conditions of the 1970s, however, its possible weaknesses became more evident, and debate also arose over whether a similar system should be pursued by the state of Alaska for the exploitation of its vast but remote oil reserves.

The Institutional Framework

Leasing

In the 1880s, courts in various states upheld what has come to be called the 'rule of capture', the principle that oil that is brought to the surface on a piece of land belongs to the owner of the land (Epple, 1975:15). This is an extension to fluid deposits of the American view that the owner of land is presumed to be owner of the minerals lying beneath it. The original owner may thus be a private person or a government. In this respect the USA is in contrast with most other countries with respect to its onshore mineral resources. Generally, governments, both state and federal, have not aspired to extract the rent from on-share minerals unless the government concerned happens to own the land under which they lie. Private owners make their own arrangements with intending extractors and, if wells turn out to be productive, they, and not the government, extract the royalties. Normally the private owner requires a fixed payment ('cash bonus') for a lease on the oil property, as well as a rent payable over the duration of the lease unless and until it becomes productive. If oil is found on the block leased, the rent is normally replaced by a royalty and the lease is extended for the life of the well (Epple, 1975:7; citing McLean and Haigh, 1954:397).

Apart from any mineral rights that they happen to possess through landowning, the state governments have the rights to offshore lands within three miles of the Atlantic and Pacific coasts and within nine natucial miles of the coast on the Gulf of Mexico. This principle was established by Congress in 1953 (Nash, 1968:194). Their methods of managing these rights have not been identical, but have tended to follow similar patterns. For the most part, the states require applications, accompanied by modest fees, for rights to explore; they determine the lease of the lands by some form of competitive bidding, either by cash bonus or by annual rental; they offer leases normally for terms of three to five years, extended automatically if a deposit is discovered; and they charge a royalty generally between $12\frac{1}{2}$ and $16\frac{2}{3}$ per cent on the value of the product (Epple, 1975:19). Recently some states, notably California and Alaska, have released substantial areas under profit share bidding.

The rules under which federal lands in the outer continental shelf should be leased were laid down in the Outer Continental Shelf Act. This empowered the Secretary of the Interior to issue leases for lots not to exceed 5,760 acres each, after competitive bidding (Nash, 1968:194). Bidding may be either for a cash bonus, in which case the royalty rate is fixed, or for a rate of royalty, in which case there is a fixed cash bonus (Dam, 1976:148). The royalty may be set at any rate not less than $12\frac{1}{2}$ per cent *ad valorem*; where bidding is by cash bonus, it has been set routinely at $16\frac{2}{3}$ per cent (Dam, 1976:148). For onshore leases on federal land, competitive bidding is required whenever the leased land is 'within any known geological structure of a producing oil or gas field' (quoted Dam, 1976:147). Otherwise (that is, on a so-called wildcat field) a petroleum lease put up for application must be given, on a $12\frac{1}{2}$ per cent royalty, to the first bidder (Dam, 1976:147).

At least until the late 1970s, bidding for OCS leases had been consistently by cash bonus (the 'auctioned fixed fee' referred to in chapter 6), except for an experimental period in 1974, when bidding was by rates of royalty and the fixed cash bonus was $25 per acre. Winning royalty bids ranged from 56.8 to 82.2 per cent (Dam, 1976: 149). Dam mentions (1976:148) that royalty bidding has also been used in the states of California, Louisiana, and Pennsylvania and in the Canadian provinces of Saskatchewan and Alberta.

Epple (1975:22) observes that the areas of federal lands released have varied erratically from year to year. The Secretary of the Interior may withdraw blocks even though bidding has occurred, and in about 5 per cent of cases the highest bid has been rejected, and the block has been withdrawn without explanation (Epple, 1975:23).

Cash bonus bids realize considerable sums. Dam (1976:153) mentions that in 1973 alone the amounts paid for OCS leases totalled $9 billion and that the highest bid in the 1969 Alaska state sales was $72 million. Bids over $150 million were not uncommon in the early 1980s. Jones *et al.* (1979:887) record that, from 1954 to 1977, the total value of oil and gas produced from the OCS was $33.4 billion, of which $25.3 billion (roughly three-quarters) went to the government as revenue, $19.7 billion of this being proceeds of cash bonus bids.

Federal leasing rules remained unchanged until 1978. Then, amendments obliged the authorities to apply other leasing systems on a proportion of federal offshore lands coming up for release. These changes are described later in the chapter. Before that time, however, the context in which they operated was greatly altered by large changes in the real price of the product and in other fiscal provisions.

Other Fiscal Provisions

US companies are subject to federal income tax. Income on mining operations has, as elsewhere, been given special concessions. These have taken the form of depletion allowances which have varied according to the product mined. They have applied to income from mining, both at home and abroad. Before the law was changed in March 1975, oil and natural gas were treated more favourably than most other minerals (McDonald, 1976:27, 29), although the arrangements for all were similar in form.

In the early years of the oil industry the Internal Revenue Bureau made a valuation of each oil discovery individually, and this formed the base on which the reduction in value each year through depletion was estimated for purposes of deduction from taxable income. This procedure was held to be administratively costly and to lead to disputes between the producers and the authority, and in 1926 legislation was passed by Congress permitting what is called 'percentage depletion' (Nash, 1968:86). This allowed oil producers to deduct from taxable income each year an amount equal to a certain percentage of the gross value of sales. The percentage was initially set at 27.5 per cent, and this applied to oil and gas until 1969 when it was reduced to 22 per cent (Epple, 1975:21). As the system developed, depletion deduction for any year by this method was limited by the rule that it could not exceed 50 per cent of net income; for this purpose net income was interpreted somewhat more generously than in assessment for actual calculation of tax, since costs associated with unsuccessful properties were deducted for the latter purpose but not for the former (McDonald,

1976:28). This is because depletion is based on the individual property, while the tax itself is based on all the taxpayer's operations, and for this purpose such expenses as drilling of dry holes can be treated as part of costs (Epple, 1975:21).

The taxpayer could choose, however, to use instead a method called 'cost depletion' and could shift from one to the other. Cost depletion involves estimation of the value of the property, and deduction in the year concerned of a proportion of that value equal to the proportion that the year's production bears to the total estimated quantity to be produced from the property (McDonald, 1976:27).

The total amount that could be deducted under percentage depletion was not limited by any estimate of the initial value of the property, and a taxpayer who changed from percentage depletion to cost depletion, or who chose to use cost depletion on a property bought from someone who had used percentage depletion, was not required, for purposes of the latter method, to take any account of depletion already deducted under the former (Epple, 1975:21).

The generosity of these provisions was increased by the fact that a number of 'intangible' expenses of a capital nature, such as those for labour and fuel involved in well development, which undoubtedly contributed to the capital value of the property, could be expensed against taxable income as they were incurred. Thus it could reasonably be claimed that there was double counting of expenses. Epple (1975: 22) cites two estimates, both suggesting that the system allowed producers to deduct as depletion roughly twice the amount that they could have deducted if costs not already expensed had alone been eligible for depreciation.

This favourable treatment applied to all those having ownership rights in oil or gas production, including those receiving royalties, which were thus deducted from a lessee's sales revenue in order to determine the gross sales revenue on which his percentage depletion would be calculated (McDonald, 1976:28).

Under the impact of the oil price rises, the system was considerably tightened in March 1975. The major integrated firms (which were generally those most likely to have overseas production) were debarred from using percentage depletion (McDonald, 1976:27). Smaller firms could apply the percentage depletion allowance to no more than 2,000 barrels per day each, an amount gradually to be reduced to 1,000 barrels per day by 1980, and the rate of percentage depletion was to be reduced gradually from 22 to 15 per cent by 1984. In addition, the limit of the allowance in relation to net income was modified, and a property switched from percentage depletion to cost depletion would have the amount already allowed deducted from its value for the purposes of the latter method (Epple, 1975:21). The Act (P.L.

94–12) did not affect the tax treatment of minerals other than oil and gas (McDonald, 1976:27). The immediate expensing of 'intangibles' remained and continued to entail, rightly or wrongly, greatly accelerated depreciation for the majority of development costs and the great majority of exploration costs.

The other special tax provision applying to oil was again a result of the price disturbances of the 1970s. This was the Crude Oil Windfall Profits Tax enacted on 2 April 1980. Such a tax had been proposed repeatedly by successive Presidents since late 1973 (Keesing:30355). It was introduced with the progressive decontrol of domestic oil prices (outlined below) and in the wake of the second major bout (1979–80) of large OPEC-directed real increases in the world oil price.

Briefly, the tax was designed to take a proportion varying from 30 to 70 per cent (and expected to average 50 per cent) of increases in producer revenue that resulted from the decontrol of prices or subsequent rises in the world price of oil. In order to provide an approximation to this result, the base of the tax was to be the difference between the actual revenue from sales and what it would have been if a set base price had ruled. The base price would be raised from its original value to take account of inflation. This price would depend on the type of oil, and the rate of tax or base price or both would vary according to a number of circumstances, of which one was the date at which the well was first tapped. The application of the tax was to be limited, however, in that it would cease to operate from January 1988 or when the sum of $227,300 million had been collected, whichever was the later. Revenues would be earmarked for certain purposes, one of which was energy development (Keesing: 30355). In addition to federal income tax and (recently) the Windfall Profit Tax, extractors operating within a state are subject to that state's taxes. These include usually an income tax, a severance tax, and sometimes property taxes, and there are also local property taxes. Where the state is also the landowner, as in much of Alaska, a very large part of the total fiscal charges may go to it. A recent study (A.G. Kemp, personal communication) has estimated that in Alaska well over 40 per cent of the fiscal total is currently accruing to the state.

Controls Directly Affecting Price of Output: Oil

Apart from wartime intervention, direct federal and state controls date mainly from the period of low prices and industrial distress in the inter-war period. A tariff on oil of $1 per barrel was imposed by the USA in 1932; the rate of tariff was later changed, and it stood at 10.5 cents per barrel when it was abolished (to be replaced by another form of duty) in 1973 (Mead, 1976:153). The various states at the same period began the system of quantitative control of each well's output known as 'prorationing', which roughly served the purposes of conservation and price stabilization. They were supported in these activities by the federal authorities under the Connally Hot Oil Act of 1935, which banned the inter-state movement of oil produced in excess of amounts allowed by the states. Prorationing typically involved the setting for each well of a maximum allowable rate of output determined by certain objective criteria such as the depth of the well and its proximity to others. This maximum was fixed in a crude way so as to avoid the over-rapid and wasteful extraction that could easily follow from the rule of capture and the fact that there were no rules obliging owners of land above the same deposit to agree on its exploitation. For any period a certain proportion of this maximum was allowed. This proportion was set on considerable market information and was generally designed to balance supply and demand within the state at the current price. It thus acted in effect as a means of stabilizing price. Apparently no very firm mechanism existed for allocating production among the producing states, and there is evidence that those states that were the largest producers tended to bear the main brunt of the restrictions on output (Epple, 1975:17). By 1972 all states were permitting the 'maximum allowable' (Epple, 1975:18).

After the Second World War, during which there had been comprehensive control of supply by federal authorities, there was a withdrawal to a large extent of federal intervention, but prorationing by the states continued with federal support. Imports provided a small proportion of domestic consumption in the 1930s and 1940s, and no further action was taken over imports until an attempt at their voluntary restraint was made in 1957. This was followed by mandatory quantitative controls (the Mandatory Oil Import Program) in 1959, with a quota that grew steadily as a proportion of domestic consumption until 1973. Mead quotes an estimate that for most of the period from 1959 to 1973 a quota allocation was worth about \$1.25 per barrel (Mead, 1976:151).

In May 1973 quantitative restrictions on imports and the existing 10.5 per cent tariff were removed. A licence fee of 21 cents per barrel on imports of crude and 63 cents per barrel on imports of product was imposed instead, but firms were allowed to import free of fee the amounts that they had previously imported under quota. This concession was to be phased out by 1980 (McAvoy, 1977:5). In early 1975 the fee was raised, with some haggling between President Ford and Congress, by \$2 per barrel on crude. The \$2 fee was removed at the end of 1975 after being declared unconstitutional by the Supreme Court (Mead, 1976:153).

More important in their effects of the disposition of economic rents in the industry were the controls on domestic prices first introduced as part of President Nixon's Economic Stabilization Program of August 1971. With most other prices, those of oil and natural gas and products were frozen for three months and subject thereafter for a period of just over a year to a maximum limit of a 3 per cent per year price rise. Then, from early in 1973, special rules were imposed on the twenty-four largest firms in the oil industry, introducing mandatory controls on their oil prices, while other firms, including the smaller ones in the oil industry, were for about five months subject to voluntary controls.

This unsatisfactory discrimination between large and small companies was removed by regulations propounded in August–October 1973, which established a two-tier system of domestic crude oil prices. As the system emerged by the end of November, what was classified as 'new oil' or 'released oil' or 'stripper well oil' would be free of controls. (Stripper well oil was oil from wells producing less than 10 barrels a day.) What was classified as 'old oil' (other than 'released oil' or 'stripper well oil') would have its price fixed (initially at 35 cents, later at $1, per barrel above its price level of 15 May 1973). 'Old oil' for any property producing in 1972 would be an amount equal to its output in 1972 (or its total current production if that were smaller). The rest of the oil from such properties, and all oil from properties not producing in 1972, would be treated as 'new oil'. However, as a property that had been producing in 1972 generated new oil, an amount of old oil equal to the new oil produced would become 'released oil' and would be free of controls (McAvoy, 1977:6–8).

Under the same rules, the prices charged by refiners and resellers would be those of 15 May 1973, increased only by rises in costs with a profit margin rate subject to a limit laid down. For certain products, sellers were permitted to allocate costs as they wished between products (McAvoy, 1977:8). A provision setting specified limits to the retail prices of certain products lasted for a few months only.

Legislation effective from 30 April 1973 gave the President authority to allocate crude oil and petroleum products. This appears to have been designed mainly to ensure sources of supply for small, independent refiners. Though voluntary controls on allocation were imposed shortly afterward, a mandatory system was not established until 15 January 1974. These were introduced under the Emergency Petroleum Allocation Act, passed by Congress on 27 November 1973 in the wake of the Arab countries' oil embargo (McAvoy, 1977:9–10). The details of these rules are peripheral to our concerns here. They tended to freeze buyer–seller relationships within the industry. Over and above this effect, further rules (the 'buy/sell program') were designed to ensure that all refiners were able to operate to similar proportions of their capacity. This was modified in May in such a way as to oblige only the fifteen largest refiners to sell 'excess' supplies of crude and then to sell such supplies only to small refiners (McAvoy, 1977:10–11).

The two-tier system for crude oil pricing made it desirable to find some device that would allow any crude oil to be sold to the most convenient buyer while at the same time not leaving those refiners who would be buying a high proportion of upper-tier oil to face higher average prices. World price changes occurring at about the end of 1973 had produced a large gap (of nearly $5 per barrel in mid-1974) between lower-tier and upper-tier prices (Mead, 1976:155). The device applied was the entitlement system, introduced in December 1974. This allowed refiners to use only the national average proportion of lower-tier oil (or in fact somewhat less if they were large refiners, and somewhat more if they were small). Each refiner was given the appropriate 'entitlement' indicating the amount of lower-tier oil he was allowed to use. Entitlements could then be sold in a free market, so that those who found it convenient to use more than the average of upper-tier oil (and thus use less of their entitlement) might be compensated, by the sale value of their excess entitlements, for the higher average crude oil prices that they would have to pay (McAvoy, 1977:12).

The Energy Policy and Conservation Act of December 1975 was aimed at gradually decontrolling crude oil prices. This was to be done by allowing the Federal Energy Administration (FEA) to raise, at the rate of up to 10 per cent a year, the national average crude price set by the Act. The FEA was allowed to achieve this average price by establishing various price tiers. The authority responded by creating a third, middle, tier for the domestic oil that had previously fallen into the upper tier. Its price was now set below the price of imported oil, which formed the top tier. Later, in August 1976, stripper well oil was moved in effect to the top tier by being freed of controls. At the same time, presumably in recognition of the fact that the 'old oil' price could provide quite inadequate incentives for marginal production in wells with declining output, the FEA was required to establish greater price incentives for tertiary extraction (McAvoy, 1977:13–14).

The three-tier system continued in essence until 1980, the further large real-price rises in imported oil making decontrol by the route proposed in 1975 seem more and more remote. Then in April 1980, with the passage of the Crude Oil Windall Profit Tax Act, more radical moves were made towards a free market. The presidential proposal for these moves and for the Windfall Profit Tax had been made a year earlier, in April 1979 (Keesing:30355). At this time the 'old oil' price had been about $6, and the 'new oil' price about $13, per barrel, while OPEC prices on average had risen from $16.75 in January, to $23.15 in February. It was clear that the marginal price received even for 'new oil' was far below its opportunity cost as measured by the value of imports replaced (Keesing:22987). The discrepancy continued to increase in early 1980.

The 1980 rules for gradual decontrol were complex, but they relied less than their predecessors on assumed stability in the world price. They provided that the middle-tier price should reach the world price by 1 October 1981, and that lower-tier oil should gradually be shifted to the upper tier at certain percentage rates per month, so that the lower tier would disappear in the early 1980s (Keesing:30355). The Reagan administration in early 1981 completed decontrol before these deadlines were reached.

Controls Directly Affecting Price or Output: Natural Gas

Price controls on natural gas have been much more deeply ingrained than those on oil. Federal regulation of natural gas that moved across state borders was authorized by the Natural Gas Act of 1938. This Act appears to have been passed principally with the aim of controlling monopolistic behaviour on the part of the pipeline companies (Helms, 1974:3). Because natural gas has been produced mainly in six states in the south and southern Mid-West, while consumption is naturally concentrated in the main centres of population, a large part of total output has moved between states and has thus come under federal jurisdiction. Because pipelines have hitherto been the main means of transport for natural gas, and pipelines are subject to considerable economies of scale, there was reasonable ground for fearing monopolistic pricing.

Under the Supreme Court's ruling in the Phillips case of 1954, the Act was held to require the Federal Power Commission (FPC) to regulate not only the prices at which the pipeline companies sold gas but also the prices at which they bought it from producers (Helms, 1974:17–18). The FPC was unwilling to undertake such regulation of producer prices and expected Congress to amend the law accordingly, but President Eisenhower vetoed a bill to effect the change, and from 1958 the authority made a serious attempt to control field prices. Intra-state sales of gas remained largely unregulated.

The Commission's original attempts to control prices of properties individually proved administratively unworkable and in any case seemed to do little to stop price rises; hence from 1960 it practised what was called 'area rate regulation', initially freezing prices in each area by setting a ceiling price on new contracts equal to the previous highest price in that area and a ceiling price on old contracts equal to the average for the area. Hearings were at the same time to be held in each area to consider changes in these limits, but the first of these, on the Permian Basin covering adjacent parts of New Mexico and Texas, set ceilings even lower than those of 1960 and established the principle of a two-tier system of gas prices, with 'old' gas priced lower than 'new' (Helms, 1974:22–3).

Because of pipeline companies' need to ensure continuity of supply, there are generally long-term contracts between them and producers, with prices that may be fixed in money terms or be subject to various forms of escalation and indexation (Helms, 1974:12–16). FPC regulation of producer prices began after a period of over a decade in which new contract prices had been rising substantially and fairly consistently in real terms, with a consequent, but less rapid, rise in average well-head prices (Helms, 1974:20–2). It was this that had encouraged the view that producer price regulation was necessary. Where an industry is substantially competitive, as most studies have found the producer section of the gas industry to be (Helms, 1974: 9–10), there is of course no presumption that unregulated prices will tend to be excessive in the sense of exceeding marginal cost. Long-term contracts, unless their prices are indexed to some such indicator as new-contract prices, also tend to remove the presumption that rising prices will create windfall rents for already established producers. The FPC, indeed, proceeded to forbid new contracts concluded after 3 April 1961 from having most kinds of 'indefinite pricing' clauses that might in effect link prices in old contracts to the current market situation (Helms, 1974:15).

In his study of gas regulation, Helms concludes that the first eight years of area rate regulation led to the emergence of an excess of demand in the deficit states over inter-state supply as new rates were allowed to respond slowly at most, and old rates scarcely at all, to changes in market conditions. Since any change in the climate of decisions to explore, or even to extract, gas takes a fairly long time to have its effect on output, this shortage was not immediately evident. Helms (1974:26) shows how between 1968 and 1971 the Commission came gradually to the view that a shortage was impending and took steps to reverse the direction of its price policies. Eventually in 1974 a nationwide ceiling price for inter-state sales of 'new gas' and later one for 'old gas', were declared in place of the area rate ceilings, but Helms notes that the 'new gas' ceiling so established was 'substantially below the intra-state prices for most unregulated gas' (1974:27).

Thus by the mid-1970s the pricing of natural gas had reached, by a somewhat different route, the three-tier system established for oil. Domestic gas sold inter-state provided the two lowest tiers, while intra-state gas in the producing states, and sometimes imported liquefied natural gas in the non-producer states (Helms, 1974:48), formed a free market top tier with considerable differentials between that and the other two. The system, unlike that for oil, remains in force at the time of writing, though with some measure of decontrol introduced by the Carter administration in 1979. It may not survive.

The Optimality of the US Federal System of Rent Taxes on Oil and Gas

The US federal system of distributing rents from the extraction of hydrocarbons has been in a state of flux for about ten years. In order to consider the main questions in a coherent way, we shall first discuss briefly the complicating side-issue of

(i) depletion allowances;

and then consider each of the three sets of devices that have been tried since 1954:

(ii) cash bonus bidding, with fixed *ad valorem* royalty, and income tax (a reference being given also to the alternative of royalty bidding);
(iii) cash bonus bidding, with fixed *ad valorem* royalty, income tax, and controlled multi-tier pricing (modified at the end of the period by experiments with other bid variables and fixed variables);
(iv) cash bonus bidding, with fixed *ad valorem* royalty, income tax, and 'windfall profit tax' (still modified by experiments with other bid variables and fixed variables).

State controls on individual well output will not be considered, since the need for them springs from the lack of 'unitization' (a rule that each reservoir should be expoited as a unit). Nor will the fact that much rent from onshore properties goes to private landowners. These are characteristics following from aspects of American legal institutions that are not widely followed elsewhere.

Depletion Allowances

As explained above, the very generous system of depletion allowances has been much modified by the legislation of March 1975. While it operated, it often allowed investment expenses to be deducted in effect twice over for income tax purposes, and in fact the deductions bore no close relationship to the investment expenses.

These arrangements must have acted as a subsidy to oil and gas production in the sense of directing investment into it rather than into other industries. The relatively favourable tax treatment would render eligible oil and gas projects that would not have been so under equal treatment. Since this subsidy did not compensate for any clearly recognized externality that might otherwise have led too little investment to enter the industry, the presumption must be that in itself it tended to misallocate resources.

It is worth stressing that this objection is one about the allocation of resources rather than about the distribution of income. If the cash bonus bidding system of lease allocation, or indeed any other competitive system, operated as it was supposed to do, any excess profit that would have accrued to oil and gas producers as a result of the allowances would be taken into account by bidders and thus transferred to the governmental or other owners of the property in advance. Distributionally speaking, it would then be a subsidy to landowners rather than to operators.

Cash Bonus Bidding, With Fixed Ad Valorem *Royalty and Income Tax*

Without the quantitative import restrictions, the small tariff, and the over-generous depletion allowances, would the simple system applying from 1954 to operators on federal lands in and near known oil reservoirs have been an ideal one? For the present we ignore the *ad valorem* royalty, the general effect of which on mineral operations has been discussed elsewhere in this book, and concentrate on whether cash bonus bidding alone is an efficient method for collecting oil and gas rents.

This question has been much discussed during the 1970s, and two main kinds of study beyond the purely *a priori* have been used to help answer it. One is a study of actual evidence relating to such matters as competitiveness in bidding, ease of entry to the industry, and rates of return after taxes and charges within it. Comparisons can be made with the experience of states and other landowners that have applied other bidding systems, and (mainly after 1978) they will be available from the federal authorities' own experiments. Another is simulation, under a range of cost and price conditions, of allocation effects and government revenue accrual with cash bonus bidding and with various possible alternatives. The case for cash bonus bidding as a system of rent collection is that it does not distort any decision on investment or production; that, if there is sufficient competition for leases, it must realize an amount approximating to the total rent on a set of tracts as the investor most optimistic about each of them conceives it before actually drilling; and that, with minimal safeguards, it avoids the possibility of official favouritism or corruption in the distribution of rights to extract.

There are several *a priori* objections to reliance on cash bonus bidding, however. One, stressed earlier in this book, is that, because of uncertainty and risk aversion, the investor's bid must generally understate the true rent even if he has a correct view of the probability distribution of the outcomes. A second objection is that, because of the large sum that needs to be found in advance of production, and in advance even of any assurance that the property will be productive, the system tends to restrict entry to firms with large resources. It will thus tend to reduce the number of operators in the industry and perhaps lead towards oligopoly. This is more likely to be true after the large increases in real oil prices, when individual bids have exceeded $100 million. A third is that the bidder will discount the rent to the time of his bid in determining how much he should offer, and that, if his discount rate is higher than that of the government, he will bid less from the latter's viewpoint than the full rent, even if from his own viewpoint he does offer the full rent (Mead, 1977:50–1). It is not obvious that this should be likely, unless the difference is on account of sovereign risk, which may be regarded by the 'subject' as a risk that increases with time from the present, while to the 'sovereign' is no risk at all; but this point may be held to be covered by the first objection. Alternatively, those who make decisions for oil companies may perhaps have shorter time horizons than officials acting on behalf of the state ought ideally to have; the former may be anxious to see results of investments during their own working lifetimes, and on this account may make demands on their investments that exceed the productivity that the resources would have if applied to other uses.

Aside from the appeal to direct evidence or simulation studies, the answers plausibly given to these arguments run roughly like this.

To the first objection, it may be maintained that risk aversion in bidding would be irrational among large firms each bidding for a number of leases in a business in which the probabilities of successful finds in given geological conditions and in proximity to existing reservoirs are fairly well known. It might be added that what we have called above (chapter 3) the 'managerial safety margin' would also be unlikely to apply in this case. Complete failure on a particular well happens about four times out of five, even with the best technical information; so that no manager can regard a dry hole as a disaster or a serious slur on his judgement.

This answer might well have seemed reasonable for the USA in the 1950s and 1960s. Its weaknesses, however, would always have been apparent for the postwar developing countries and would have stood out clearly in the USA itself during the 1970s. In most developing countries political risk must be taken seriously by private foreign investors. However much they know about the characteristics of their leases, or even about the markets for their products, they can not assume that the rules will remain unchanged. Any investor in a developing country who was prepared to bid a sum over a group of leases equal to the present value of their expected rent for the next twenty years *on the assumption that he would remain in possession under a constant set of fiscal laws* would be extraordinarily sanguine. The horizon of foreign investors in developing countries must be short.

Even in America tax laws change, but bidders for oil and gas leases have doubtless assumed that laws will not move too drastically to their disfavour. There, however, the flaw in relying on cash bonus bidding has lain on the other side. Investors may have become increasingly knowledgeable about petroleum geology, but no one could have been expected to predict the cataclysmic leaps in the real international price of oil that occurred in the 1970s or the response of the American authorities to them. An oil company bidding for leases in the late 1950s or early 1960s could not, on any reasonable rule, allow for the price movements that were in fact to occur. If the cash bonus system had been left alone during the 1970s, without price regulation or additional taxes, oil properties on average would have earned far more than their operators could reasonably have foreseen. The economic rent received by the producers would have far exceeded their winning bids. This revealed inadequacy played a large part in stimulating first multi-tier pricing and then the tax on windfall profits. By contrast, a rent collection system based on realized profits might have reduced the apparent need for changes in the rules.

A standard answer to the second objection is that oil drilling in conditions such as those of the OCS or the Alaskan North Slope, where much of the development is now taking place, would in any case be well outside the capacity of independent small firms acting individually, and that on the other hand bidding is frequently done by consortia, so that any difficulty experienced by an individual firm in raising all the capital required for a high bid on top of developmental expenses need not prevent it from participating in oil or gas production. It may be added that the typical American concern over the smaller home-based independents does not have the same relevance for much of the rest of the world. Elsewhere independents, if they showed interest at all, would probably be American independents. Outside North America, small private oil firms, foreign or indigenous, are rare; yet it seems that since the 1970s competition has not thereby been excluded. As recent British experience has shown, the number of firms showing interest may still be considerable, which suggests that there is at least potential competition for rights to exploit. The two dozen or so larger American companies, plus a handful from Europe, provide a large selection. It is quite possible that auctioning in Britain would have made greater use of this potential competition.

The third objection may or may not be of great quantitative importance. In so far as it is significant, it points to the desirability of a device for collecting economic rents as they are realized rather than as they are foreseen.

A useful summary of evidence, as it could be read in the early 1970s, generally in favour of the cash bonus bidding system, is given by Mead (1977:53–6). Though the number of bids per US tract receiving bids was small (3.6 over the twenty years), no less than 132 firms took part as winning bidders. The eight largest buyers admittedly bought a high proportion of the OCS tracts, but this proportion declined from 85.5 per cent in 1954–66 to 62.0 per cent in 1967–early 1974. The government rejected 78 top bids from among those in the 1,377 OCS sales to early 1974; a third of these 78, which were offered again, received their top bids on average nearly ten times as high as on the former occasions, which suggests that the federal authorities exercised sensible discretion in avoiding the release of tracts for which the top bids were unreasonably low. Regression on 184 OCS tracts off the cost of Louisiana released in 1954 and 1955 showed no significant relationship between size of firm and the amount of the winning bid (the coefficient being in fact negative) or between the amount of the winning bid and whether firms bid solo or jointly. Most interestingly of all for the present purpose, these same 184 early OCS leases had shown (apparently over about thirteen years) rates of return before income tax of only 7.5 per cent. (Mead notes that after-tax rates of return would be only slightly lower.) A later statement by Mead and others (Jones *et al.*, 1979:896), referring to all 839 federal OCS leases issued between 1954 and 1962, gives a rate of return before income tax of 9.5 per cent, while they note that the average return in all US manufacturing from 1954 to 1975 before income tax was 19.2 per cent.

In total, this evidence suggests that smaller firms were not excluded; that potential competition was considerable; that, within the size range of the firms bidding, smallness did not inhibit a firm from participating in a bid for any tract, however promising; that the effects of occasional weak bidding could be avoided by official discretion; and that the federal authorities collected something approaching, or even exceeding, the economic rents of the early OCS tracts considered as a whole. The last of these inferences, if correct, would suggest either that risk aversion on the part of bidders had negligible effects, or that winning bidders (who were presumably the ones taking the most favourable views about the respective tracts that they won) made on the whole an over-favourable assessment of prospects. If the former explanation is correct and risk aversion on the part of the bidders had no impact, then it would be hazardous to transfer the experience to other situations, especially to those where bidders had less knowledge about structures or where there was significant sovereign risk.

Leland (1977:59–60), in commenting on this last bit of Mead's evidence and its interpretation, notes that, given a plausible distribution of oil yields, the result does not imply that all the rent was collected on the successful leases, since on most tracts rates of return could be expected to be subnormal. This criticism may not be entirely relevant. If 7.5 per cent before tax were a rate of return that investors would expect on all their oil-producing operations taken together, then it follows that a much higher rate of return must be required of those tracts that were successful. On these tracts taken alone, rent might be regarded as not accruing until a much higher rate of return had been reached. A tax of RRT type, applied tract by tract, would need to set a threshold rate substantially above 7.5 per cent (or whatever the overall required rate of return appeared to be). Leland's other main criticism is that Mead's early sample may not be typical, and that the tracts released in 1964–7 on the Alaska North Slope would have shown much higher returns. This may have been in part because uncertainty over the costs of extraction and delivery were greater in Alaska than in the Gulf of Mexico, though in 1954–5 offshore drilling must also have been a fairly recent development with technical problems still to be solved. The extent of reserves in Alaska may also have been less accurately guessed. Quite apart from any particular characteristics of the Alaska North Slope, however, we might expect that tracts won by bidding in the 1960s and producing into the 1970s and 1980s would perform much more profitably than the bidders could have foreseen.

The Alaskan experience has been surveyed, less favourably to cash bonus bidding, by Gaffney (1977). While recognizing the useful role that cash bonus bidding can play as part of the fiscal regime, Gaffney notes a number of problems that arise when heavy reliance is placed on it. First, a high capital barrier to entry, which reduces the number of bidders, might cause reduced competition to prevent the state's receiving fair value. Second, the internal cost of capital to the State of Alaska is probably lower than to the large oil companies. Third, smaller firms may be drained of cash by a successful bid, which may affect the allocation of funds to exploration. Fourth, risk aversion on the part of investors can reduce the size of bids, and the differential importance of risk across firms of varying size introduces a further screening device that favours large firms. These and other possible problems of cash bonus bidding are especially important when the value of leases is very large, as in Alaska.

These suspicions about bonus bidding in Alaska were thought by Gaffney to have been confirmed by Noorgaard (1977), who estimates that bonuses captured only from 9 to 16 per cent of the surplus DCF above costs and royalties in Cook Inlet. Noorgaard argues that in these valuable areas there were too few bidders to ensure effective competition, and that bidders successfully used advanced gaming strategies and statistical analysis to take advantage of the competitive environment and of the lack of expertise available to the state.

Another class of evidence comes from simulation studies. One by Kalter, Stevens, and Bloom (1975) compares the then-existing system with various alternatives. This takes account of the complex effect that changes in the *ad valorem* royalty level may have on the initial rate, and hence on the ultimate extent, of extraction of petroleum: an effect modifying the usual high-grading that would apply in the case of any mineral. Precise details of the various alternative systems examined are not made clear in the paper; technical characteristics, costs and prices are those officially estimated for the batch of MAFLA (Mississippi, Alabama, and Florida) tracts released in 1973; there is assumed to be a known probability distribution of outcomes; and the cash bonus of the winning bidder is assumed to be exactly equal to the expected present value of each lease with no allowance for risk aversion. On these assumptions the cash bonus bid, plus existing royalty, plus (pre-March-1975) depletion allowances, give as high a revenue as the highest given by any alternative. Substituting a profit share (based on the definition of profit used for income tax) for the royalty gives roughly equal revenue. Removing or reducing the depletion allowance tends to reduce revenue (presumably by making certain leases ineligible to the investor while shifting revenue from cash bonus to income tax), but for the standard reasons improves allocation in the economy. Total recovery is highest with the higher royalty (50 per cent) plus cash bonus bid and (even more) with the royalty bid alone. This is because slower recovery can increase the amount taken. It does not appear to be implied, however, that these solutions are on balance more socially efficient. (The possibility that the royalty bidder might 'walk away' if he found the well to be less productive than he had estimated is not taken into account in the simulation.)

The simulation in itself is not a conclusive test of the superiority of the cash bonus bid for revenue purposes since the study assumes that the cash bonus will absorb all the remaining present value without any deduction for risk. Yet the fact that the actual sum ($1.49 billion) realized in cash bonus bids from the MAFLA sales is only about 5 per cent below the amount ($1.57 billion) projected on those assumptions does fit with the *ex post* evidence cited by Mead. What it suggests at least is that the bidders may have made assumptions rather similar to those of the federal authorities about technical characteristics and costs, and made something like the price assumption adopted by Kalter *et al.*: $8 per barrel increasing at 8 cents a year. On these suppositions, with unchanged taxes and time discounted at 10 per cent, what they bid *would have been* about 95 per cent of the remaining rent. But the course of producer prices was *in fact* far different, and different again from what it would have been without forms of official intervention which could not have been foreseen in 1973. An unexpected rise in real product price would presumably tip the balance of revenue advantage in favour of profit-share systems.

A more elaborate simulation study conducted by Kalter, Tyner, and Hughes (1975) assumes risk aversion on the part of bidders and considers a range of possible fiscal systems, some with and some without a cash bonus bid. The systems are applied to estimated characteristics of OCS reserves and price–cost relationships considered reasonable at the time. The systems are evaluated on five criteria: government revenue, total expected production, chance of a less-than-normal profit, bonus ratio (ratio of after-tax NPV of each system to that of the existing system), and ratio of the mean after-tax NPV to its standard deviation. The last three criteria are used to assess the comparative effectiveness of the various systems in transferring risk from private investors to the government. (Presumably, low is good on the third, and high is good on all of the other four.)

The outcome of the study is that, on these five criteria together, five systems are good as, or better than, the existing cash bonus system. All these five, however, include the cash bonus as the bid variable. The five include one in which the royalty rate varies directly with the value of production in each year and four described as 'capital recovery profit-share' systems (respectively 'fixed-rate annuity', 'variable-rate annuity', 'fixed-rate British-type', and 'variable-rate British-type'). These capital recovery profit-share systems are distinguished from Internal Revenue Service-based profit-share systems (presumably HRIT in our terminology). The 'annuity method is what we have called a PPT (see chapter 6), and the 'British type' is one in which a multiple of capital expenses is deducted without interest before tax liability begins. The Internal Revenue Service-based profit-share systems come out inferior to, or no better than, the existing one.

These results appear to be generally in line with the *a priori* conclusions derived earlier in this book, except that we have omitted any consideration of the special considerations affecting the extent of extraction in the case of oil and gas.

Broadly, these two simulation studies by Kalter and others tend to vindicate the usefulness (in American conditions) of cash bonus bidding but they are consistent with the view that, if bidders are subject to risk aversion, a better outcome for government revenue or private risk taking or both may be achieved by an additional profitability-based tax than by the existing fixed-rate royalty.

Royalty bidding is subject to the objection that, the higher the rate of royalty, the greater the misallocation owing to high grading, and (what is really an extension of the same effect) that the successful bidder may 'walk away' and not exploit the lease at all if on further investigation he decides that it will not justify his investment. Dam

(1976) records that, in the royalty bidding experiment for OCS leases, winning bids ranged from 56.8 to 82.2 per cent. This is extremely high. Though the high grading effects would partly be offset by the fact that slower initial rates of extraction resulting from the higher royalty would make it possible to extract a larger quantity by primary extraction methods, with rational present-value-maximizing investors the effect must be socially wasteful unless the investor's discount rate for time was different from the social discount rate.

The evidence about the systems has been supplemented since 1978 by the amendments of that year to the OCS Lands Act. These will be considered in their chronological position under the next section.

Cash Bonus Bidding, With Fixed Ad Valorem *Royalty, Income Tax, and Controlled Multi-Tier Pricing*

The introduction of two-tier, and then three-tier, pricing can be seen as a response to the desire to prevent oil companies from receiving, at the expense of the consuming public, profits that they had never expected to enjoy, while at the same time leaving them with an incentive to extract commensurate with the new, higher, marginal value of petroleum to the USA. As we have seen, a similar device was adopted for the same reason in Australia. The earlier and longer-lasting three-tier pricing of natural gas, less justified and less a matter of coherent policy, owed its existence to the same general set of objectives. Admittedly, legislators may have been influenced more by the desire to keep down consumer prices than by the need to prevent the oil companies from reaping excess profits, but one consideration supported the other.

Multi-tier pricing, however, was an unsatisfactory solution for a number of reasons. In the first place, the user would be buying petroleum or natural gas products at below their marginal social opportunity cost and would thus not be basing his consumer decisions on the relative scarcity of the products. If the authorities had wished to avoid a transfer of income from the rest of the community to the oil and gas producers as free market prices of the products rose, the more efficient method in principle would have been to extract the windfall from the oil and gas producers and to pay it to the rest of the community by some method that did not involve a subsidy on the use of oil and gas.

Second, the distinction between 'old oil' and 'new oil' prices could not discriminate satisfactorily between oil for which a rise of price would, and oil for which it would not, affect production and investment decisions. To take the most obvious reason, oil output from any given reservoir tends to decline over the extraction period, so long as primary techniques of extraction are used. Hence oil from the sole reservoir on a particular property that had been producing in 1972 would be 'old oil', unless secondary methods were used in such a way as to raise output above its rate in the base year. Yet is is important that such a reservoir should not be abandoned so long as its marginal costs of output are less than the import price. By March 1979, the price paid for 'old oil' was little more than a quarter of the import price. Thus, unless secondary methods could actually raise output above its 1972 level, the reservoir would be abandoned well before its marginal costs of production reached anywhere near the true social marginal value of the oil to the nation. As mentioned above, an attempt was made to correct this deficiency by providing incentives for tertiary extraction techniques. It is doubtful, however, whether any complete solution on these lines would be possible.

A third deficiency was perhaps an accidental feature of the system actually adopted rather than an intrinsic fault of multi-tier pricing. This is that 'new oil' and 'new gas' were both priced in the late 1970s well below the free market price. Even if, in the example just given, a property producing throughout 1972 could have managed by secondary techniques to exceed its 1972 rate of output in 1979, the 'new oil' price that it would have received on marginal output would still have been only something like three-fifths (as viewed in March 1979) of the import price; and that price itself would apply not to the whole of its output but only to the part corresponding to twice the excess over 1972 output. If two tiers provided a second-best solution, three tiers provided a third-best that could be justified only by an irresistible political constraint.

Fourth, it is worth mentioning that multi-tier pricing of natural gas not only dissipated some rent by misallocation and diverted much of the remainder to consumers of the product, but also clearly caused a transfer of income to some consumers (who were lucky enough to get inter-state gas) from others (who, because of the disincentives to production, could not receive natural gas at all, or had to pay more for intra-state gas than they otherwise would, or who had to buy the much more expensive product imported in liquefied form). This last effect was partly, but perhaps not wholly, avoided with oil through the system of allocation.

Amendments passed in 1978 to the Outer Continental Shelf Lands Act introduced radically new elements into the system by allowing the Interior Secretary to choose a combination of any one of four bid variables with any one of a number of fixed charges and combinations of fixed charges, or indeed any other system that has no more than one bid variable (but in this last 'unspecified' case the system would be subject to veto by either House of Congress). The four specified bid variables are cash bonus, royalty, net profit share, and work commitment. The specified fixed items include the following and certain combinations of pairs of them: fixed royalty of at least $12\frac{1}{2}$ per cent, sliding-scale royalty of at least $12\frac{1}{2}$ per cent at beginning of lease period, fixed share of net profit of at least 30 per cent, specified work commitment expressed in terms of dollars spent on exploration, and fixed cash bonus. Eight combinations of bid variable and fixed element or elements were in fact specified in the legislation, but this list of combinations was not to be exclusive. In each year of the first five after passage of the amendment, not less than 20 per cent or more than 60 per cent of the area offered for leasing would be leased by systems other than the cash bonus bid and fixed royalty. The Secretary could defer payment of any part of a cash bonus, whether fixed or variable, for up to five years from the date of sale (Jones *et al.*, 1979:888–9). Recent OCS sales in the Alaska area ran on three different systems: the standard cash bonus bid plus $16\frac{2}{3}$ per cent royalty; bonus bidding plus fixed net profit share of 40 per cent; and fixed cash bonus of $2 million plus net profit share bidding (A.G. Kemp, personal communication). The federal Alaskan sale in December 1979 employed a cash bonus bid and a variable royalty, calculated on a sliding logarithmic scale, between a minimum of $16\frac{2}{3}$ per cent when the quarterly value of production adjusted for inflation was $13.2 million or less, and a maximum of 65 per cent when it was $166.8 million or more (A.G. Kemp, personal communication).

Apart from these changes, the legislation provided for a range of planning controls (designed in part for environmental protection) and provisions for compensation over environmental damage and also a measure to favour independent refiners. Lessees were also required to provide the Interior Secretary in confidence with all information obtained from exploration, development, or production on federal leases (Jones *et al.*, 1979:889–92). This last provision was apparently designed to help the government to judge the value of leases.

The variation permitted and enjoined in fiscal systems for leasing was clearly designed for purposes of experiment. The Secretary might choose bidding systems at random over a series of leases, apparently to provide a statistical test of their differential effect (Jones *et al.*, 1979:889). Jones *et al.* make the point that the experiment is not necessary since plenty of evidence is there to be analysed from the variety of systems applied by various states, and that some of the bidding systems are manifestly bad, either because of their intrinsic character or because of the high rates sometimes applied as a result of competitive bidding. They mention (p. 903) that a civic lease sale by the city of Long Beach realized a bid of 95.56 per cent of accounting profit, which, added to certain other charges including income tax, would leave the operator with only 0.75 per cent of profit. The highest net profit share bid in the sales, made by the state of Alaska in the

Beaufort Sea in December 1979, was 93.2 per cent, and several in that sale exceeded 85 per cent (A.G. Kemp, personal communication). At rates such as that of the Long Beach lease the incentive to commercially efficient management can be presumed to disappear, as experience in that case suggested that it did. The system of experiment itself might also, as they suggest, lead to distortion of decisions in the interest of avoidance since different regimes might be applied on neighbouring leases of the one operator.

The new scheme would appear to have offered an opportunity of a kind rare in economic policy studies for something approaching a controlled experiment, and the Department of the Interior has attempted to use it in this way. Actual results, as compared with simulated results, with clear implications for assessment clearly can not be expected for some years. As we have seen, both simulation studies and some pieces of comparative experience (albeit less systematically elicited) have been available in the past. It is perhaps unfortunate that the system of profit-sharing so far adopted is the one described as fixed capital recovery, which is similar to that in what we have called the 'first tax' of the British PRT. Profit is shared after a certain multiple of capital expenses has been recovered, without regard to its timing. Recent federal sales in Alaskan waters used a 0.75 capital recovery factor (identical, as it happens, with that used by the UK for the first years of its Petroleum Revenue Tax) for both fixed and bid-variable profit shares (A.G. Kemp, personal communication). Lower factors tend to be used for less difficult or deep sites. It appears that recovery factors up to 1.5 have been used; this maximum would imply that 250 per cent of capital would be recovered before the profit began to be shared. It would depend entirely on the timing of cash flow from a particular lease whether any figure chosen was about right or unduly high or low. If, to take a hypothetical and extreme example, the whole figure were recovered in a year after the capital expenses were incurred, the effective threshold rate of return would in effect be 150 per cent with the 1.5 recovery factor. This, being presumably far above the operator's discount rate, would give considerable incentives to over-investment. At the other extreme, a recovery factor of 0.25, which might be reasonable for a project with high and quick returns, would provide a very low threshold, and a corresponding deterrent effect, for a marginal project, and especially so if the recovery would be spread over a long period. This method of profit-sharing thus requires *ad hoc* judgement about each lease area to determine what the capital recovery factor should be. This increases the demands on administrative expertise and decision-making. Moreover, because the authorities cannot be confident that they have got it right for a particular area, the rate of profit share may need to be kept fairly low to reduce the distorting effects in either direction. There seems so far to have been no opportunity to examine the actual working of other profit share systems such as PPT or RRT, which offer less risk of distortion, and therefore justify higher rates, though other methods have been discussed and advocated by the Interior Department.

Cash Bonus Bidding, With Fixed Ad Valorem *Royalty, Income Tax, and 'Windfall Profit Tax'*

This is the system for crude oil designed in 1979, enacted in 1980, and supposed to be fully operative over a few years in the mid-1980s. It was to involve fully decontrolled oil prices and an extra tax to rake off, for the benefit of the public, some of the excess profits that the producers would otherwise have received.

This set of measures would overcome the first of the three deficiencies of multi-tier pricing mentioned above. It would expose the consumer, whatever compensation he received, to the current scarcity price of oil (though not yet of gas). The real fault of the tax authorized under the Crude Oil Windfall Profit Tax Act, 1980 (P.L. 96–223), is that, despite its name, it is designed as a windfall *price* tax rather than a windfall *profit* tax. Indeed, it was described as an excise. It was to take a proportion, of a size depending on circumstances, of the excess of the actual price received for each unit over an indexed base-level price. Thus the authorities, having abolished a distortion on consumption, introduced one on production. In addition to the $16\frac{2}{3}$ per cent or other royalty where that was payable, the price received by the most favoured (presumably, most price-responsive) producer would be reduced by 30 per cent of the excess of the free market price over an indexed price approximating that of 1979. There were to be three tiers of oil subject to different rates of tax, and within the two most highly taxed tiers there was to be a difference in rate between oil from independent producers and other oil. The lowest taxed tier, tier 3, consisted of newly discovered oil, heavy oil, and incremental tertiary oil. Base prices were fixed differently across the three tiers, and indexing was to be applied more generously to tier 3. Clearly, some of the difficulties of distinguishing between 'new' and 'old' oil would remain under the system, since some producers were to pay as much as 70 per cent of the excess and some only 30 per cent. The main point, however, is that the tax would leave price incentives to investment and production too low. There was also some oil exempt from the tax altogether, notably some tertiary oil. The Economic Recovery Act of August 1981 introduced in effect a fourth tier at lower rates, which would decline progressively from $27\frac{1}{2}$ to 15 per cent, for newly discovered oil; and further reductions are possible.

The Windfall Profit Tax was to be phased out over a period of thirty-three months after the end of December 1987, or after the end of the month in which the aggregate revenue from the tax minus the resulting aggregate reduction in income tax would be reported to have exceeded $227,300 million, whichever was the earlier. (Details of the Act are in 26 *U.S. Code Annotated*, Internal Revenue Code, chapter 45, paras. 4986–98).

During his first few days in office, on 29 January 1981, President Reagan lifted all federal price and allocation controls on gasoline and fuel oil (*International Herald Tribune*). A number of the price controls were in any case due to expire at the end of September 1981. Controls on natural gas, however, which were due to run out in 1985, were not affected immediately by these changes.

Is There a Solution?

If history could be re-run from some time in the early 1950s, the obvious device would have been to impose an additional profitability-related tax, such as the RRT, as a supplement to the cash-bonus bid and the income tax, on all new oil and gas properties. The proceeds of the additional tax could have been used in the 1970s to mitigate the effect on living costs of the oil price rises, if that had been considered the use of highest priority. The tax would have avoided the uncovenanted transfer of immense rents to the oil producers: the prospect to which multi-tier pricing and the 'windfall profit tax' were unsatisfactory responses. In addition, it would doubtless have reduced the size of cash bonus bids, so allaying the suspicion, justified or not, that cash bonus bidding restricts entry to the industry. Conversely, the cash bonus bidding system would remove much of the force of one objection to a profitability tax such as the RRT: that it takes only an arbitrary, if large, fraction of the economic rent. If applied with an RRT, the cash bonus bid would also have shortened the period for any tract over which RRT was actually collected, and would probably have been held to justify lower rates for an RRT, or indeed for any other profit-related tax, so reducing any possible disincentive effect that such a tax might have.

History can not be re-run. In any case, the idea of an extra profit tax on oil and gas would probably have appeared unnecessary and inequitable twenty years before the Yom Kippur war. Can anything be done now? In the first place, it would seem urgent to introduce now a tax or charge of RRT or similar form, combined with cash bonus bidding and dealing with the operation of each new oil or gas property individually. The opportunity to do this now, without further legislation, exists for up to 60 per cent of new federal offshore leases under the 1978 Act. The decision to apply a fixed net profit share in RRT form under that Act would be a purely administrative one. To make the system universal on federal OCS lands, however, or to introduce it to federal lands on-shore, legislation would need to be altered. A sufficiently high profit share under such a system would remove all the case for continuing the Windfall Profit Tax on oil from the leases affected, and legislation might be presented to suspend the operation of that tax on those leases.

Real oil and gas prices in the world market may well have further large rises to make; the 'new' oil of today may be the embarrassingly profitable 'old' oil of tomorrow. An efficient tax closely related to profitability and applying only to new property could not face plausible charges of inequity; future bids for leases could take it into account. The cash flows of the RRT applied in this way would need to be indexed; the cash bonus might or might not be treated as a deductible expense, but all other taxes, domestic and foreign, applied before this tax would be so treated; and product price might need to be fixed for purposes of the tax at a declared level over any given tax period in order to forestall avoidance by transfer pricing. Such a charge, applying only to new properties, would infringe no understandings over existing leases; it would be no more discriminatory than the Crude Oil Windfall Profit Tax. We suggest that strong arguments of equity and efficiency could be used in its favour against most of the alternatives. If such a tax were introduced as a leasing charge by the federal authorities, *ad valorem* royalties would naturally be abolished on new federal leases with a probable improvement in allocative efficiency and no presumption of any loss in total federal revenue.

The application of such a tax or charge to oil from outside the federal lands raises much more difficult questions. If it is to replace the Windfall Profit Tax and to obviate the need for similar charges in future, it must apply to all oil produced within the USA. In view of the number of other agents involved as landowners and the considerable attention given by some of the states to leasing systems, any notion of bargaining with them to adjust to a common scheme, as suggested above in chapter 15 for Australia, must probably be ruled out. US income tax law and practice would probably make it necessary to introduce any new charge 'between' the lease payments on the one hand and (federal and state) income tax on the other. If the charge applied in this way were not to risk being both inequitable and inefficient, it would need to be based very closely on the economic rent of each lease before federal and state income tax had been deducted. The RRT would seem to offer the best approximation to this objective, and, if there were an RRT in the federal leasing conditions, it would seem an obvious move to apply a tax of exactly the same form here with the same thresholds and rates. State and private leasing changes and royalties should then be not deductible but creditable, with the possible exception of any cash bonus bid, which should not be creditable and should be deductible only if similar bids were deductible under the federal leasing charge. For creditability purposes, the state and private charges might be carried forward. It would then be open to a state or any other landowner to reframe its own leasing arrangements for the future to conform to the pattern of this federal tax or charge. If it did so, it would in effect deprive the federal authorities of any revenue from this imposition and relieve its own lease-holders of any continuing charge additional to what they would have borne as federal leaseholders. At the same time, the charge itself would ensure that such leaseholders faced *no less* a levy on windfall and other rent than they would as federal leaseholders. The nature of the tax as an RRT, with contingent leasing charges and royalties creditable, would largely prevent any inefficiency from arising through a double imposition.

Further problems again, and more serious ones, would arise however if an attempt were made to apply such an RRT-type charge or tax to oil from pre-existing leases as a substitute for the Windfall Profit Tax. We leave aside any legal and constitutional questions, on which we could not offer an opinion, and concentrate on problems of administration, equity, and economic effects. The RRT-type charge in this case would be not a leasing condition but a tax imposed by specific legislation. The fact that it would be an imposition of which the leaseholders had not taken account when bidding for their leases would in principle make it no more objectionable on grounds of equity than the Windfall Profit Tax itself. The justification of either imposition is that they are collecting rents that the leaseholders could not reasonably have expected and of which they could not therefore have taken account. Federal leaseholders could, if the government so decided, be relieved of other continuing financial obligations of their leases (which in most cases would mean the royalty). State and other royalty recipients could presumably negotiate with their leaseholders to modify any continuing lease obligations in the light of the federal charge, if that were thought to be the fair thing to do, or in the interests of both parties, so that some of the revenue from the federal imposition might in that case be effectively diverted to the landowner. There would be administrative problems to be overcome. Probably the best way would be to impose the tax as, say, a 60 per cent charge on current positive net cash flow after deduction of contingent leasing charges and royalties and indirect taxes, but to allow remissions on the RRT principle where the producer could give evidence of past outlays on the producing property that had not been recovered at the threshold rate of interest (augmented by price inflation). If clear records applying to the past expenses of the particular lease claimed for purposes of deduction were not available, claims could be made, and the federal authorities would have to judge by various rules of thumb whether those claims were reasonable. Such administrative arrangements would be less than ideal but might be workable if a government were determined to make them so.

The 'lease' or 'project' basis of the tax would demand a fairly high threshold rate for the carrying forward of past expenses in order to make up for the unsuccessful tracts, on which the RRT gives no bonuses to match its imposition on successful ones. (Alternatively, unsuccessful exploration costs might be deductible from successful projects' taxable receipts, as they are in the UK. A related question is whether cash bonus payments on unsuccessful properties not otherwise deducted should be credited against the leaseholder's fiscal obligations, or alternatively should be treated as a deductible expense. Deduction of unsuccessful exploration expenses, and deduction or crediting of cash bonus payments on unsuccessful properties, would make it reasonable to reduce the threshold rate, with resulting advantages because a lower discount rate for time on subsequent less risky outlays would introduce less incentive to over-investment. Nevertheless, the need for information on unsuccessful, as well as successful, projects would greatly complicate administration if the tax were to be imposed on properties in operation at the time of its introduction, and it is probably not advisable in this case. The threshold for new expenses might instead be set lower than that for old ones.) The threshold should be indexed for inflation or alternatively set for each year of discounting by some named standard long-term rate; the former method seems preferable for retrospective application. The rate of tax should be fixed with recognition of the fact that a cash bonus bid would often have 'preceded' this tax and that federal and sometimes state income tax would 'follow' it. A 60 per cent rate before income taxes would mean that after those taxes the effective rate on marginal income might be 80 per cent or more. In addition, much or some of the rent would have been 'prepaid' through the bid variable in the many cases subject to bonus bidding; some of this would indeed have been recovered on *unsuccessful* leases, which must be regarded as contributing part of the industry's dues on successful ones. If exploration costs and cash bonus payments on an unsuccessful project were to be deductible on the next (or a chosen) successful one, then a real threshold of 15 per cent (or alternatively, say, a threshold 10 percentage points nominal above a standard long-term interest rate) might be adequate. Without either of these deductions, thresholds of 20 per cent real (or, say, 15 percentage points nominal above a standard long-term rate) might be appropriate.

Though the area is a minefield of complications, we believe that there exists a course that, if it surmounts legal and political barriers, appears capable of being administered; avoids the inefficiencies of multi-tiered pricing and windfall excises together with some of those in the standard form of the federal leasing system; can be plausibly presented as equitable to producers; collects a substantial share of the windfall rents when they exist and does not apply when they do not; and influences the leasing activities of states and other landowners in the direction of greater general efficiency but does so by giving them opportunities for raising extra revenue rather than by taking the revenue away.

Summary

Federal authorities in the USA have played several roles in fiscal policy towards the domestic production of oil and gas: extraction projects throughout the nation have been subject to federal income tax on their earnings, which is often added to state and local direct taxes; nationally, too, there has been direct federal intervention over imports, prices, and briefly allocation, while the states have for a period in the past controlled production within their jurisdictions; and finally, over federal lands alone, principally areas off-shore, the federal government has applied leasing systems designed to extract residual economic rent, a role played by other landowners (state, civic, and private) for the rest of the nation's territory. Though many projects are subject to both federal and state tax, there is not quite the degree of potential conflict between tiers of government present in Australia, where both sets of authorities may collect special taxes directed at mineral rent, or the actual conflict present in Canada. Until recently one authority, the landowner, has applied the special mineral tax in each case. This rule has perhaps been mildly breached, however, by the Crude Oil Windfall Profit Tax of 1980.

The US federal income tax system has discriminated in favour of oil and gas production in a way that in itself has possibly introduced distortions towards relative over-investment in those industries. This discrimination, however, has been greatly reduced since 1975.

The *ad valorem* royalty in the US federal system of oil and gas leasing probably deterred investment in some properties where exploitation would have been socially desirable, and possibly tended to cut short production prematurely. Because of the special characteristics of oil and gas drilling, however—namely that the extent of extraction is inversely related to its initial pace, and the possibility that investors may apply a different discount rate for time from that of society—the royalty may sometimes have shifted extraction towards, rather than away from, the optimum.

Whatever these possible distortions, the federal system of cash bonus bidding for leases was probably an efficient way of collecting the residual economic rent so long as bidders were on familiar territory and so long as product prices did not depart radically from historical patterns. Since prices did precisely this in the 1970s, the inadequacy of the cash bonus bidding system and fixed royalty alone (and of similar state and private methods) to collect the rents was then recognized in practice, first by the introduction of multi-tier pricing for oil, and then by a so-called 'windfall profit tax' designed to subject consumers to the world price but to retain a self-adjusting multi-tier price system for producers. A longer-standing multi-tier price system has also been applied to natural gas. From the late 1970s federal experiments also began on more profit-related fixed elements in the leasing system as well (perhaps mistakenly) as different bidding methods.

Neither multi-tier pricing nor the Windfall Profit Tax was satisfactory. Multi-tier pricing introduced distortions of allocation to both consumption and production, thus dissipating some of the economic rent as well as diverting some of it to consumers and (especially in the case of gas) effecting a transfer of income from some consumers to others. The inaptly named 'windfall profit tax' will retain some of the distortions on the side of production and almost certainly will fail to deliver all of the remaining windfall rent to the federal authorities. More hopeful is the authorized experimentation since 1978 with a variety of other leasing systems, including the combination of cash bonus bidding with a fixed profit share, though we believe that the form of profit share typically used is less than ideal.

The solution of choice would appear to involve introduction of an additional tax or charge closely related to profitability on the production of all US oil and gas, ideally on 'old' as well as 'new' properties and on the output of state and private, as well as federal, lands. The Windfall Profit Tax or any similar levy that might be contemplated for the future would then become unnecessary. The new imposition could be introduced as part of the leasing system for new federal leases, as existing legislation already permits on up to 60 per cent of those on OCS lands. Further legislation would be required to cover new oil from all of the federal lands, and a tax in a different legal category, but of the same fiscal form, would need to be applied to make it likely that the same profit share would be taken by one authority or other on oil from new non-federal leases and from all pre-existing leases. We have set out the requirements that might make such a system of charges workable and have argued that, with appropriate tax thresholds, it would be unimpeachable on grounds of fairness to producers. The addition of such a tax could make possible a complete deregulation of gas, as well as oil, with minimal windfall profits to producers (whether in response to past, or to any future, large real price rises). Some form of compensation could thereby be paid, if appropriate, to the consuming public for the losses that such price rises would otherwise entail. The pricing, fiscal, and federal leasing system could be stable again (as it was from 1954 to 1971) and without the frenzied changes that have occupied so much political time over the last decade.

19
CONCLUSION

We have attempted in this study to clarify some of the issues relating to the extraction of revenue by governments from non-renewable natural resources. The conclusions drawn have depended on a combination of *a priori* reasoning and appeal to public experience. We have not undertaken simulation studies of the effects of various tax systems. This has exposed the element of judgement in evaluation where a number of possibly conflicting considerations have a bearing on deciding which of a set of policies is optimal by some criterion. This element of judgement over the relevance of differing models or the balance of diverse forces is inevitable where an ultimate practical recommendation in social matters is required.

On the other hand, we have tried to be as realistic as possible about the circumstances in which fiscal arrangements for mining are actually made. In particular, we have stressed what seems to be an overwhelmingly important element: the ignorance of all parties, especially governments, about the outcome of mining investments before they are actually undertaken. Given this ignorance, certain simple and otherwise attractive solutions to the problem are not of much practical use.

With the uncertainty of the outcome of mining investment that follows from the ignorance of all parties, we assume also that investors are subject to what in technical language is called 'risk aversion', namely that they will expect rewards for taking risks. This has the important implication that, if a government by its fiscal arrangements can reduce the risk involved in the investment, the investors will settle for less.

The third major supposition that we have made about the world is that there is usually a potential for substantial competition in the market for mineral investment opportunities, in other words, that a government normally does not need to put itself in the position of having to negotiate with one investing firm and one alone. (This question is explored in chapter 6, and some evidence on the matter can be drawn from the case studies, especially from the Papua New Guinea experience outlined in chapter 14.) If the investor enjoys the power of a 'sole buyer' in his relationship with the government, this is generally, we suggest, because of the government's ignorance or the way in which it has chosen to play the rules. The implication of this belief is that the government does not usually need to enter into a game of bargaining and bluff with the investor but can rely on something in the nature of a 'market price'. Because of the government's ignorance, however, the market price can not satisfactorily be expressed for any project in advance as a price of the resource in terms of a sum of cash or an *ad valorem* rate of duty on the product, but should rather be expressed as a price of investment in terms such as that of the discount rate that at the time investors will apply to returns on their investments. This statement may be qualified in certain cases in which rights can effectively be auctioned to applicants with sufficient knowledge and capacity to spread their risks, but such fixing of a price in advance by auction cannot be a complete substitute for a valuation based on retrospective knowledge. The possibility of competition among investors means that a small Third World country need not, if it adopts sensible policies, accept a position of weakness over the exploitation of its minerals. Examples such as those of Botswana and Papua New Guinea support this conclusion.

These three important elements (ignorance, risk aversion, and potential competition) suggest that, where other institutions permit, the extraction of economic rent from mining undertaken by private companies should be effected, wholly or in large part, through taxes whose yields are closely related to realized profitability as the investor perceives it, and that the fiscal rules that an investor is to face should be made clear to him before he begins any outlays even on exploration. The main levies that might be combined with taxes of this kind are (for reasons given below): (i) company profit tax where it already exists, and (ii) a flat fee determined by auction, either where exploration has been carried out at public expense before mining rights are allocated, or in the case of petroleum or natural gas. We have tried to elaborate various systems of taxation that might fit the requirement of tailoring yields closely to profitability, and we have compared them principally on the interlocking grounds of neutrality, capacity for raising revenue, and administrative feasibility.

We have concentrated on the extraction of rent from mining undertaken by private investors, because in most of the mining that actually takes place a private firm both manages the operation and has a substantial equity. There are generally good grounds for this, especially in small Third World countries. There are, however, at least two possible reasons of a strictly 'economic' kind why a government might sensibly buy some equity at the market value in a mine within its territory: (i) because its participation would significantly reduce the risk to the private investor and hence lower the 'supply price' of the investment, with resulting benefits to government revenue which exceeded the social cost of the extra risk that the government thereby accepted; in other words, because the government's own proper discount rate would otherwise be lower at the margin that that of the investor; or (ii) because the government could borrow for the purpose more cheaply than the investor. In addition, however, government equity may be desired for what can broadly be called 'political' reasons: because the idea of ownership is attractive, or because it is thought to imply a greater degree of control. Where equity is desired for reasons such as these and its purchase 'at cost' cannot be justified on economic grounds, it may reasonably be taken by one of the methods outlined at the end of chapter 7 as a substitute for tax that would otherwise be due. Equity may be acquired by the government in such a way that the effect of taking it corresponds closely to that of one or other of the profit-related taxes. Any disadvantage of such acquisition of equity over simple collection of the corresponding taxes lies mainly in possible differences in tax treatment on the part of the investors' home governments (as mentioned near the end of chapter 10).

As fiscal instability is an important matter related to the extraction of mineral rent, we have in chapter 12 briefly considered ways in which the fluctuations in economic activity and wage employment that might otherwise be generated by mining projects and by government spending based on mining revenue might be mitigated. Most important among such methods is the introduction of a form of fiscal discipline whereby mining revenues are spent by governments at rates determined by longer-run expectations of their availability, and government reserves and debt are correspondingly varied from year to year. Such discipline might be achieved, for example, by instituting a statutory obligation to pay mining revenues into a special fund which could be drawn on for purposes of government spending only under certain rules, and by simultaneously agreeing to plan for a rate of real growth in government spending from year to year which seemed likely to be sustainable over a period such as a decade ahead.

On the question of the particular form of additional profit tax to be used for the extraction of mineral rent, the balance of considerations favours either what we have described as the resource rent tax (RRT) or what we have described as the Brown tax (BT), unless administrative difficulties peculiar to the country exclude them. These two forms of tax are especially likely to be administratively advantageous (by comparison with the 'higher rate of income tax' (HRIT) and the 'progressive profit tax' (PPT), which are the two main alternatives considered) where a corporate income tax does not already exist.

Overseas creditability, administrative ease, or neutrality among industries: one or more of these considerations will often lead to the conclusion that, where a corporate income tax already exists, any additional profit tax should be applied on top of it rather than replacing it in its application to mining investments; but in a country dependent largely on foreign investment the regular income tax may advantageously be replaced entirely by one of the forms proposed for additional profit tax unless overseas creditability clearly favours the income tax.

US experience with the use of the 'cash bonus bid' (the auctioned fixed fee) for both allocating rights to, and extracting rents from, petroleum and natural gas suggests that it may make an important contribution to the collection of rents in appropriate circumstances, though it has clearly been recognized as inadequate on its own in times when product prices are undergoing vast unexpected changes. The cases in which it appears to be useful are (i) those in which (because of prior government exploration) much is known about deposits before rights are allocated, and (ii) those in which rights to petroleum or natural gas are released in the form of numerous small blocks in areas of known geological character. In these circumstances the addition of an auctioned fixed fee to a profitability-based tax relieves the latter's dilemma between charging so high a rate that incentives to economizing diminish and missing a substantial share of the rent.

Indexation for exchange rate changes should normally be applied to the RRT or the PPT. Either indexation for inflation in the investor's home currency or attachment of the 'threshold' to some listed international interest rate should also be applied to the RRT or the PPT. (These statements are explained and qualified in chapter 10, and the forms of tax themselves are described in chapter 7.) Though inconsistent in principle with the possibility of strict neutrality, the use of more than one 'threshold' in the case of RRT or the PPT may in practice make it possible to approximate more closely to neutrality and to revenue maximization (a point argued, though not conclusively, in chapter 10). No simple universal statement can be made on whether it is better to base the additional profit tax on the project, on the company's operations within national boundaries, or on some compromise between these two bases, but, where exploration is carried out privately and is unsubsidized, it is probably best to go at least as far in the direction of a 'company' basis as to allow the deduction of the cost of unsuccessful exploration against positive cash flow on the next project that would otherwise be taxable. The rules for creditability against US tax may also dictate at least some approach to the use of a 'company' basis.

A Model for Fiscal Arrangements on Mining

The typical pattern of fiscal arrangements for mining to which these recommendations lead might be something like this. There would be an additional profit tax, of resource rent tax or Brown tax form, with its provisions uniform within each mineral industry, though the threshold applied on the former type of tax might vary over time according to fixed rules designed to keep it in a coherent relationship with the discount rate applied by investors in the industry concerned. Normal company profit tax might or might not be charged in addition. The maximum marginal rate of tax on net cash flow should generally be between 70 and 80 per cent, or possibly rather higher on petroleum, except where an auctioned fixed fee is operated under favourable circumstances, in which case the maximum might be 60 to 70 per cent for hard minerals and 75 per cent for petroleum. Auctioning of rights for a fixed fee would be applied in the case of petroleum and natural gas, with rights allocated in numerous small blocks, though this should not exclude the requirement for lessees with rights in a single field to operate together. Exploration for hard minerals might be undertaken by the government (probably with help from international agencies in the case of a small Third World country) or subsidized. Generally the case for government involvement would be stronger in relation to preliminary work of the form commonly described as 'geological survey', and weaker for the testing of mineral deposits requiring specialized expertise. If exploration were undertaken by the government, mining or operating rights would be auctioned to investors for a once-for-all cash fee, the government first indicating, beside the statutory fiscal terms, roughly what capital contribution if any it would propose to make to any resulting project. If exploration were not undertaken by the government and there were no auction of extracting rights, an exploration licence would carry some near-automatic rights to exploitation under existing law, subject only to a time limit and to agreement on such matters as environmental protection and the respective capital contributions of government and investor. Conditions other than strictly fiscal requirements set on mining investments should be limited to (i) those embodied in general laws; (ii) those required to prevent, or to compensate for, some external social cost; and (iii) those that could reasonably be held to provide for the host country quantifiable benefits whose value would exceed the social cost entailed by any loss of government revenue through reduction of the investor's gross profit. Government investment in infrastructure associated with a mining project would be rewarded with equity on the same terms as the investor's own capital expenditure, unless the facility produced a readily marketable input (such as electricity) for the mine, in which case this input might be sold to the mining company on tariffs that were likely to cover its full cost. In general, all capital expenditure by government that was solely directed to the needs of a mine would be rewarded with equity on the same terms as the private investor's expenditure, but the *amount* of the government's capital contribution would be subject to negotiation. If the requirement on 'political' grounds for equity in excess of what the government considered it prudent to buy 'at cost' were irresistible, the government would adopt one of the methods for substituting the receipt of equity for the additional profit tax. There would be legislation to make this possible and to define its maximum extent.

In the text of the book we have provided a number of qualifications to the recommendations implied in this model. The principal foreseeable circumstances in which it might need to be seriously modified are the unavoidable existence of monopoly in the supply of investment and insuperable administrative difficulties in devising and administering a new form of profit tax. We do not believe, however, that such ineluctable obstacles are typical. Hence we propose the set of arrangements outlined in table 19.1, as a standard with a sound rationale behind its provisions, and submit that good reasons should be required for departures from it.

Table 19.1

Decisions on Fiscal System

The starting point of this table is the presumption that the desirable norm for a fiscal system on minerals will be to include some tax on rent that is closely related to profitability, probably an RRT, with the total top marginal rate of taxes on cash flow not more than about 80 per cent. There will normally be a company income tax with the rent tax. There may or may not be an auctioned fixed fee to allocate the lease and to contribute to the extraction of rent. The rules are assumed to be fixed across each mining industry. The table indicates possible variants that are appropriate under various circumstances.

	Circumstance	*Response*						
		(A) Type of rent tax	(B) Allocation by auctioned fixed fee (cash bonus bid)	(C) Combination with income tax	(D) Top effective marginal rate of tax on cash flow	(E) Multiple rates and thresholds on RRT and PPT	(F) Equity participation by government	(G) Fiscal terms negotiated by project
1	Exploration undertaken by government		Yes					
2	Value of each lease small in relation to world total		Yes					
3	Low risk perceived in industry	Favours (relatively) HRIT as against PPT, RRT and BT	Yes					

4	Costs low in relation to value of output				Top rate *may* be above 80%, say to 90%, if no cash bonus bid		
5	Lack of government knowledge of investors' discount rate	Tends to favour HRIT and BT as against PPT and RRT				Favours multiple rates and thresholds	Tends to favour equity as against PPT or RRT
6	Most mining investment from overseas			*May* favour applying rent tax without income tax			
7	Much US investment expected			Probably demands addition of income tax			
8	Government averse to risk in project revenue (e.g. because individual project is very large in relation to government budget)	Favours HRIT over PPT over RRT over BT					

Table (*cont.*)

	Circumstance	*Response* (A) Type of rent tax	(B) Allocation by auctioned fixed fee (cash bonus bid)	(C) Combination with income tax	(D) Top effective marginal rate of tax on cash flow	(E) Multiple rates and thresholds on RRT and PPT	(F) Equity participation by government	(G) Fiscal terms negotiated by project
9	Government with very high discount rate for time (impatience for revenue)	Favours HRIT over PPT over RRT over BT						
10	Poor capacity for tax enforcement	*May* favour *ad valorem* duty, preferably on sliding scale						
11	No existing company income tax	Favours (relatively) RRT and BT against HRIT and PPT		May not be necessary to devise income tax				
12	One producer or cartel has effective monopoly of industry							

	(a) which is secure					Fixed terms may still be preferable
	(b) which may be challenged					Negotiation may give better deal
13	Desire to prolong oil or gas extraction	*Ad valorem* duty on sliding scale included				
14	High possibility of political pressure for renegotiation			High rate, if no great government equity participation	Favours equity participation e.g. by substituting equity for rent tax (RRTE or CI)	
15	Addition of cash bonus bid in appropriate conditions			May be lower than otherwise (e.g. 65–75% instead of 80–90%)		
16	Oil, natural gas		Yes (as under 2 above)	Top rate may be up to say 90% if no cash bonus bid; otherwise perhaps 75%		

Table (*cont.*)

Circumstance	*Response* (A) Type of rent tax	(B) Allocation by auctioned fixed fee (cash bonus bid)	(C) Combination with income tax	(D) Top effective marginal rate of tax on cash flow	(E) Multiple rates and thresholds on RRT and PPT	(F) Equity participation by government	(G) Fiscal terms negotiated by project
17 Hard minerals		No (except as under 1 above)		Top rate probably not above 80%			

REFERENCES

Aharoni, Y. (1966). *The Foreign Investment Decision Process*. Cambridge, Mass.: Harvard University.

Allan, B. and Hinchliffe, K. (1982). *Planning, Policy Analysis and Public Spending: Theory and the Papua New Guinean Practice*. Aldershot: Gower Press.

American Society of International Law: Proceedings (1967).

Arndt, H. W. *et al.* (1980). *The World Economic Crisis: A Commonwealth Perspective*. London: Commonwealth Secretariat.

Arsjad, Anwar (1980). Trade strategies and industrial development in Indonesia. *In* R. Garnaut (ed.), *ASEAN in a Changing Pacific and World Economy*. Canberra: Australian National University Press.

Averch, H. and Johnson, L. L. (1962). Behaviour of the firm under regulatory constraint. *American Economic Review*, 52, 5, December, pp. 1053–69.

Booth, A. and McCawley, P. (eds.) (1981). *The Indonesian Economy During the Soeharto Era*. Kuala Lumpur: Oxford University Press.

Bostock, M. and Harvey, C. (eds.) (1972). *Economic Independence and Zambian Copper*. London: Praeger.

Brandt, W. *et al.* (1980). *North–South: A Programme for Survival*. London: Pan Books.

Brigden, J. B. *et al.* (1929). *The Australian Tariff: an Economic Enquiry*. Melbourne University Press.

Brown, E. Cary (1948). Business income taxation and investment incentives. *In Income, Employment and Public Policy: Essays in Honor of Alvin H. Hansen*. New York: Norton.

Brown, R. (1977). The choice of law provisions in concession and related contracts. *In* Brown and Faber (1977).

Brown, R. and Faber, M. (1977). *Some Policy and Legal Issues Affecting Mining Legislation and Agreements in African Commonwealth Countries*. London and Brighton: Commonwealth Secretariat and Institute of Development Studies.

Campbell, H. and Scott, A. (1979). Attenuation vs postponement. Unpublished paper.

Campbell, H. and Scott, A. (1980). Cost of learning about the environmental damage. *Economic Record*, 56, 152, March, pp. 36–53.

Caves, R. E. (1978). Policies toward trade in raw materials. *In* Crawford and Okita (1978), pp. 288–311.

Commonwealth Secretariat (n.d.). *Compendium of Mining Legislation and Agreements in African Commonwealth Countries*. London: Commonwealth Secretariat.

Copland, D. (1931). A neglected phase of tariff controversy. *Quarterly Journal of Economics*, 45, February, pp. 289–308.

Crawford, J. G. and Okita, S. (eds.) (1978). *Raw Materials and Pacific Economic Integration*. London: Croom Helm.

Crommelin, M. and Thompson, A. R. (1977). *Mineral Leasing as an Instrument of Public Policy*. Vancouver: University of British Columbia Press.

Dafter, R. (1981). Mutiny on the oil bounty. *Financial Times*, 11 August 1981.

Dafter, R. (1982). Government fails to please offshore oil companies. *Financial Times*, 10 March 1982.

Dam, K. W. (1976). *Oil Resources: Who Gets What How?* University of Chicago Press.

Department of Energy (UK) (1977, 1978, 1979). *Development of the Oil and Gas Resources of the United Kingdom*. London: HMSO.

Dowell, R. (1980). Profit based royalties and productive efficiency. *Resources and Energy*, February, pp. 103–30.

Drysdale, P., Nakatani, I. and Smith, B. (n.d.). Market structure, ownership ties and long-term contracts in Australia's resources trade. Unpublished paper, Japan–Australia Project, Australian National University, Canberra.

Economist (1980). Loose change. *The Economist*, 29 March 1980, pp. 34–5.

Epple, D. N. (1975). *Petroleum Discoveries and Government Policy*. Cambridge, Mass.: Ballinger.

Erickson, G. K. (1977). Work commitment bidding. *In* Crommelin and Thompson (1977), pp. 61–77.

Faber, M. L. O. (1977). The fiscal regime. *In* Brown and Faber (1977).

Faber, M. L. O. and Brown, R. (1980). Changing the rules of the game: political risk, instability and fair play in mineral concession contracts. *Third World Quarterly*, 2, 1, January, pp. 100–19.

Faber, M. L. O. and Potter, J. G. (1971). *Towards Economic Independence: Papers on the Nationalization of the Copper Industry in Zambia*. Cambridge University Press.

Financial Times (1980). UK to license 90 new North Sea blocks. *The Financial Times*, 7 May 1980.

Fitzgerald, T. M. (1974). *The Contribution of the Mineral Industry to Australian Welfare*. Canberra: Australian Government Publishing Service.

Fritzsche, M. and Stockmayer, A. (1978). Mining agreements in developing countries. *Natural Resources Forum*, 2, 3, April, pp. 215–27.

Gaffney, M. (1977) Oil and gas leasing policy: alternatives for Alaska in 1977. A report to the State of Alaska and to the Alaska State Legislature Interim Committee on Oil and Gas Taxation and Leasing Policy. Anchorage, Alaska: State of Alaska.

Garnaut, R. G. (1979a). Survey of recent developments. *Bulletin of Indonesian Economic Studies*, 15, 3, November, pp. 1–42.

Garnaut, R. G. (1979b). Australia's shrinking markets. *In* L. B. Krause and Sueo Sekiguchi (eds.), *Economic Interaction in the Pacific Basin*. Washington, DC: Brookings Institution.

Garnaut, R. G. (1980). General repercussions of the minerals boom in the segmented Indonesian economy. *In* Garnaut and McCawley, 1980, pp. 413–23.

Garnaut, R. G. (1981). The framework of economic policy-making. *In* J. A. Ballard (ed.), *Policy Making in a New State: Papua New Guinea, 1972–1977*. Brisbane: University of Queensland Press, pp. 157–211.

Garnaut, R. G. and Anderson, K. (1980). ASEAN export specialisation and the evolution of comparative advantage in the western Pacific region. *In* R. G. Garnaut (ed.), *ASEAN in a Changing Pacific and World Economy*. Canberra: Australian National University Press, pp. 374–412.

Garnaut, R. G. and Clunies Ross, A. I. (1975). Uncertainty risk aversion and the taxing of natural resource projects. *Economic Journal*, 85, 338, June, pp. 272–87.

Garnaut, R. G. and Clunies Ross, A. I. (1979). The neutrality of the resource rent tax. *Economic Record*, 55, 150, September, pp. 193–201.

Garnaut, R. G. and McCawley, P. T. (eds.) (1980). *Indonesia: Dualism, Growth and Poverty*. Canberra: Research School of Pacific Studies, Australian National University.

Gillis, M. *et al.* (1978). *Taxation and Mining*. Cambridge, Mass.: Ballinger.

Gillis, M. *et al.* (1980). *Tax and Investment Policies for Hard Minerals: Public and Multinational Enterprises in Indonesia*. Cambridge, Mass.: Ballinger.

Gordon, A. (1972). The prospects for new mine investment. *In* Bostock and Harvey (1972), pp. 189ff.

Gordon, H. Scott (1954). The economic theory of a common property resource: the fishery. *Journal of Political Economy*, 42, 2, April, pp. 124–42.

Gutman, G. O. (1973). Objectives, strategy and tactics in negotiations for mining projects. *In* J. G. Zorn and P. Bayne (eds.). *Foreign Investment, International Law and National Development*, Papers of the Seventh Waigani Seminar. Sydney: Butterworth.

Gray, L. C. (1914). Rent under the assumption of exhaustibility. *Quarterly Journal of Economics*, 28, May, pp. 466–89.

Gregory, R. (1976). Some implications of the growth of the mining sector. *Australian Journal of Agricultural Economics*, 20, 2, August, pp. 71–91.

HMSO (1974). *United Kingdom Offshore Oil and Gas Policy*. Cmnd 5696, 11 July 1974. London: HMSO.

Hamilton, A. (1978). *North Sea Impact*. London: International Institute for Economic Research.

Harrigan, F. J. (1977). Economic rent and government policy (1964–1976) in the North Sea province. Unpublished MA dissertation, University of East Anglia, Norwich.

Healy, A. M. (1967). *Bulolo, a History of the Development of the Bulolo Region, New Guinea*. New Guinea Research Bulletin no. 15. Canberra and Port Moresby: New Guinea Research Unit, Austrialian National University.

Helms, R. B. (1974). *Natural Gas Regulation*. Washington, DC: American Enterprise Institute for Public Policy Research.

Herfindahl, O. C. (1974). The value of mineral surveys to economic development. *In* D. M. Brooks (ed.), *Resource Economics: Selected Works of Orvis C. Herfindahl*. Baltimore: Johns Hopkins University Press.

Herfindahl, O. C. and Kneese, A. V. (1974). *Economic Theory of Natural Resources*. Columbus, Ohio: Merrill, for Resources for the Future.

Hotelling, H. (1931). The economics of exhaustible resources. *Journal of Political Economy*, 39, 2, pp. 137–75.

Hughes, H. (1975). Economic rents: the distribution of gains from mineral exploitation, and mineral development policy. *World Development*, 3, 11 and 12, pp. 811–25.

Hunter, A. (1965). The oil industry: the 1963 agreements and after. *Bulletin of Indonesian Economic Studies*, 2, September, pp. 16–33.

Industries Assistance Commission (Australia) (1976a). *Report on the Mining and Petroleum Industries*. Canberra: Australian Government Publishing Service.

Industries Assistance Commission (Australia) (1976b). *Crude Oil Pricing*. Canberra: Australian Government Publishing Service.

Johnson, C. J. (1981). Minerals objectives, policies and strategies in Botswana–analysis and lessons. *Natural Resources Forum*, 5, 4, October, pp. 347–67.

Jones, R. O., Mead, W. J. and Sorenson, P. E. (1979). The Outer Continental Lands Act Amendments of 1978. *Natural Resources Journal*, October.

Kalter, R. J., Stevens, T. H. and Bloom, O. A. (1975). The economics of outer continental shelf leasing. *American Journal of Agricultural Economics*, 57, 2, May, pp. 251–8.

Kalter, R. J., Tyner, W. E. and Hughes, D. W. (1975). *Alternative Leasing Strategies and Schedules for the Outer Continental Shelf*. Ithaca, NY: Cornell University Press.

Keating, P. (1982). Address to the 1982 Australian Coal Conference. Brisbane, 21 April.

Keesing's Contemporary Archives.

Kemp, A. G. (1975). Fiscal policy and the profitability of North Sea oil exploitation. *Scottish Journal of Political Economy*, 22, 3, November, pp. 237–60.

Kemp, A. G. (1976). The taxation of North Sea oil. University of Aberdeen, Department of Political Economy, *North Sea Study Occasional Papers*, no. 11, August.

Kemp, A. G. and Cohen, D. (1980). The impact of the system of petroleum taxation in the UK on oil operations and government revenue. *Fraser of Allander Institute Research Monograph*, no. 8. Glasgow: University of Strathclyde Press.

Leland, H. E. (1977). Comment. *In* Crommelin and Thompson (1977), pp. 57–60.

Leland, H. E. (1978). Optimal risk sharing and the leasing of natural resources, with application to oil and gas leasing on the OCS. *Quarterly Journal of Economics*, 92, August, pp. 413–37.

Lloyd, P. J. (1981). The optimal tax on mines. Unpublished paper, Department of Economics, Research School of Pacific Studies, Australian National University, Canberra.

Lowenfield, A. F. (1967). Diplomatic intervention in investment disputes. *American Society of International Law: Proceedings*, pp. 96–101.

McAvoy, P. W. (ed.) (1977). *Federal Energy Administration: Report of the Presidential Task Force*. Washington, DC: American Enterprise Institute for Public Policy Research.

McCawley, P. (1979). Industrialization in Indonesia: Developments and prospects. *Occasional Paper* no. 13, Development Studies Centre, Australian National University, Canberra.

McCawley, P. (1980). Stability, equity, growth and devaluation: A new direction for Repelita III. *In* Garnaut and McCawley (1980), pp. 539–54.

McDonald, S. L. (1976). Taxation system and market distortion. *In* R. J. Kalter and W. A. Vogely (eds.), *Energy Supply and Government Policy*. Ithaca, NY: Cornell University Press, pp. 26–50.

McKinnon, R. I. (1976). *Money and Finance in Economic Growth and Development*. New York: Dekker.

McLean, J. G. and Haigh, R. W. (1954). *The Growth of Integrated Oil Companies*. Cambridge, Mass.: Harvard University Press.

Manning, C. (1980). Dualism in labour markets and labour market segmentation in Indonesian manufacturing. *In* Garnaut and McCawley (1980), pp. 307–20.

Marshall, A. (1920). *Principles of Economics*. London: Macmillan, 8th edn. 1952.

Martin, A. (1972). *Minding Their Own Business*. London: Hutchinson.

Mead, W. J. (1976). Petroleum: An unregulated industry? *In* R. J. Kalter and W. A. Vogely (eds.), *Energy Supply and Government Policy*. Ithaca, NY: Cornell University Press, pp. 130–60.

Mead, W. J. (1977). Cash bonus bidding for mineral resources. *In* Crommelin and Thompson (1977), pp. 46–57.

Mead, W. J. and Deacon, R. T. (1979). Proposed windfall profit tax on crude oil: Some major errors in estimation. *Journal of Energy and Development*, V, 1, Autumn, pp. 32–43.

Mikesell, R. F. (1975). *Foreign Investment in Copper Mining: Case Studies of Mines in Peru and Papua New Guinea*. Baltimore: Johns Hopkins University Press, for Resources for the Future.

Mikesell, R. F. (1979). *The World Copper Industry: Structure and Economic Analysis*. Baltimore: Johns Hopkins University Press.

Mikesell, R. F., Bartsch, W. H. and Bahrman, J. N. (eds.) (1971). *Foreign Investment in the Petroleum and Mineral Industries: Case Studies of Investor-Host Country Relations*. Baltimore: Johns Hopkins University Press.

Mikesell, R. F. (1981). Options for packaging agreements to meet host country and investor financial goals. *Materials and Society*, 5, 4, pp. 449–54.

Moran, T. H. (1975). *Multinational Corporations and the Politics of Dependence: Copper in Chile*. Princeton University Press.

Morgan, J. D. Jr (1974). The mineral position in the United States. *In* R. W. Marsden (ed.) *Politics, Minerals and Survival*. Madison, Wisconsin: University of Wisconsin Press, pp. 3–26.

Nash, G. D. (1968). *United States Oil Policy, 1890–1964*. University of Pittsburgh Press; reprinted Greenwood Press, Westport, Conn., 1976.

Noorgaard, R. B. (1977). Uncertainty, competition, and leasing policy, Appendix E. *In* Gaffney (1977).

O'Faircheallaigh, C. (1980). The role of foreign investment in mineral development: A comparative analysis. Unpublished PhD thesis, Research School of Pacific Studies, Australian National University, Canberra.

Orchard, J. E. (1922). The rent of mineral lands. *Quarterly Journal of Economics*, 36, pp. 290–318.

Palmer, K. F. (1980). Mineral taxation policies in developing countries: An application of resource rent tax. *IMF Staff Papers*, 27, 3, September, pp. 517–41.

Penrose, E. T. (1959). *The Theory of the Growth of the Firm*. Oxford: Basil Blackwell.

Radetzki, M. (1976). The potential for monopolistic commodity pricing by developing countries. *In* G. K. Helleiner (ed.), *A World Divided*. Cambridge University Press, pp. 53–76.

Ralph, J. T. (1971). Feasibility and financing. Paper delivered to Australian Society of Accountants and Institute of Chartered Accountants Joint Seminar: Special Accounting and Financial Considerations Relating to Mining and Exploration Companies, Sydney, 27 September 1971.

Reddaway, W. B. (1937). Some effects of the Australian tariff. *Economic Record*, 13, 1, June, pp. 20–30.

Ricardo, D. (1911). *The Principles of Political Economy and Taxation*. London: Dent (Everyman), 1962 edn.

Sibthorp, M. M. (ed.) (1975). *The North Sea: Challenge and Opportunity*. London: Europa, for the Institute of International Studies.

Smiley, A. K. (1979). *Competitive Bidding Under Uncertainty: The Case of Offshore Oil*. Cambridge, Mass.: Ballinger.

Smith, B. (1978). Australia's minerals production and trade. *In* L. Krause and H. Patrick (eds.), *Mineral Resources in the Pacific Area*. San Francisco: Federal Reserve Bank of San Francisco.

Smith, B. (ed.) (1979). *Taxation of the Mining Industry*. Canberra: Centre for Resource and Environmental Studies, Australian National University.

Smith, D. and Wells, L. J. (1969). *Report to the Indonesian Government*. Jakarta: Government of Indonesia.

Smith, D. and Wells, L. J. (1975). *Negotiating Third World Mineral Agreements*. Cambridge, Mass.: Ballinger.

Snape, R. H. (1977). Effects of mineral development on the economy. *Australian Journal of Agricultural Economics*, 21, 3, pp. 147–56.

Solow, R. M. (1974). The economics of resources or the resources of economics. *American Economic Review*, 64, 2, May, pp. 1–14.

Sumner, M. (1978). Progressive taxation of natural resource rents. *Manchester School*, 46, March, pp. 1–16.

Sundrum, R. M. and Booth, A. E. (1980). *In* Garnaut and McCawley (1980), pp. 455–86.

Swan, P. (1976). Income taxes, profit taxes and neutrality of optimizing decisions. *Economic Record*, 52, 138, June, pp. 166–81.

Taussig, F. W. (1933). *Principles of Economics*. London: Macmillan, 3rd edn 1934.

Treadgold, M. (1974). Recent economic performance in Papua New Guinea. *Economic Record*, 50, 129, March, pp. 39–56.

United Kingdom (1975). Oil Taxation Act 1975, c.22.

United Kingdom (1975). Petroleum and Submarine Pipelines Act 1975, c.74.

United Kingdom (1981). Finance Act 1981, c.35.

United Nations (1972). *Guidelines to Project Evaluation* (by A. K. Sen, S. Marglin, and P. Dasgupta). New York: United Nations.

Vernon, R. (1967). Long-run trends in concession contracts. *American Society of International Law: Proceedings*, pp. 81–9.

Waelde, T. W. (1977). Lifting the veil from transnational mineral contracts: A review of the recent literature. *Natural Resources Forum*, 1, 3, April, pp. 277–84.

Wells, L. T. (1973). *A Review of the Bougainville Copper Agreement*. Report to the Papua New Guinea Minister of Mines, Port Moresby.

Zambia Mining Yearbook (1978).

Zorn, A. S. (1977). New development in Third World mining agreements. *Natural Resources Forum*, 1, 3, April, pp. 239–50.